International Association of Fire Chiefs

 National Fire Protection Association

Fundamentals of Fire Fighter Skills

SECOND EDITION

Bridgeport Public Library
925 Broad St.
Bridgeport, CT 06604

JONES AND BARTLETT PUBLISHERS
Sudbury, Massachusetts
BOSTON   TORONTO   LONDON   SINGAPORE

Jones and Bartlett Publishers
World Headquarters
40 Tall Pine Drive
Sudbury, MA 01776
978-443-5000
info@jbpub.com
www.jbpub.com

National Fire Protection Association
1 Batterymarch Park
Quincy, MA 02169-7471
www.NFPA.org

International Association of Fire Chiefs
4025 Fair Ridge Drive
Fairfax, VA 2
www.IAFC.org

Jones and Bartlett Publishers Canada
6339 Ormindale Way
Mississauga, Ontario L5V 1J2
Canada

Jones and Bartlett Publishers International
Barb House, Barb Mews
London W6 7PA
United Kingdom

Jones and Bartlett's books and products are available through most bookstores and online booksellers. To contact Jones and Bartlett Publishers directly, call 800-832-0034, fax 978-443-8000, or visit our website, www.jbpub.com.

Substantial discounts on bulk quantities of Jones and Bartlett's publications are available to corporations, professional associations, and other qualified organizations. For details and specific discount information, contact the special sales department at Jones and Bartlett via the above contact information or send an email to specialsales@jbpub.com.

Editorial Credits
Author: Douglas C. Ott

Production Credits
Chief Executive Officer: Clayton E. Jones
Chief Operating Officer: Donald W. Jones, Jr.
President, Higher Education and Professional Publishing: Robert W. Holland, Jr.
V.P., Sales and Marketing: William J. Kane
V.P., Production and Design: Anne Spencer
V.P., Manufacturing and Inventory Control: Therese Connell
Publisher, Public Safety Group: Kimberly Brophy
Senior Acquisitions Editor, Fire: William Larkin

Associate Editor: Lindsay Murdock
Production Supervisor: Jenny L. Corriveau
Production Assistant: Sarah Bayle
Photo Research Manager/Photographer: Kimberly Potvin
Photo Researcher: Lee Michelsen
Director of Marketing: Alisha Weisman
Marketing Manager: Brian Rooney
Cover Image: © Daniel Templeton/Alamy Images
Text Design: Anne Spencer
Cover Design: Kristin E. Ohlin
Composition: Spoke & Wheel/Jason Miranda
Text Printing and Binding: Courier
Cover Printing: Courier

Additional photo credits appear on page 543 which constitutes a continuation of the copyright page.

ISBN: 978-0-7637-5749-6

Copyright © 2009 by Jones and Bartlett Publishers, LLC, and the National Fire Protection Association®.

All rights reserved. No part of the material protected by this copyright notice may be reproduced or utilized in any form, electronic or mechanical, including photocopying, recording, or by any information storage and retrieval system, without written permission from the copyright owner.

The procedures and protocols in this book are based on the most current recommendations of responsible sources. The International Association of Fire Chiefs (IAFC), National Fire Protection Association (NFPA®), and the publisher, however, make no guarantee as to, and assume no responsibility for, the correctness, sufficiency, or completeness of such information or recommendations. Other or additional safety measures may be required under particular circumstances.

Notice: The individuals described in "Fire Alarms" throughout this text are fictitious.

6048
Printed in the United States of America
13 12 11 10   10 9 8 7 6 5 4

Table of Contents

> Note to the student: Exercises indicated with a 🎩 are specific to the Fire Fighter II level.

Chapter 1 1
The History and Orientation of the Fire Service

Chapter 2 8
Fire Fighter Safety

Chapter 3 24
Fire Service Communications

Chapter 4 34
Incident Command System

Chapter 5 44
Fire Behavior

Chapter 6 54
Building Construction

Chapter 7 62
Portable Fire Extinguishers

Chapter 8 76
Fire Fighter Tools and Equipment

Chapter 9 86
Ropes and Knots

Chapter 10 104
Response and Size-Up

Chapter 11 112
Forcible Entry

Chapter 12 124
Ladders

Chapter 13 138
Search and Rescue

Chapter 14 150
Ventilation

Chapter 15 162
Water Supply

Chapter 16 172
Fire Hose, Nozzles, Streams, and Foam

Chapter 17 198
Fire Fighter Survival

Chapter 18 210
Salvage and Overhaul

Chapter 19 222
Fire Fighter Rehabilitation

Chapter 20 230
Wildland and Ground Fires

Chapter 21 236
Fire Suppression

Chapter 22 246
Preincident Planning

Chapter 23 256
Fire and Emergency Medical Care

Chapter 24 264
Emergency Medical Care

Chapter 25 282
Vehicle Rescue and Extrication

Chapter 26 294
Assisting Special Rescue Teams

Chapter 27 304
Hazardous Materials: Overview

Chapter 28 312
Hazardous Materials: Properties and Effects

Chapter 29 322
Hazardous Materials: Recognizing and Identifying the Hazards

Chapter 30 332
Hazardous Materials: Implementing a Response

Chapter 31 338
Hazardous Materials: Personal Protective Equipment, Scene Safety, and Scene Control

Chapter 32 350
Hazardous Materials: Response Priorities and Actions

Chapter 33 358
Hazardous Materials: Decontamination Techniques

Chapter 34 366
Terrorism Awareness

Chapter 35 374
Fire Prevention and Public Education

Chapter 36 380
Fire Detection, Protection, and Suppression Systems

Chapter 37 390
Fire Cause Determination

Answer Key

Chapter 1 399
The History and Orientation of the Fire Service

Chapter 2 403
Fire Fighter Safety

Chapter 3 410
Fire Service Communications

Chapter 4 412
Incident Command System

Chapter 5.................................. 415
Fire Behavior

Chapter 6.................................. 418
Building Construction

Chapter 7.................................. 420
Portable Fire Extinguishers

Chapter 8.................................. 426
Fire Fighter Tools and Equipment

Chapter 9.................................. 429
Ropes and Knots

Chapter 10................................. 436
Response and Size-Up

Chapter 11................................. 438
Forcible Entry

Chapter 12................................. 442
Ladders

Chapter 13................................. 448
Search and Rescue

Chapter 14................................. 452
Ventilation

Chapter 15................................. 457
Water Supply

Chapter 16................................. 460
Fire Hose, Nozzles, Streams, and Foam

Chapter 17................................. 471
Fire Fighter Survival

Chapter 18................................. 476
Salvage and Overhaul

Chapter 19................................. 481
Fire Fighter Rehabilitation

Chapter 20................................. 483
Wildland and Ground Fires

Chapter 21................................. 485
Fire Suppression

Chapter 22................................. 489
Preincident Planning

Chapter 23................................. 493
Fire and Emergency Medical Care

Chapter 24................................. 496
Emergency Medical Care

Chapter 25................................. 503
Vehicle Rescue and Extrication

Chapter 26................................. 508
Assisting Special Rescue Teams

Chapter 27................................. 511
Hazardous Materials: Overview

Chapter 28..**514**
Hazardous Materials: Properties and Effects

Chapter 29..**517**
Hazardous Materials: Recognizing and Identifying the Hazards

Chapter 30..**521**
Hazardous Materials: Implementing a Response

Chapter 31..**523**
Hazardous Materials: Personal Protective Equipment, Scene Safety, and Scene Control

Chapter 32..**528**
Hazardous Materials: Response Priorities and Actions

Chapter 33..**529**
Hazardous Materials: Decontamination Techniques

Chapter 34..**533**
Terrorism Awareness

Chapter 35..**535**
Fire Prevention and Public Education

Chapter 36..**537**
Fire Detection, Protection, and Suppression Systems

Chapter 37..**539**
Fire Cause Determination

Student Resources

To help students retain the most important information and assist them in preparing for examinations, Jones and Bartlett Publishers has developed a complete set of student resources.

Skills Evaluation Workbook
ISBN-13: 978-0-7637-6040-3

The *Skills Evaluation Workbook* offers step-by-step instruction on the most vital and fundamental Fire Fighter I and II skills. Addressing more than 265 skills, it covers every job performance requirement within the 2008 edition of NFPA 1001 and NFPA 472. This vital resource was developed to help both fire fighter students and instructors review, document, and evaluate each candidate's progress through hands-on skills demonstrations. Each customizable skill sheet provides the following items:

- Evaluator instructions
- Tasks to be performed
- Performance outcomes
- Candidate directives
- Numbered, step-by-step instructions checklist
- Area for evaluator and candidate comments

Student Review Manual
Print—ISBN-13: 978-0-7637-5750-2
Online—ISBN-13: 978-0-7637-5751-9

The *Student Review Manual* is designed to prepare students for exams by asking them the same types of questions that they are likely to see on classroom and certification examinations. The manual contains multiple-choice questions along with an answer key and appropriate page references.

Also of interest:

Fire Fighter Health Evaluation and Training Workout Manual
ISBN-13: 978-07637-6634-4

The *Fire Fighter Health Evaluation and Training Workout Manual* is the only health and fitness manual developed exclusively for fire fighters. Fire fighters now have a comprehensive training workout program designed with their specific job functions in mind. The program takes into account the energy systems that fire fighters need to develop and the stabilization muscles that help keep them injury free. All of this can be done without spending hours in the gym. The *Fire Fighter Health Evaluation and Training Workout Manual* encompasses a complete and comprehensive 12-month health and fitness regimen. Each month the exercises build in complexity and rigor so that by the end of 12 months fire fighters will be in peak physical performance.

Fire Fighter Field Guide
ISBN-13: 978-0-7637-4143-3

A quick reference to essential information for fire fighters, *Fire Fighter Field Guide* includes charts and tables to provide easy access to key topics. Designed to withstand the elements, this field guide is pocket sized, spiral bound, and water resistant. This handy reference provides instant access to information about the following subject areas:

- Response and size-up
- Forcible entry, search, and rescue
- Ladders and ventilation
- Fire suppression
- Salvage, overhaul, and fire cause determination
- Vehicle rescue and extrication
- Tools and equipment
- Special operations
- Safety

Technology Resources

Key components of the teaching and learning system that accompany *Fundamentals of Fire Fighter Skills, Second Edition* are activities and simulations that help students become great fire fighters.

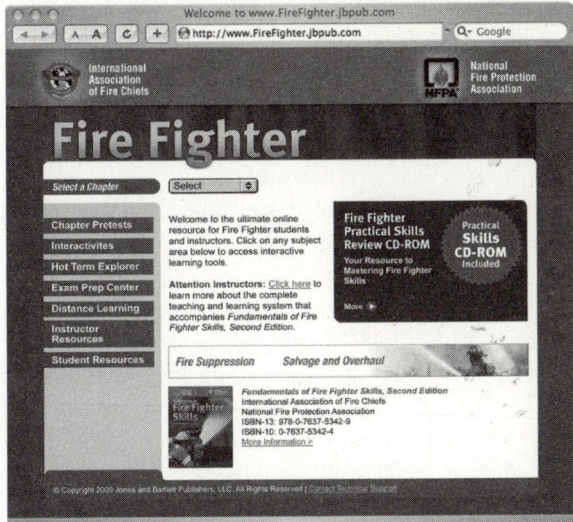

www.FireFighter.jbpub.com

This site has been specifically designed to complement *Fundamentals of Fire Fighter Skills, Second Edition* and is regularly updated. Resources available include:

- **Chapter Pretests** that prepare students for training. Each chapter has a pretest and provides instant results, feedback on incorrect answers, and page references for further study.
- **Interactivities** that allow students to reinforce their understanding of the most important concepts in each chapter.
- **Hot Term Explorer**, a virtual dictionary that allows students to review key terms, test their knowledge of key terms through quizzes and flashcards, and complete crossword puzzles.
- **Review Manual** that allows students to prepare for classroom and certification examinations and provides instant results, feedback, and page references.

FireLearn
ISBN-13: 978-0-7637-3145-8

FireLearn consists of a series of online modules that bring Fire Fighter I and II material to life with interactive audio, video, Flash™ animation, photographs, illustrations, and case-based scenarios. *FireLearn* is designed to enhance students' learning experience by allowing them to experiment with key concepts and skills in the safety of a virtual environment.

FireLearn can be used in a variety of ways. Use either the individual chapters or the complete course as follows:

- As a distance-learning alternative to the traditional classroom setting
- As a review tool to assess students' progress and reinforce important concepts presented in the classroom
- In a modular format to clarify more challenging topics and allow students to experiment with skills in the safety of a virtual environment

The History and Orientation of the Fire Service

Workbook Activities

The following activities have been designed to help you. Your instructor may require you to complete some or all of these activities as a regular part of your fire fighter training program. You are encouraged to complete any activity that your instructor does not assign as a way to enhance your learning in the classroom.

Chapter Review

The following exercises provide an opportunity to refresh your knowledge of this chapter.

Matching

Match each of the terms in the left column to the appropriate definition in the right column.

_____ 1. Battalion chief   A. To take off
_____ 2. Captain   B. A tool used to pull down burning structures
_____ 3. Discipline   C. The guidelines that a department sets for fire fighters to work within
_____ 4. Don   D. A valve installed to control water accessed from wooden pipes
_____ 5. Driver/operator   E. The position responsible for a fire company and for coordinating activities of that company among the shifts
_____ 6. Fireplug   F. The position responsible for operating the fire apparatus
_____ 7. Safety officer   G. The position often in charge of running calls and supervising multiple stations or districts within a city
_____ 8. Doff   H. To put on
_____ 9. Fire hook   I. The position with the authority to stop any firefighting activity until it can be done safely and correctly
_____ 10. SCBA   J. Respirator with independent air supply

Multiple Choice

Read each item carefully, and then select the best response.

_____ 1. Augustus Caesar created what was probably the first fire department, called the Familia Publica, in
   A. 100 B.C.
   B. 24 B.C.
   C. 1 B.C.
   D. 10 A.D.

_____ 2. The first fire insurance company in the United States was established in 1736
   A. by George Washington.
   B. by Benjamin Franklin.
   C. in Charleston, South Carolina.
   D. by the Alexandria Fire Department.

CHAPTER 1

_____ 3. In 1871, a historic fire, which was believed to be started by a cow, burned for three days, destroyed more than 2,000 acres and 17,000 homes, and killed 300 people. This was the
   A. Great Chicago Fire.
   B. Peshtigo Fire.
   C. Green Bay Burn.
   D. Alexandria Fire.

_____ 4. Colonial fire fighters had limited equipment; most departments had only buckets, ladders, and
   A. hand-powered pumpers.
   B. horse-drawn water carriages.
   C. fire hooks.
   D. hoses.

_____ 5. The first water system valves or fire hydrants used by fire fighters were called
   A. fire taps.
   B. water valves.
   C. water boxes.
   D. fireplugs.

_____ 6. Before radios or bullhorns, the _____ allowed communication during an incident. Today it serves as a symbol of authority.
   A. chief's trumpet
   B. call box
   C. monitor
   D. commander's horn

_____ 7. The fire service draws its authority from the governing entity, and the head of the department is accountable to the
   A. fire chief.
   B. insurance companies.
   C. leader of the governing body.
   D. civil servants.

_____ 8. The theory that each fire fighter answers to only one supervisor is referred to as
   A. unity of command.
   B. span of control.
   C. division of labor.
   D. discipline.

_____ 9. New fire fighters usually report to a
   A. lieutenant.
   B. captain.
   C. battalion chief.
   D. division chief.

_____ 10. The overall responsibility for the administration and operations of the department belongs to the
   A. battalion chief.
   B. chief of the department.
   C. incident commander.
   D. government.

_____ 11. The company responsible for securing a water source, deploying handlines, and putting water on the fire is the
   A. truck company.
   B. brush company.
   C. water company.
   D. engine company.

_____ 12. To provide a uniform way to deal with emergency situations, departments develop and follow
   A. laws.
   B. regulations.
   C. standard operating procedures (SOPs).
   D. policies.

_____ 13. The company that specializes in forcible entry, ventilation, roof operations, search and rescue, and ground ladders is the
   A. truck company.
   B. brush company.
   C. water company.
   D. engine company.

_____ 14. In some states, the preferred terminology for standard operating procedures is
   A. policies.
   B. regulations.
   C. rules.
   D. suggested operating guidelines.

_____ 15. The organizational structure of a fire department consists of a(n)
   A. chain of custody.
   B. incident management system.
   C. chain of command.
   D. division of labor.

_____ 16. Which of the following is not a form of discipline?
   A. Regulations
   B. Policies
   C. Span of control

_____ 17. Most experts believe that span of control should not exceed how many people in a complex or rapidly changing environment?
   A. 3
   B. 4
   C. 5
   D. 6

_____ 18. What are the most important "machines" on the fire scene?
   A. Hand tools
   B. Engines
   C. Well-trained fire fighters
   D. Ladders

_____ 19. The majority of fire departments consist of
   A. all career fire fighters.
   B. mostly career fire fighters.
   C. mostly volunteer fire fighters.
   D. all volunteer fire fighters.

_____ 20. In the 1700s, a fire mark indicated
   A. the homeowner had fire insurance.
   B. the homeowner was a career fire fighter.
   C. the homeowner was a volunteer fire fighter.
   D. the home had a previous fire.

Vocabulary

Define the following terms using the space provided.

1. Incident commander (IC):

   _____
   _____
   _____

2. Training officer:

   _____
   _____
   _____

3. Company officer:

   _____
   _____
   _____

4. Safety officer:

   _____
   _____
   _____

5. Emergency Medical Technician–Paramedic:

   _____
   _____
   _____

Fill-in

Read each item carefully, and then complete the statement by filling in the missing word(s).

1. The first fire regulations in North America were established in Boston, Massachusetts, when the city banned _____ and _____.

2. The first volunteer fire company began in Philadelphia in 1735, under the leadership of _____.

3. _____ developed the first municipal water systems.

4. George Smith, a fire fighter in New York City, developed the first _____ in 1817.

5. In Washington, D.C., _____ were introduced as the first communication tool used to send coded telegraph signals to the fire departments.

6. _____ provide specific information on the actions that should be taken to accomplish a certain task.

7. Each fire department is responsible for a specific _____ area.

8. _____ personnel administer prehospital care to people who are sick and injured.

9. When multiple agencies work together, a unified command system must be established. This system is referred to as the _____.

True/False

If you believe the statement to be more true than false, write the letter "T" in the space provided.
If you believe the statement to be more false than true, write the letter "F."

1. _____ George Washington established one of the first fire departments in Alexandria, Virginia, in 1765.
2. _____ The Peshtigo fire storm jumped the 60-mile-wide Green Bay and continued to burn on Wisconsin's northeast peninsula.
3. _____ Today almost all fire protection in the United States is funded directly or indirectly through tax dollars.
4. _____ The organizational structure of a fire department consists of a division of labor.
5. _____ Captains report directly to chiefs.
6. _____ The fire fighter is responsible for dispatching units to an incident.
7. _____ "Info techs" are responsible for communicating information to the news media.
8. _____ Consensus documents are developed through agreement between people representing different organizations and interests.
9. _____ Covering a fire to ensure a low burn is called "banking."
10. _____ The battalion chief is the second rank of promotion, responsible for managing a fire company.

Short Answer

Complete this section with short written answers using the space provided.

1. Identify the four basic management principles utilized in most fire departments.

   _____
   _____
   _____
   _____

2. Identify and describe the role of five companies common to most fire departments.

   _____
   _____
   _____
   _____
   _____

3. Identify and describe 10 of the common and/or specialist positions a fire fighter may assume in his or her career as a fire fighter.

   _____
   _____
   _____
   _____
   _____
   _____
   _____
   _____
   _____
   _____

4. Describe the differences between regulations, policies, and standard operating procedures.

   _____
   _____
   _____

CLUES

Across

1. A valve installed to control water accessed from wooden pipes.
5. The second rank of promotion, responsible for a fire company.
9. Rules, usually issued by a government, that dictate how something must be done.
11. Usually the officer in charge of a single-alarm working fire.
12. Covering a fire to ensure low burning.
13. A tool used to pull down burning structures.
14. To take off.
15. The procedures that provide information on the actions that should be taken.
16. Formal statements that provide guidelines for present and future actions.

Down

1. Officials responsible for protecting fire fighters by controlling traffic and securing the scene from public access during an incident.
2. A team usually charged with the task of rescuing victims from confined spaces, trenches, and high-angle situations.
3. A company officer who is usually responsible for a single fire company on a single shift.
4. A person who is tasked with anything from hose line placement to extinguishing fires.
6. A symbol showing that a building was insured by a company that would pay fire fighters for extinguishing the fire.
7. Responsible for updating the training of current employees and for training new fire fighters in the current techniques of firefighting and EMS.
8. To put on.
10. An officer who has the authority to stop any firefighting activity until it can be done safely and correctly.

Word Fun

The following crossword puzzle is an activity provided to reinforce correct spelling and understanding of terminology associated with firefighting. Use the clues provided to complete the puzzle.

Fire Alarms

The following real case scenarios will give you an opportunity to explore the concerns associated with the history and orientation of the fire service. Read each scenario, then answer each question in detail.

1. You are outside the fire station washing the fire truck when you are approached by three children on bicycles. None of the children is wearing a helmet. They ask you if you will show them the truck and the station. How will you proceed?

   _____
   _____
   _____
   _____
   _____
   _____

2. You have chosen the fire services as a career, and you have worked hard to get to this point. You have successfully completed the entry requirement, and you have been issued your bunker gear and uniform. You need to keep yourself on target to become a proud and accomplished fire fighter. What do you need to do to succeed?

   _____
   _____
   _____
   _____
   _____
   _____
   _____

Fire Fighter Safety

Workbook Activities

The following activities have been designed to help you. Your instructor may require you to complete some or all of these activities as a regular part of your fire fighter training program. You are encouraged to complete any activity that your instructor does not assign as a way to enhance your learning in the classroom.

Chapter Review

The following exercises provide an opportunity to refresh your knowledge of this chapter.

Matching
Match each of the terms in the left column to the appropriate definition in the right column.

_____ 1. Compressor   A. To take off an item of clothing or equipment
_____ 2. Face piece   B. A process to provide periods of rest and recovery for emergency workers during an incident
_____ 3. Two-way radio   C. A fire-resistant synthetic material
_____ 4. Doff   D. To put on an item of clothing or equipment
_____ 5. Rehabilitation   E. A portable communication device used by fire fighters
_____ 6. Don   F. A component of SCBA that fits over the face
_____ 7. Cascade system   G. A device used to refill SCBA cylinders
_____ 8. Nomex   H. An apparatus consisting of multiple tanks used to store compressed air
_____ 9. Freelancing   I. The dangerous practice of acting independently of command instructions
_____ 10. SCBA   J. A respirator with an independent air supply that is used by fire fighters

Multiple Choice
Read each item carefully, and then select the best response.

_____ 1. The system in which two fire fighters work as a team for safety purposes is referred to as the
   A. incident management system.
   B. buddy system.
   C. personnel accountability system.
   D. personal alert system.

_____ 2. Putting on an item of clothing or equipment is called
   A. doffing.
   B. freelancing.
   C. prepping.
   D. donning.

CHAPTER 2

_____ 3. An electronic semiconductor that emits a single-color light when activated is an LED, which is an acronym for
   A. light-emergent device.
   B. light-exiting device.
   C. light-emitting diode.
   D. laser-emitting diode.

_____ 4. Oxygen deficiency occurs when the atmosphere's oxygen level drops below
   A. 19.5 percent.
   B. 21 percent.
   C. 9 percent.
   D. 6 percent.

_____ 5. An electronic device that sounds a loud audible signal when a fire fighter becomes trapped or injured is a(n)
   A. personal alert safety system (PASS).
   B. cascade system.
   C. emergency bypass mode.
   D. SCBA regulator.

_____ 6. PBI, Kevlar, and Nomex are materials used in the construction of
   A. personal protective clothing.
   B. firefighting ropes.
   C. communications equipment.
   D. the buddy system.

_____ 7. An SCBA in which exhaled air is released into the atmosphere and is not reused is a(n)
   A. closed-circuit breathing apparatus.
   B. cascade system.
   C. open-circuit breathing apparatus.
   D. supplied-air apparatus.

_____ 8. The written rules and procedures that outline how to perform various functions and operations are the
   A. general operating guidelines.
   B. NIOSH Codes.
   C. incident management system (IMS).
   D. standard operating procedures (SOPs).

_____ 9. In general, an SCBA weighs at least
   A. 25 pounds.
   B. 30 pounds.
   C. 40 pounds.
   D. 45 pounds.

_____ 10. The U.S. Department of Transportation requires _____ testing on a periodic basis to ensure that SCBA cylinders are in good working condition.
   A. thread
   B. hydrostatic
   C. NIOSH
   D. standard operating procedure

Fundamentals of Fire Fighter Skills

_____ 11. The device on an SCBA that measures and displays the amount of pressure currently in the cylinder is the
   A. personal safety gauge.
   B. pressure gauge.
   C. SCBA regulator.
   D. air line.

_____ 12. An SCBA designed to recycle the user's exhaled air is a(n)
   A. closed-circuit breathing apparatus.
   B. cascade system.
   C. open-circuit breathing apparatus.
   D. supplied-air apparatus.

_____ 13. Which part of the personal protective equipment is worn over the head and under the helmet to protect the neck and ears?
   A. Helmet shell
   B. Protective hood
   C. Bunker hood
   D. Face piece

_____ 14. During operating mode, if the SCBA regulator fails to function properly, what releases a constant flow of breathing air into the face piece?
   A. Respirator
   B. Air line
   C. Emergency bypass mode
   D. PBI

_____ 15. What is the leading cause of death among fire fighters?
   A. Vehicle accidents
   B. Smoke inhalation
   C. Heart attack
   D. Cancer

_____ 16. When plastic products burn, one of the most dangerous gases produced is
   A. hydrogen cyanide.
   B. phosgene gas.
   C. hydrochloric gas.
   D. phosphorous gas.

_____ 17. The straps and fasteners used to attach the SCBA to the fire fighter are part of the
   A. face piece.
   B. backpack.
   C. bunker coat.
   D. harness.

_____ 18. If you need assistance during an emergency situation, you should activate your
   A. respirator.
   B. PASS device.
   C. SCBA regulator.
   D. LED.

_____ 19. Donning protective clothing must be done in a specific order, and quickly. Fire fighters should be able to don protective clothing in
   A. 30 seconds.
   B. 45 seconds.
   C. 60 seconds.
   D. 120 seconds.

Vocabulary

Define the following terms using the space provided.

1. Employee assistance program (EAP):
   _____
   _____
   _____

2. National Institute for Occupational Safety and Health (NIOSH):
   _____
   _____
   _____

3. Personnel accountability system:
   _____
   _____
   _____

4. Safety officer:
   _____
   _____
   _____

5. Standard operating procedures (SOPs):
   _____
   _____
   _____

6. Buddy system:
   _____
   _____
   _____

7. Immediately dangerous to life and health (IDLH):
   _____
   _____
   _____

Fill-in

Read each item carefully, and then complete the statement by filling in the missing word(s).

1. When the fuel is not completely consumed during the burning process, _____ occurs.

2. An atmosphere is described as _____ when the oxygen level is 19.5 percent or less.

3. The _____ officer has the authority to stop any activity that is judged to be unsafe.

4. When fire fighters make independent decisions or do not follow command instructions, they are taking part in the dangerous practice of _____.

5. A(n) _____ is a breathing apparatus that uses an external source for the breathing air.

6. The _____ has developed programs with the goal of reducing line-of-duty deaths by _____ in 5 years.

7. _____ provide confidential help with a wide range of problems that might affect performance.

8. Emergency vehicle operators are subject to all _____, unless a specific exemption is made.

9. _____ will prevent most vehicle collisions.

True/False

If you believe the statement to be more true than false, write the letter "T" in the space provided.
If you believe the statement to be more false than true, write the letter "F."

1. _____ Carbon monoxide is a toxic gas produced through incomplete combustion.
2. _____ Members of rapid intervention teams are the first fire fighters to enter a structure in an emergency operation.
3. _____ SCBA regulators control the flow of air by increasing the low pressure in the cylinder to a usable higher pressure for the user.
4. _____ Phosgene gas is a dangerous by-product of incomplete combustion.
5. _____ A prompt response is a higher priority than a safe response.
6. _____ On the fire ground, the company officer must always know where his or her teams are and what they are doing.
7. _____ Most fire departments have employee assistance programs to provide counseling services to support fire fighters.
8. _____ Freelancing is a good method of discovering new firefighting techniques.
9. _____ Even with an emergency driving exemption, the operator can be found criminally or civilly liable if involved in a crash.
10. _____ Every fire department must have a personnel accountability system.

Short Answer

Complete this section with short written answers using the space provided.

1. Identify the four major components of a successful safety program.

2. Identify five limitations of personal protective equipment (PPE).

3. What are the seven types of protection provided by PPE?

4. Discuss three reasons why fire fighters need respiratory protection during fire incidents.

5. Describe the purpose of a critical incident stress debriefing (CISD).

Word Fun

The following crossword puzzle is an activity provided to reinforce correct spelling and understanding of terminology associated with firefighting. Use the clues provided to complete the puzzle.

CLUES

Across

1. The individual in overall command of an emergency incident.
6. A designated individual who oversees safety practices at an emergency scene and during training.
9. To put on an item of clothing or equipment.
10. The dangerous activity of acting independently of command instructions.
12. The hose through which air flows within an SCBA.
13. A system in which two fire fighters always work as a team for safety purposes.
14. The protective coat worn by a fire fighter for interior structural firefighting.
15. Part of SCBA that allows fire fighters to wear it as a "backpack."
16. A protective device used to provide safe breathing air to a user in a dangerous atmosphere.
17. A respirator that gets its air through a hose from a remote source.

Down

2. An insert inside the face piece of an SCBA that fits over the user's mouth and nose.
3. The operating mode that allows an SCBA to be used even if part of the regulator fails to function properly.
4. The component of the SCBA that stores the compressed air supply.
5. Airborne solid material consisting of ash and unburned or partially burned fuel released by a fire.
7. A dangerous by-product of incomplete combustion.
8. A mechanical device that is used to refill SCBA cylinders.
9. To take off an item of clothing or equipment.
11. Protective head covering worn by fire fighters to protect the head from falling objects, blunt trauma, and heat.
14. The harness of the SCBA that supports the components worn by a fire fighter.

Fundamentals of Fire Fighter Skills

Fire Alarms

The following real case scenarios will give you an opportunity to explore the concerns associated with fire fighter qualifications and safety. Read each scenario, and then answer each question in detail.

1. You have just returned from a commercial structure fire at a mattress store. Your PPE is soiled with the products of combustion. Although you were sprayed off with water at the scene, your gear is still very dirty. How will you proceed?

   _____
   _____
   _____
   _____
   _____
   _____

2. It is just after dinner when your ladder truck company is dispatched to an apartment fire. Upon arrival, you are assigned to search and rescue on the second floor. You have just completed searching the first apartment when your SCBA regulator malfunctions. How should you proceed?

   _____
   _____
   _____
   _____
   _____
   _____

Skill Drills

Skill Drill 2-1: Donning Personal Protective Clothing
Test your knowledge of this skill drill by filling in the correct words in the photo captions.

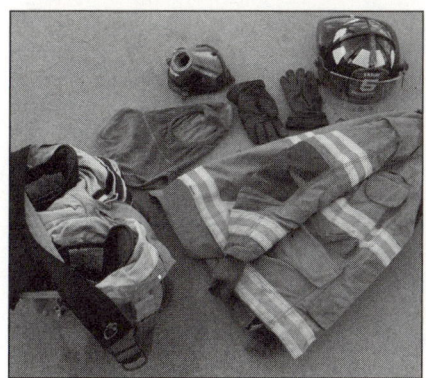

1. Place your equipment in a(n) _____ order for donning.

2. Place your protective _____ over your head and down around your neck.

3. Put on boots and pull up bunker pants. Place the _____ over your shoulders and secure the front of the pants.

4. Put on your _____ coat and close the front of the coat.

5. Place the _____ on your head and adjust the chin strap securely. Turn up your coat collar and secure it in front.

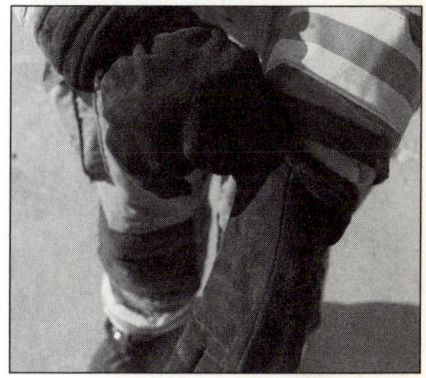

6. Put on your _____.

7. Have your partner _____ your clothing.

Skill Drill 2-2: Doffing Personal Protective Clothing

Test your knowledge of this skill drill by placing the photos below in the correct order. Number the first step with a "1," the second step with a "2," and so on.

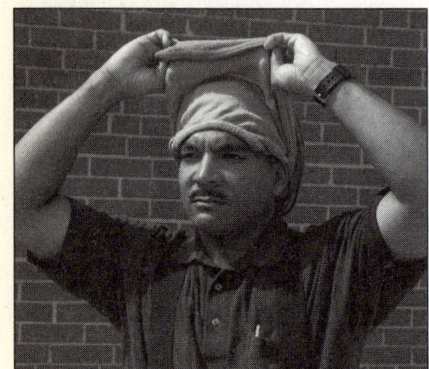

_____ Remove your protective hood.

_____ Remove your bunker pants and boots.

_____ Remove your turnout coat.

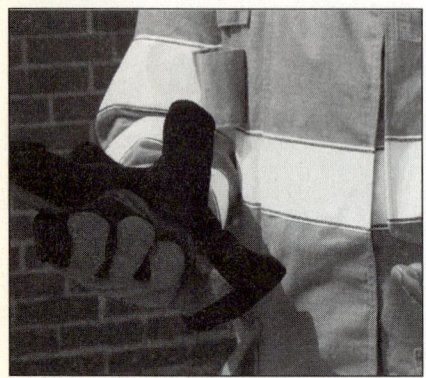

_____ Remove your gloves.

_____ Release the helmet chin strap and remove your helmet.

_____ Open the collar of your turnout coat.

Skill Drill 2-3: Donning SCBA from an Apparatus Seat Mount
Test your knowledge of this skill drill by filling in the correct words in the photo captions.

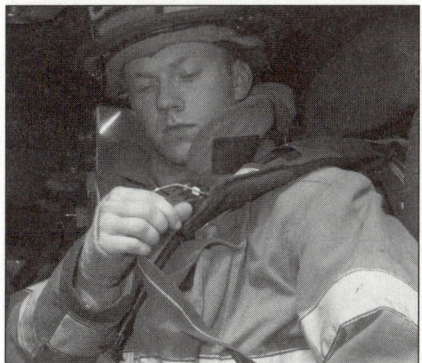

1. Don full PPE ensemble prior to _____ fire apparatus. Safely mount the apparatus and sit in the seat, placing your arms through the SCBA shoulder straps.

2. Fasten your seat belt. Partially tighten the _____ straps. When the apparatus stops, release the seat belt and release the SCBA from its brackets. Exit the apparatus.

3. Attach the _____ belt and cinch down.

4. _____ shoulder straps until they are tight.

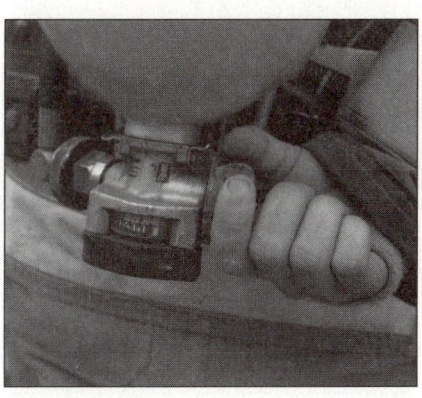

5. Open the main _____ valve.

6. Loosen or remove your helmet and pull the _____ back. Don the face piece and check for leaks. Replace the protective hood and helmet, and secure the chin strap.

7. If necessary, connect the _____ to the face piece.

8. Activate the air flow and _____ alarm.

Skill Drill 2-5: Donning SCBA Using the Over-the-Head Method

Test your knowledge of this skill drill by placing the photos below in the correct order. Number the first step with a "1," the second step with a "2," and so on.

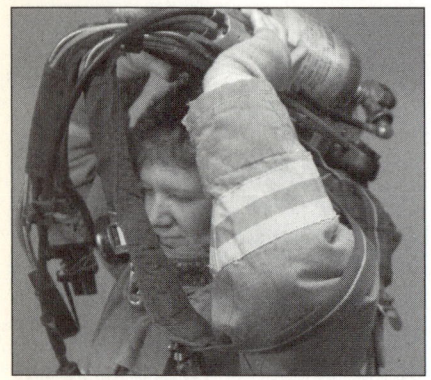

_____ Bend down and grasp the SCBA backplate with both hands. Using your legs, lift the SCBA over your head. Rotate the SCBA 180 degrees, so the waist straps are pointed to the ground.

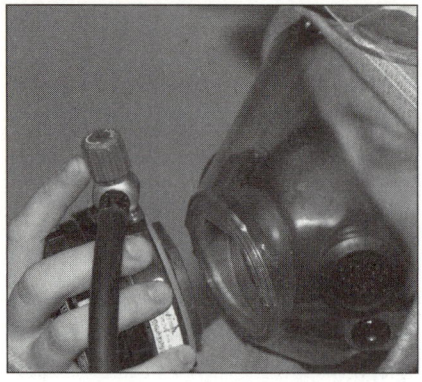

_____ If necessary, connect the regulator to the face piece. Activate the air flow and PASS alarm.

_____ Remove your helmet and pull the hood back. Don your face piece and check for an adequate seal. Pull your protective hood into position, replace your helmet, and secure the chin strap.

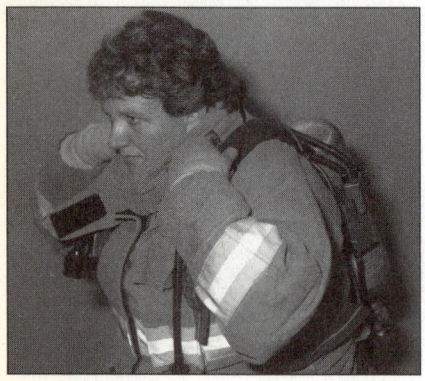

_____ Slide the SCBA down your back while your arms slide into the shoulder straps. Tighten the shoulder straps and secure the waist belt.

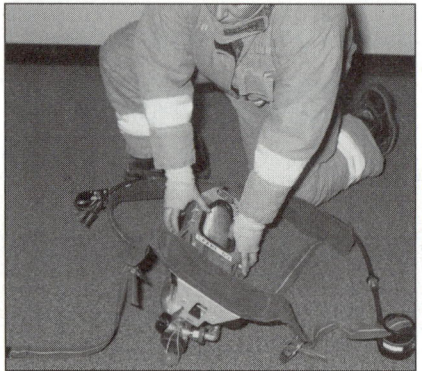

_____ Open the case and lay out the SCBA with the cylinder valve away from you and the shoulder straps out to the sides.

_____ Fully open the main cylinder valve.

Skill Drill 2-6: Donning SCBA Using the Coat Method
Test your knowledge of this skill drill by filling in the correct words in the photo captions.

1. Open the case and lay out the SCBA with the cylinder valve away from you and the shoulder straps out to the sides. Fully open the main cylinder valve. Place your _____ hand on the opposite shoulder strap.

2. Lift the SCBA and _____ it over your dominant shoulder.

3. Slide your other hand between the SCBA cylinder and the _____ shoulder strap.

4. Tighten the _____ straps.

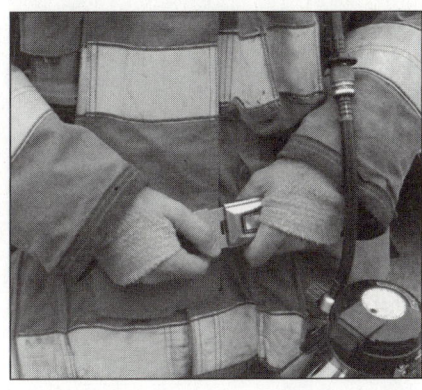

5. Attach the _____ and adjust tightness.

6. Remove your helmet and pull your hood back. Don the _____ and check for an adequate seal.

7. Pull the hood into position, replace the helmet, and secure the chin strap. If necessary, connect the regulator to the face piece. Activate the _____ and PASS alarm.

Skill Drill 2-8: Donning a Face Piece

Test your knowledge of this skill drill by placing the photos below in the correct order. Number the first step with a "1," the second step with a "2," and so on.

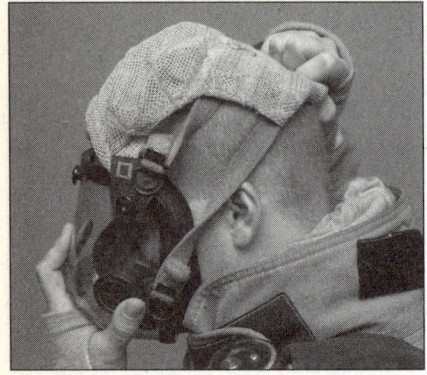

_____ Fit the face piece to your face, bringing the straps or webbing over your head.

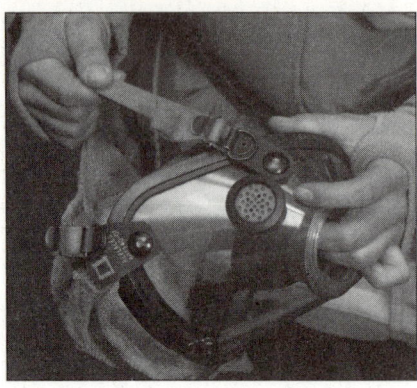

_____ Fully extend the straps on the face piece.

_____ Pull your protective hood so it covers all bare skin. Don your helmet and secure the chin strap.

_____ If there are more straps, tighten the top straps last.

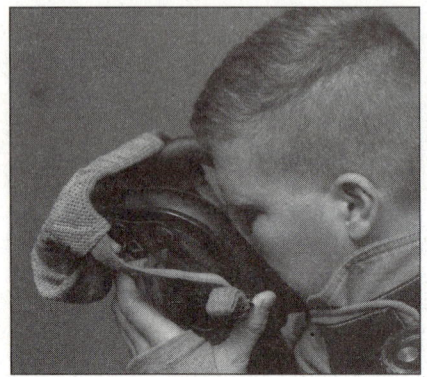

_____ Place your chin in the chin pocket.

_____ Install the regulator on your face piece or attach the low-pressure air supply hose to the regulator.

_____ Tighten the lowest two straps.

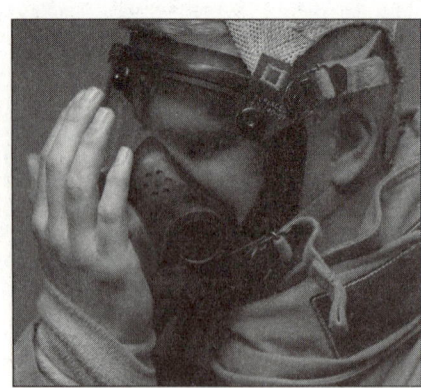

_____ Check for a proper seal.

Skill Drill 2-13: Replacing an SCBA Cylinder
Test your knowledge of this skill drill by placing the photos below in the correct order. Number the first step with a "1," the second step with a "2," and so on.

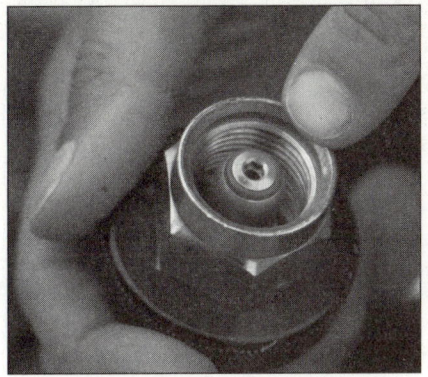

_____ Check that the "O" ring is present and in good shape.

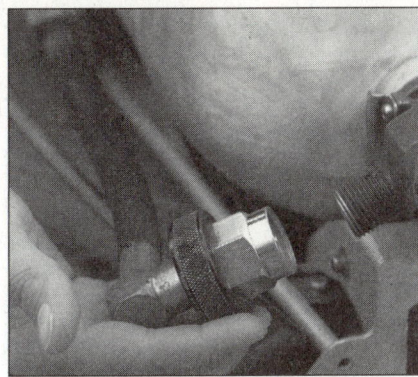

_____ Disconnect the high-pressure supply hose.

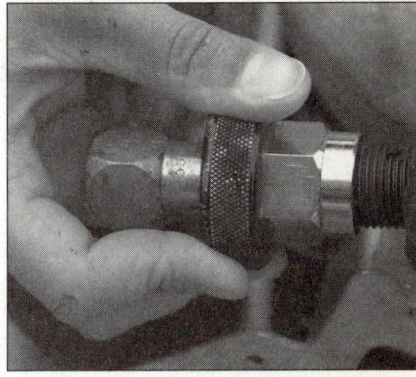

_____ Connect the high-pressure hose to the air cylinder.

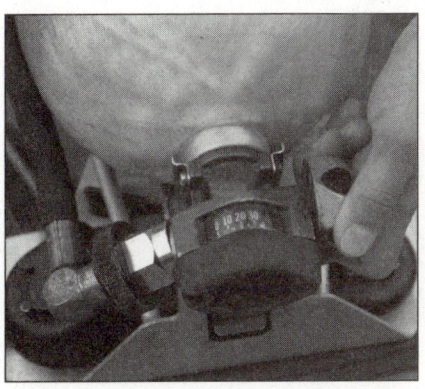

_____ Turn off the cylinder valve.

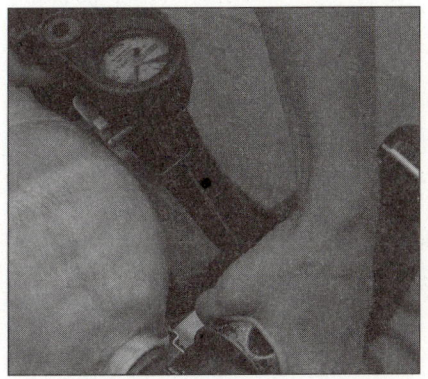

_____ Open the cylinder valve. Check the gauge reading.

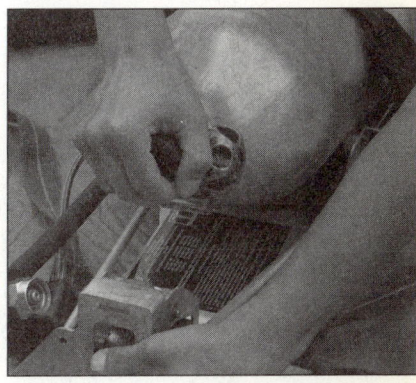

_____ Release the cylinder from the backpack.

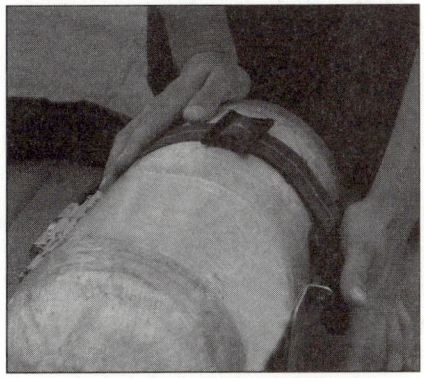

_____ Slide a full cylinder into the backpack. Align the outlet to the supply hose. Lock the cylinder in place.

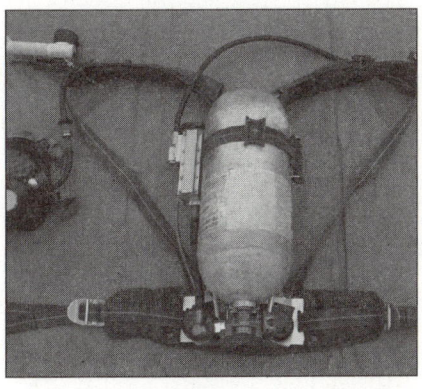

_____ Place the SCBA on the floor or a bench.

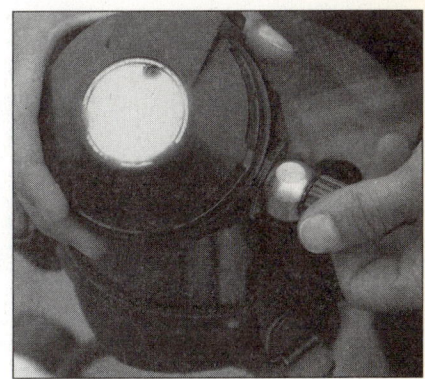

_____ Open the bypass valve to bleed off pressure.

Fire Service Communications

Workbook Activities

The following activities have been designed to help you. Your instructor may require you to complete some or all of these activities as a regular part of your fire fighter training program. You are encouraged to complete any activity that your instructor does not assign as a way to enhance your learning in the classroom.

Chapter Review

The following exercises provide an opportunity to refresh your knowledge of this chapter.

Matching

Match each of the terms in the left column to the appropriate definition in the right column.

_____ 1. Portable radio
_____ 2. Radio repeater system
_____ 3. Mobile data terminals
_____ 4. Mayday
_____ 5. CAD (computer-aided dispatch)
_____ 6. Time marks
_____ 7. Duplex channel
_____ 8. Talk-around channel
_____ 9. Simplex channel
_____ 10. ANI

A. Code indicating that a fire fighter is lost, is missing, or requires immediate assistance
B. Allow fire fighters to receive information in the apparatus or at the station
C. Automated systems used by telecommunicators to obtain and assess dispatch information
D. A battery-operated, hand-held transceiver
E. A radio system that automatically retransmits a radio signal on a different frequency
F. Update that should include the type of operation, the progress of the incident, the anticipated actions, and the need for additional resources
G. Automatic number identification
H. A radio channel using two frequencies
I. A radio channel that bypasses a repeater system.
J. A radio channel using one frequency

Multiple Choice

Read each item carefully, and then select the best response.

_____ 1. The central processing point for all information relating to an emergency incident is the
   A. incident commander.
   B. communications center.
   C. fire department.
   D. computer-aided dispatch.

_____ 2. A CAD system helps meet the most important objective in processing an emergency call, which is
   A. recording communications messages.
   B. documenting the incident.
   C. sending the appropriate units to the correct location as quickly as possible.
   D. identifying the potential casualties.

CHAPTER 3

_____ 3. The telecommunicator's first responsibility is to obtain
   A. the caller's identification.
   B. time marks.
   C. the location and nature of the emergency.
   D. communication with the responding units.

_____ 4. The process of assigning a response category is based on the nature of the reported problem or
   A. classification and prioritization.
   B. location validation.
   C. unit selection.
   D. dispatch directive.

_____ 5. Unit selection is the process of determining exactly which
   A. radio frequency to assign.
   B. equipment will be needed in the response.
   C. attack strategy will be assigned.
   D. unit(s) to dispatch.

_____ 6. Telecommunicators must follow standard operating procedures (SOPs) and use
   A. the incident management system.
   B. active listening.
   C. FCC guidelines.
   D. talk-around channels.

_____ 7. The telecommunicator can initiate a response after determining the
   A. location and nature of the problem.
   B. time of the communication and nature of the problem.
   C. urgency of the response.
   D. fire department location.

_____ 8. To assist with pinpointing calls from wireless phones, phones are being updated to include
   A. GPS technology.
   B. CAD systems.
   C. FCC regulated technology.
   D. TDD systems.

_____ 9. A call box connects a person directly to a(n)
   A. fire department.
   B. police station.
   C. incident commander.
   D. telecommunicator.

_____ 10. Call classification determines the
   A. incident management system.
   B. equipment to transport to the incident.
   C. record documentation format.
   D. number and types of units that are dispatched.

_____ 11. Someone in the communications center must remain in contact with the responding units
   A. until an incident commander is on the scene.
   B. throughout the incident.
   C. throughout the on-site scene assessment.
   D. until the incident is under control.

_____ 12. Two-way radios that are permanently mounted in vehicles are called
   A. portable radios.
   B. base stations.
   C. simplex channel radios.
   D. mobile radios.

_____ 13. A group of shared frequencies controlled by a computer is called a
   A. trunking system.
   B. mobile radio system.
   C. radio repeater system.
   D. base station system.

_____ 14. When you speak into the microphone, always speak across the microphone
   A. at a 90-degree angle.
   B. at a 45-degree angle, holding the microphone 1 to 2 inches from the mouth.
   C. as loudly as possible.
   D. without background interference.

_____ 15. The first-arriving unit at an incident should always give a brief initial report and
   A. control traffic.
   B. determine the duration of the ongoing incident.
   C. establish a command post.
   D. prepare an offensive attack unit.

_____ 16. Urgent messages that take priority over all other communications are known as
   A. time marks.
   B. dispatch information.
   C. emergency traffic.
   D. ten-code communications.

_____ 17. One of the first things you should learn when assigned to a fire station is how to use the
   A. personal protective equipment (PPE).
   B. response vehicles.
   C. incident management system.
   D. telephone and intercom system.

_____ 18. Radio codes, such as "ten codes,"
   A. are widely used and popular.
   B. are understood by all radio operators.
   C. can be problematic.
   D. work well when responding with other jurisdictions.

_____ 19. The first-arriving unit at an incident should
   A. give a brief initial radio report.
   B. establish a command post.
   C. tell other responding units what is happening.
   D. all of the above

Chapter 3: Fire Service Communications

Vocabulary

Define the following terms using the space provided.

1. TDD/TTY/text phone:

2. Ten-codes:

3. Activity logging system:

4. Automatic location identification:

5. Evacuation signal:

6. Run cards:

7. Time marks:

8. Computer-aided dispatch (CAD):

Fill-in

Read each item carefully, then complete the statement by filling in the missing word(s).

1. The _____ is a trained individual responsible for answering requests for emergency assistance from citizens.
2. A(n) _____ signal warns all personnel to pull back to a safe location.
3. _____ is an emergency code indicating that a fire fighter is missing or requires immediate assistance.
4. _____ provide status updates to the communications center at predetermined intervals.
5. A(n) _____ is a summons to fire department units to respond to an emergency.
6. The facility that receives the emergency reports and is responsible for dispatching fire department units is the _____.
7. Radio systems that use one frequency to transmit and receive all messages are called _____ channels.
8. A(n) _____ radio is a two-way radio that is permanently mounted in a fire apparatus.
9. A(n) _____ channel radio utilizes two frequencies per channel.
10. A(n) _____ system uses a shared bank of frequencies to make the most efficient use of radio resources.

True/False

If you believe the statement to be more true than false, write the letter "T" in the space provided.
If you believe the statement to be more false than true, write the letter "F."

1. _____ All calls to 9-1-1 are directed to a designated public safety answering point for that jurisdiction.
2. _____ Size-up should be transmitted by the first-arriving unit at an incident.
3. _____ A telecommunicator can initiate a response with just two pieces of information—the location and nature of the problem.
4. _____ Automatic number identification displays the telephone number where the call originated.
5. _____ Individuals with speech or hearing impairments can access the 9-1-1 system through telephones.
6. _____ Municipal fire alarm boxes are the most reliable source of contact with the communications center to indicate an emergency.
7. _____ The police, fire, and EMS departments must always have separate communication centers.
8. _____ The telecommunicator who takes a call must conduct a "telephone interrogation."
9. _____ Most fire departments use clear speech (plain English) for radio communications.
10. _____ Before transmitting over a radio, push and hold the PTT button for at least 2 seconds.

Short Answer

Complete this section with short written answers using the space provided.

1. List five of the pieces of equipment often found in a communications center.

   _____
   _____
   _____
   _____
   _____

2. Identify the five major steps in processing an emergency incident.

   _____
   _____
   _____
   _____
   _____

3. What are five of the basic functions performed in a communications center?

   _____
   _____
   _____
   _____
   _____

4. Why is accuracy important in report documentation?

   _____
   _____
   _____

Word Fun

The following crossword puzzle is an activity provided to reinforce correct spelling and understanding of terminology associated with firefighting. Use the clues provided to complete the puzzle.

Across

1. A trained individual responsible for answering requests for emergency and nonemergency assistance from citizens.
6. An electric circuit designed to cut off weak radio transmissions that are only capable of generating noise.
8. A battery-operated, hand-held transceiver.
9. Status updates provided to the communications center every 10 to 20 minutes.
10. A summons to fire department units to respond to an emergency.
11. A radio system that uses a shared bank of frequencies to make the most efficient use of radio resources.
12. Radios that are permanently mounted in a building.
13. Telephone that connects two predetermined points.
14. Cards used to determine a predesignated response to an emergency.

Down

2. Warn all personnel to pull back to a safe location.
3. Facility that receives emergency or nonemergency reports from citizens.
4. System of connected telephones that allows one to communicate with a communications center or fire department.
5. Code that indicates that a fire fighter is lost, missing, or requires immediate assistance.
7. An urgent message transmitted over a radio that takes precedence over all normal radio traffic.
11. System of predetermined coded messages used by responders over the radio.

Fire Alarms

The following real case scenarios will give you an opportunity to explore the concerns associated with fire service communications. Read each scenario, and then answer each question in detail.

1. Your engine company is dispatched to a commercial structure fire in a large grocery store. This is the second fire to which you have responded on your shift. You and your crew are assigned to a search and rescue team in the storage area of the store. You become disoriented and cannot find your way out. You have your radio and need to contact the IC to let him know you are lost. How will you proceed?

   _____
   _____
   _____
   _____
   _____
   _____

2. You are sitting in the report room finishing up paperwork from an earlier call when the phone rings. You answer the phone using your department's SOPs. The caller states that her neighbor's burn pile has grown out of control and that it is rapidly getting larger. The caller states that her neighbor is chasing the fire with a garden hose. How should you proceed?

   _____
   _____
   _____
   _____
   _____
   _____
   _____

Skill Drills

Skill Drill 3-3: Using a Radio
Test your knowledge of this skill drill by filling in the correct words in the photo captions.

1. Before transmitting, determine that the _____ is clear of any other traffic. Depress the "push-to-talk" (PTT) button and wait at least _____ seconds before speaking.

   Speak _____ the microphone at a(n) _____ angle and hold the microphone 1 to 2 inches from the mouth.

   Speak clearly and keep the message brief. Do not release the _____ button until you have finished speaking.

Skill Drill 3-5: Operating and Answering the Fire Station Telephone and Intercom Systems
Test your knowledge of this skill drill by filling in the correct words in the photo captions.

1. Determine immediately if the caller has an emergency. If there is an emergency, follow your department's _____.

2. If you take the information from the caller, focus on obtaining _____ information. If your station or your unit will be responding to the call, advise the _____ center immediately. Always be prepared to take accurate information or messages for emergency, nonemergency, and _____ calls. Never leave someone on hold for a long time. Always let the caller _____ first.

Incident Command System

Workbook Activities

The following activities have been designed to help you. Your instructor may require you to complete some or all of these activities as a regular part of your fire fighter training program. You are encouraged to complete any activity that your instructor does not assign as a way to enhance your learning in the classroom.

Chapter Review

The following exercises provide an opportunity to refresh your knowledge of this chapter. All questions in this chapter are Fire Fighter II level.

Matching

Match each of the terms in the left column to the appropriate definition in the right column.

_____ 1. Branch
_____ 2. Division
_____ 3. Group
_____ 4. Crew
_____ 5. Staging area
_____ 6. Company officer
_____ 7. Command
_____ 8. Passing command
_____ 9. Transfer command
_____ 10. Safety officer

A. The first component of the ICS
B. An organized group of fire fighters under the leadership of a company officer
C. The officer in charge of a fire department company
D. When the first-arriving company officer directs the next-arriving unit to assume command
E. Position responsible for identifying and evaluating hazards or unsafe conditions at an incident
F. A supervisory level established to manage the span of control above the division, group, or sector level
G. Occurs when one person relinquishes command to another individual
H. Usually refers to companies and/or crews working on the same task or objective
I. Usually refers to companies and/or crews working in the same geographic area
J. A strategically placed area where support resources are held in an organized state of readiness

Multiple Choice

Read each item carefully, and then select the best response.

_____ 1. The only position in the ICS that must always be filled is
   A. safety officer.
   B. incident commander.
   C. liaison officer.
   D. public information officer.

_____ 2. An ICS developed in the 1970s for day-to-day fire department incidents was the
   A. Unified Command System.
   B. FIRESCOPE.
   C. Fire-ground Command System.
   D. Task Force System.

_____ 3. Which position is established when the first-arriving unit arrives on the scene?
   A. Incident commander
   B. Division supervisor
   C. Operations Section Chief
   D. ICS director

_____ 4. When there are overlapping responsibilities between different agencies the ICS may employ a(n)
   A. incident command system.
   B. unity of command.
   C. fire-ground command system.
   D. unified command.

_____ 5. What is the management concept in which each person has only one direct supervisor?
   A. Integrated communications
   B. Unified command
   C. Unity of command
   D. Span of control

_____ 6. The number of subordinates who report to one supervisor at any level within the organization is the
   A. integrated communications.
   B. unified command.
   C. unity of command.
   D. span of control.

_____ 7. What is used to ensure that everyone at an emergency site can transfer information with both their supervisors and subordinates?
   A. Integrated communications
   B. Span of control
   C. Universal communications
   D. Command staff

_____ 8. A standard system of assigning and keeping track of the resources involved in the incident is the
   A. incident resources system.
   B. resource management system.
   C. unified resources system.
   D. operational resources system.

_____ 9. A location close to the incident scene where resources can be prearranged and organized for rapid response is a(n)
   A. resources area.
   B. operations area.
   C. staging area.
   D. rapid intervention area.

_____ 10. The only position that must be filled at every incident is the
   A. incident commander.
   B. safety officer.
   C. liaison officer.
   D. Operations Section Chief.

_____ 11. The incident commander (IC), all direct support staff, and command functions should always be located at the
   A. management post.
   B. incident command post.
   C. Planning Section.
   D. staging area.

_____ 12. The safety officer, liaison officer, and, public information officer are always part of the
   A. command staff.
   B. rapid intervention team.
   C. Operations Section.
   D. ICS general staff.

_____ 13. Which officer has the authority to stop or suspend operations when unsafe situations occur?
   A. Health officer
   B. Liaison officer
   C. Operations officer
   D. Safety officer

_____ 14. The IC's representative and the position responsible for exchanging information with outside agencies or directing people to the proper authority is the
   A. liaison officer.
   B. safety officer.
   C. public information officer.
   D. Planning Section Chief.

_____ 15. The officer responsible for gathering and releasing incident information to the news media is the
   A. liaison officer.
   B. public information officer.
   C. Operations Section Chief.
   D. Public Relations Chief.

_____ 16. The section of the ICS that is responsible for the management of all actions that are directly related to controlling the incident is the
   A. Operations Section.
   B. Planning Section.
   C. Logistics Section.
   D. Administration Section.

_____ 17. The section of the ICS that is responsible for the collection, evaluation, dissemination, and use of information relevant to the incident is the
   A. Operations Section.
   B. Planning Section.
   C. Logistics Section.
   D. Administration Section.

_____ 18. The section of the ICS that is responsible for providing supplies, services, facilities, and materials during the incident is the
   A. Operations Section.
   B. Planning Section.
   C. Logistics Section.
   D. Administration Section.

_____ 19. The section of the ICS that is responsible for any legal issues that may arise during an incident is the
   A. Operations Section.
   B. Planning Section.
   C. Logistics Section.
   D. Finance/Administration Section.

_____ 20. A crew is a group of fire fighters who are working
   A. on their own.
   B. as ICS Section Chiefs.
   C. without apparatus.
   D. outside the ICS.

_____ 21. Companies or crews working in the same geographic area are termed
   A. divisions.
   B. groups.
   C. sectors.
   D. teams.

_____ 22. The exterior sides of a building are generally known as sides
   A. north, east, south, and west.
   B. A, B, C, and D.
   C. one, two, three, and four.
   D. right, left, top, and bottom.

_____ 23. Five units of the same type with an assigned leader is referred to as a(n)
   A. operating unit.
   B. strike team.
   C. branch.
   D. sector.

_____ 24. When one person relinquishes command of an incident and another individual becomes the IC, there is a(n)
   A. transfer of command.
   B. passing command.
   C. operational transfer.
   D. resource transfer.

Fundamentals of Fire Fighter Skills

Labeling
Label the following diagrams with the correct terms.

1. The ICS organization chart.

ICS Organizational Chart

A. _____

B. ____ Officer

C. ____ Officer

D. ____ Officer

E. ____ Section
F. ____ Section
G. ____ Section
H. ____ Section

A. _____
B. _____
C. _____
D. _____
E. _____
F. _____
G. _____
H. _____

Figure 4-5

2. Major functional components of the ICS.

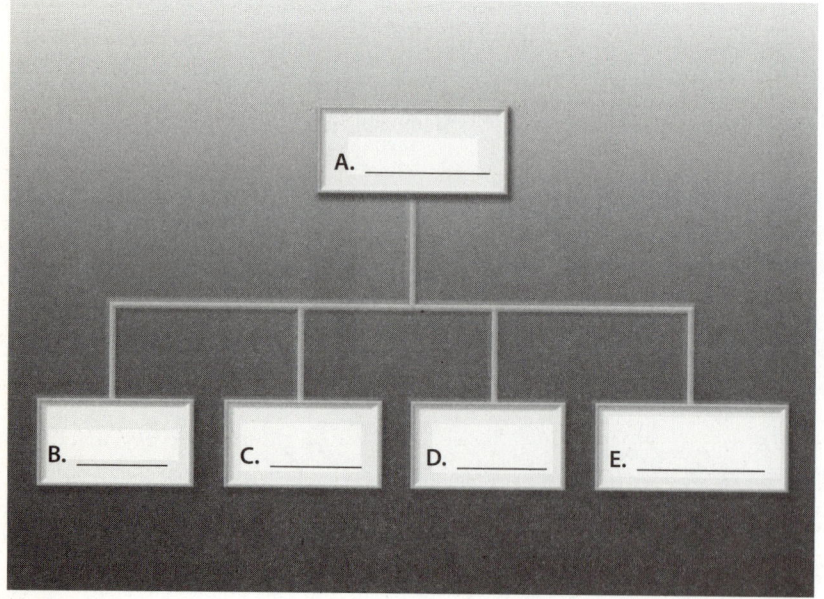

A. _____

B. ____ C. ____ D. ____ E. ____

A. _____
B. _____
C. _____
D. _____
E. _____

Figure 4-9

Vocabulary
Define the following terms using the space provided.

1. Command staff:

   _____
   _____
   _____

2. Incident command system:

   _____
   _____
   _____

3. Incident action plan:

   _____
   _____
   _____

4. Unified command:

   _____
   _____
   _____

5. Division:

   _____
   _____
   _____

6. National Incident Management System:

   _____
   _____
   _____

7. Resource management:

   _____
   _____
   _____

Fill-in

Read each item carefully, and then complete the statement by filling in the missing word(s).

1. An ICS provides a standard _____, _____ and _____ procedure to organize and manage any operation.
2. Planning, supervision, and _____ are key components of an ICS.
3. In the ICS structure, _____ is ultimately responsible for managing an incident and has the _____ to direct all activities.
4. The modular design of ICS allows the _____ to expand based on the needs of the incident.
5. The staging, rehabilitation, and treatment areas are _____ where particular functions take place.
6. The IC is the only position in the ICS that must _____ be filled.
7. A(n) _____ is a fire fighter who serves as a direct assistant to a command officer.
8. The chiefs of the four sections within the ICS are known as the ICS _____.
9. Standard _____ is a strength of the ICS, as it assists in better communication.
10. The primary reason for establishing divisions and groups is to maintain a(n) _____.

True/False

If you believe the statement to be more true than false, write the letter "T" in the space provided.
If you believe the statement to be more false than true, write the letter "F."

1. _____ Prior to the ICS, the organization established to direct operations often varied based on the chief on duty.
2. _____ An ICS provides a standard, professional, organized approach to managing emergency incidents.
3. _____ The ICS is designed to be flexible and modular.
4. _____ The different components of the ICS have different goals and objectives.
5. _____ Each block on an ICS organizational chart refers to an individual.
6. _____ The incident action plan outlines the overall strategy for an emergency incident.
7. _____ The Operations Section is responsible for updating the incident action plan.
8. _____ The Logistics Section Chief reports directly to the Planning Section Chief.
9. _____ The issues addressed by the Finance/Administration Section are usually addressed after the incident.
10. _____ An engine and its crew are considered to be a single resource.

Short Answer

Complete this section with short written answers using the space provided.

1. Identify five important characteristics of an ICS.

2. List the five major components of an ICS organization.

3. Identify the four major functional components within the ICS.

4. Identify the three basic components that apply in all ICS incident and training operations.

5. When an officer relinquishes command, he or she needs to give the new IC a current status report that includes which seven pieces of information?

… CLUES

Across

3. ICS option that allows representatives from multiple jurisdictions and/or agencies to share command authority and responsibility.
5. A prearranged, strategically placed area where support personnel and equipment are in a state of readiness.
8. The only position in the ICS that must always be staffed.
9. The position within the ICS responsible for identifying and evaluating hazardous or unsafe conditions at the scene of the incident.
10. Any combination of single resources assembled for a particular tactical need.
11. The location at the scene where command, coordination, control, and communications are centralized.
14. The section within the ICS responsible for providing facilities, services, and materials for the incident.
16. The section within the ICS responsible for all tactical operations at the incident.
17. The section within the ICS responsible for the collection, evaluation, and dissemination of tactical information related to the incident.

Down

1. The ability of all appropriate personnel at the emergency scene to communicate with their supervisor and their subordinates.
2. An option that can be used by the first-arriving company officer to direct the next-arriving unit to assume command.
4. An organizational level within ICS that divides an incident into geographic areas of operational responsibility.
6. An organized group of fire fighters under the leadership of a company officer.
7. The position within ICS that establishes a point of contact with outside agency representatives.
8. The officer in charge of a fire department company or station.

Word Fun

The following crossword puzzle is an activity provided to reinforce correct spelling and understanding of terminology associated with firefighting. Use the clues provided to complete the puzzle.

12. Alternate terminology used for an organizational level defined by either a geographic or functional assignment.
13. A supervisory level established to manage the span of control above the division, group, or sector level.
15. An organization level within ICS that divides an incident according to functional areas of operation.

Fire Alarms

The following real case scenarios will give you an opportunity to explore the concerns associated with ICS. Read each scenario, and then answer each question in detail.

1. It is a quiet Sunday afternoon when you are dispatched to an old four-story furniture warehouse located in the historical district of your city. Dispatch stated the fire started on the second floor. Upon arrival, you find fire pouring out of the A/D corner of the building, lapping into the third-floor windows. Your Lieutenant states that the preincident plan calls for the first engine to charge the standpipe. The Battalion Chief arrives on the scene, and your Lieutenant prepares to transfer command. How will she proceed?

   _____
   _____
   _____
   _____
   _____
   _____

2. You are the team leader for an exposure crew on a fire in an apartment complex under construction. You are assigned to Division C, and your task is to protect an adjacent apartment building labeled exposure C. The Division D Supervisor directs your crew to move your line and protect exposure D. How should you proceed?

   _____
   _____
   _____
   _____
   _____
   _____

Fire Behavior

Workbook Activities

The following activities have been designed to help you. Your instructor may require you to complete some or all of these activities as a regular part of your fire fighter training program. You are encouraged to complete any activity that your instructor does not assign as a way to enhance your learning in the classroom.

Chapter Review

The following exercises provide an opportunity to refresh your knowledge of this chapter.

Matching

Match each of the terms in the left column to the appropriate definition in the right column.

_____ 1. Decay
_____ 2. Conduction
_____ 3. Endothermic
_____ 4. Plume
_____ 5. Hypoxia
_____ 6. Convection
_____ 7. Gas
_____ 8. Nuclear fission
_____ 9. Oxidation
_____ 10. Radiation

A. A state of inadequate oxygenation of the blood and tissue
B. The process in which oxygen combines chemically with another substance to create a new compound
C. The phase of fire where the fire is running out of fuel
D. Created by splitting the nucleus of an atom
E. Heat transfer by circulation within a medium
F. Transfer of heat through the emission of energy in the form of invisible waves
G. The process of transferring heat through matter by movement of the kinetic energy from one particle to another
H. Reactions that absorb heat or require heat to be added
I. One of the three phases of matter
J. A thermal column

Multiple Choice

Read each item carefully, and then select the best response.

_____ 1. The basic units of measure in the United States are the
   A. American system of units.
   B. British system of units.
   C. scientific system of units.
   D. International System of Units.

_____ 2. The basic units of measure in Canada are the
   A. American system of units.
   B. British system of units.
   C. scientific system of units.
   D. International System of Units.

_____ 3. The four conditions that must be present for fire to take place are represented in the
   A. fire tetrahedron.
   B. fire square.
   C. fire triangle.
   D. fire rectangle.

_____ 4. The actual material that is consumed by a fire is
   A. oxygen.
   B. carbon.
   C. fuel.
   D. base.

_____ 5. When substances are heated, they tend to change from a
   A. solid to a liquid state, and then to a gas state.
   B. gas to a liquid state, and then to a solid state.
   C. solid to a liquid state, and then to a smoke state.
   D. liquid to a gas state, and then to a vapor state.

_____ 6. The decomposition of a material brought about by heat in the absence of oxygen is called
   A. evaporation.
   B. dehydration.
   C. decomposition.
   D. pyrolysis.

_____ 7. A thin piece of wood burns quickly due to its
   A. mass.
   B. composition.
   C. weight-to-mass ratio.
   D. large surface area.

_____ 8. Flammable limits will vary
   A. from one fuel to another.
   B. with the amount of energy present.
   C. with the vapor pressure.
   D. with the vapor density.

_____ 9. Incomplete combustion produces
   A. pure air.
   B. solids.
   C. smoke.
   D. oxidizers.

_____ 10. The key to preventing a BLEVE is to
   A. flush the spill.
   B. ventilate the area.
   C. cool the top of the tank.
   D. apply an oxidizing agent.

_____ 11. When two materials rub together and produce friction, they create
   A. light energy.
   B. chemical energy.
   C. mechanical energy.
   D. chemical energy.

_____ 12. A very rapid chemical process that combines oxygen with another substance and results in the release of heat and light is called
   A. oxidization.
   B. combustion.
   C. pyrolysis.
   D. decomposition.

_____ 13. Particles, vapors, and gases are the three major components of
   A. fumes.
   B. silt.
   C. smoke.
   D. exhaust.

_____ 14. The movement of heat through a fluid medium such as air or a liquid is
   A. convection.
   B. endothermic.
   C. exothermic.
   D. conduction.

_____ 15. The transfer of heat energy in the form of invisible waves is called
   A. radiation.
   B. oxidization.
   C. volatility.
   D. transpiration.

_____ 16. The lowest temperature at which a liquid produces a flammable vapor is the
   A. flame point.
   B. fire point.
   C. ignition temperature.
   D. flash point.

_____ 17. The weight of a gaseous fuel is the
   A. gas mass.
   B. vapor density.
   C. explosive limit.
   D. BLEVE.

_____ 18. Which class of fires involve flammable or combustible liquids such as gasoline?
   A. Class A fires
   B. Class B fires
   C. Class C fires
   D. Class D fires

_____ 19. Which class of fires involve burning metals?
   A. Class A fires
   B. Class B fires
   C. Class C fires
   D. Class D fires

_____ 20. When hot gases flow across a ceiling and cooler air stays closer to the floor. there is
   A. thermal differentiation.
   B. thermal division.
   C. thermal layering.
   D. thermal balance.

_____ 21. In which phase of fire does additional fuel become involved in the fire?
   A. Ignition phase
   B. Growth phase
   C. Fully developed phase
   D. Decay phase

_____ 22. What is the primary factor in upward fire spread?
   A. Conduction
   B. Convection
   C. Radiation
   D. Hypoxia

Labeling

Label the following diagram with the correct terms.

1. The fire tetrahedron.

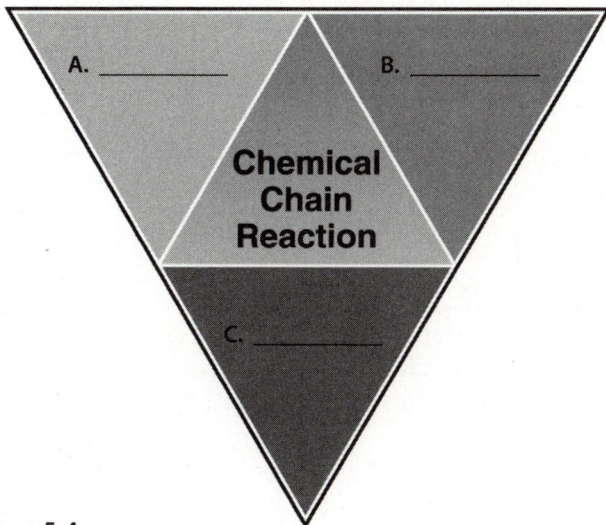

Figure 5-4

A. _____

B. _____

C. _____

Vocabulary

Define the following terms using the space provided.

1. Lower flammable limit:
   _____
   _____
   _____

2. BLEVE:
   _____
   _____
   _____

3. Fire triangle:
   _____
   _____
   _____

4. Ignition temperature:
   _____
   _____
   _____

5. Thermal layering:
   _____
   _____
   _____

6. Flashover:
   _____
   _____
   _____

7. Flash point:
   _____
   _____
   _____

Fill-in

Read each item carefully, and then complete the statement by filling in the missing word(s).

1. In the British system, distance is measured in _____ and _____; liquid volume is measured in _____; temperature is measured in degrees _____; and pressure is measured in _____.

2. In the metric system, distance is measured in _____; liquid volume is measured in _____; temperature is measured in degrees _____; and pressure is measured in _____.

3. Matter exists in three states: _____, _____, and _____.

4. A fire involving a liquid fuel can be extinguished by shutting off the _____ of fuel, or using _____ to exclude oxygen from the fuel.

5. Reactions that produce heat are referred to as _____ reactions.

6. Carbon _____ is deadly in small quantities.

7. The amount of liquid that is vaporized when it is heated relates to the _____ of the liquid.

8. The _____ temperature is the temperature at which the fuel-air mixture produced by a liquid will spontaneously ignite.

9. When heated, most plastics will pyrolyze to _____ products that are both flammable and toxic.

10. Carpet is readily ignitable by _____, even when it is some distance away from the fire.

True/False

If you believe the statement to be more true than false, write the letter "T" in the space provided.
If you believe the statement to be more false than true, write the letter "F."

1. _____ The three basic ingredients required to create a fire are fuel, oxygen, and air.
2. _____ The thinner wood is cut, the faster it ignites and burns.
3. _____ Gas has neither independent shape nor volume and tends to expand indefinitely.
4. _____ Mechanical, electrical, and chemical energy can be converted to heat.
5. _____ Smoke often contains a wide variety of hot, toxic gases.
6. _____ The flash point is the lowest temperature at which a liquid produces enough vapor to sustain a continuous fire.
7. _____ Fire suppression efforts can greatly influence the spread of fire.
8. _____ Flashover occurs when the temperature in a room becomes high enough to ignite the room's contents.
9. _____ A backdraft can occur when oxygen is introduced into a closed, superheated room.
10. _____ The growth of a fire is limited by the amount of fuel or oxygen available.

Short Answer

Complete this section with short written answers using the space provided.

1. Identify the hazards associated with smoke.

2. Identify the four basic methods of extinguishing fires.

3. List the three conditions that must be present for a vapor-air mixture to ignite.

4. Identify the four phases of fire.

5. Describe fire conditions leading to a flashover.

6. List three signs of an impending backdraft.

52 FUNDAMENTALS OF FIRE FIGHTER SKILLS

CLUES

Across

2 The emission and propagation of energy through matter or space by means of electromagnetic disturbances that display both wavelike and particle-like behavior.
4 An airborne particulate product of incomplete combustion suspended in gases.
7 A state of inadequate oxygenation of the blood and tissue.
8 The ready ability of a substance to produce combustible vapors.
9 The minimum temperature at which a fuel will ignite in air and continue to burn.
10 The type of heat energy created by friction.

Down

1 The phase of fire development where the fire is spreading beyond the point of origin and beginning to involve other fuels in the immediate area.
3 The chemical decomposition of a compound into one or more other substances by heat alone.
5 A chemical reaction initiated by combining an element with oxygen, resulting in the form of the element or one of its compounds.
6 All combustible materials.

Word Fun

The following crossword puzzle is an activity provided to reinforce correct spelling and understanding of terminology associated with firefighting. Use the clues provided to complete the puzzle.

Fire Alarms

The following real case scenarios will give you an opportunity to explore the concerns associated with fire behavior. Read each scenario, and then answer each question in detail.

1. Your engine company is dispatched to a two-story, single-family home in a newer development. Upon arrival, you find there is no flame visible and the window glass is smoke-stained with a lot of heat inside. Upon investigation, you see smoke emanating under pressure from cracks. The smoke is puffing and being drawn back like it is breathing. How will you proceed?

   _____
   _____
   _____
   _____
   _____
   _____

2. It is 3:00 in the afternoon when your engine company is dispatched to a kitchen fire in a multifamily condominium unit. You and your Lieutenant enter the unit, and it appears that the kitchen has flashed over. You are on the nozzle and your Lieutenant tells you that this is a hot fire and not to disrupt the thermal balance. How should you proceed?

   _____
   _____
   _____
   _____
   _____
   _____

Building Construction

Workbook Activities

The following activities have been designed to help you. Your instructor may require you to complete some or all of these activities as a regular part of your fire fighter training program. You are encouraged to complete any activity that your instructor does not assign as a way to enhance your learning in the classroom.

Chapter Review

The following exercises provide an opportunity to refresh your knowledge of this chapter.

Matching

Match each of the terms in the left column to the appropriate definition in the right column.

_____ 1. Fire partition   A. Chipping or pitting of concrete or masonry surfaces
_____ 2. Live load   B. Describes how quickly a material will conduct heat
_____ 3. Combustibility   C. A natural material composed of calcium sulfate and water molecules
_____ 4. Gypsum   D. How a building is used
_____ 5. Fire window   E. Built-up unit of construction materials set in mortar
_____ 6. Load-bearing walls   F. Interior walls extending from the floor to the underside of the floor above
_____ 7. Masonry   G. Determines whether a material will burn
_____ 8. Spalling   H. Walls designed for structural support
_____ 9. Thermal conductivity   I. The weight of the building contents
_____ 10. Occupancy   J. A glass block assembly with a fire-resistive rating

Multiple Choice

Read each item carefully, and then select the best response.

_____ 1. When selecting materials for building construction, architects most often place a priority on
   A. price and ease of construction.
   B. functionality and aesthetics.
   C. availability of materials and price.
   D. durability and maintenance expenses.

_____ 2. Which of the following materials will expand at extremely high temperatures, conducts heat well, and loses its strength as the temperature increases?
   A. Steel
   B. Concrete
   C. Masonry
   D. Gypsum

_____ 3. Which type of glass consists of a thin sheet of plastic between two sheets of glass?
   A. Tempered glass
   B. Wired glass
   C. Laminated glass
   D. Glass blocks

_____ 4. What is the most commonly used building material?
   A. Steel
   B. Concrete
   C. Aluminum
   D. All of the above

_____ 5. Thin sheets of wood that are glued together are called
   A. wood panels.
   B. laminated wood.
   C. wood trusses.
   D. wooden beams.

_____ 6. Which synthetic material is found in many products and may be transparent or opaque, stiff or flexible, tough or brittle?
   A. Glass
   B. Plastic
   C. Aluminum
   D. Copper

_____ 7. Thermoplastic materials melt and drip when exposed to high temperatures, some even as low as
   A. 100 °F.
   B. 250 °F.
   C. 500 °F.
   D. 650 °F.

_____ 8. Thermoset materials are fused by heat and
   A. will melt in low temperatures.
   B. will not burn.
   C. will always maintain their strength.
   D. will not melt.

_____ 9. Buildings having masonry exterior walls and interior walls, floors, and roofs made of wood are considered to be
   A. Type II construction.
   B. Type III construction.
   C. Type IV construction.
   D. Type V construction.

_____ 10. What is the length of time that a building or building components can withstand a fire before igniting called?
   A. Pyrolysis
   B. Thermal resistance
   C. Fire retardance
   D. Fire resistance

11. Which type of building construction provides the highest degree of safety and is usually made of reinforced concrete and protected steel-frame construction?
    A. Type I
    B. Type II
    C. Type IV
    D. Type V

12. Which type of building construction has two separate fire loads?
    A. Type II
    B. Type III
    C. Type IV
    D. Type V

13. What is another term for wood-frame construction?
    A. Type I
    B. Type II
    C. Type IV
    D. Type V

14. The weight of the building is called the
    A. live load.
    B. total load.
    C. dead load.
    D. structural load.

15. Lightweight and heavy timber construction are examples of
    A. Type I construction.
    B. window frames.
    C. wood floor structures.
    D. roofs.

16. Pitched, curved, and flat are types of
    A. awnings.
    B. roofs.
    C. stairways.
    D. rafters.

17. How many layers will a typical built-up roof covering have?
    A. 3
    B. 5
    C. 7
    D. 9

18. Trusses are used extensively in support systems for
    A. both floors and roofs.
    B. floors.
    C. roofs.
    D. roofs, with the exception of flat roofs.

19. A steel bar joist is an example of a
    A. bowstring truss.
    B. pitched chord truss.
    C. parallel chord truss.
    D. flat chord truss.

20. Walls that are constructed on the line between two properties and are shared by a building on each side of the line are called
    A. fire walls.
    B. fire partitions.
    C. curtain walls.
    D. party walls.

_____ 21. Fire doors and fire windows are rated for a particular duration of
   A. heat resistance to controlled temperatures.
   B. internal temperature compliance.
   C. standard fire resistance.
   D. fire resistance to a standard test fire.

_____ 22. The exposed interior surfaces of a building are commonly referred to as the
   A. interior finish.
   B. building surfaces.
   C. structural surfaces.
   D. structural finish.

Vocabulary

Define the following terms using the space provided.

1. Dead load:

_____
_____
_____

2. Thermoplastic materials:

_____
_____
_____

3. Interior finish:

_____
_____
_____

4. Load-bearing wall:

_____
_____
_____

5. Bowstring truss:

_____
_____
_____

6. Balloon-frame construction:

_____
_____
_____

Fill-in

Read each item carefully, and then complete the statement by filling in the missing word(s).

1. The term _____ refers to how a building is used.
2. A(n) _____ helps prevent the spread of a fire from one side to the other side of the wall.
3. When wood is exposed to high temperatures, its strength can be decreased through the process of _____.
4. Type _____ is the most fire-resistive category of building construction.
5. Fire severity in a Type II building is determined by the _____.
6. Type _____ building construction is the most commonly used type of construction today.
7. _____-frame construction is used for almost all modern wood-frame construction.
8. A(n) _____ chord truss is typically used to support a sloping roof.
9. A building with a(n) _____ will have a distinctive curved roof.

True/False

If you believe the statement to be more true than false, write the letter "T" in the space provided.
If you believe the statement to be more false than true, write the letter "F."

1. _____ Aluminum is more expensive than, and not as strong as, steel.
2. _____ Concrete is one of the most commonly used building materials.
3. _____ The structural components and building contents in Type III, Type IV, and Type V buildings will burn.
4. _____ Type V building construction provides the highest degree of safety.
5. _____ Aluminum floors are common in fire-resistive construction.
6. _____ The support systems for most flat roofs are constructed of aluminum.
7. _____ Trusses are used in support systems for both floors and roofs.
8. _____ Most doors are constructed of aluminum.
9. _____ Fire doors and fire windows are rated for a particular duration of fire resistance to a standard fire test.
10. _____ The entire structure of a manufactured (mobile) home can be destroyed by fire within a few minutes.

Short Answer

Complete this section with short written answers using the space provided.

1. Identify and briefly describe the five types of building construction.

_____
_____
_____
_____
_____

Chapter 6: Building Construction 59

2. List the four key factors that affect building materials under fire.

   _____
   _____
   _____
   _____

3. List and briefly describe the five factors that affect how fast wood ignites, burns, and decomposes.

   _____
   _____
   _____
   _____
   _____
   _____

4. List the seven major components of a building.

   _____
   _____
   _____
   _____
   _____
   _____
   _____

5. Identify and briefly describe the five NFPA 80 designations for fire doors and fire windows.

   _____
   _____
   _____
   _____
   _____

6. Briefly describe gusset plates, and how they respond when heated.

   _____
   _____
   _____

Fundamentals of Fire Fighter Skills

Clues

Word Fun

The following crossword puzzle is an activity provided to reinforce correct spelling and understanding of terminology associated with firefighting. Use the clues provided to complete the puzzle.

Across

2. Glass manufactured with a thin vinyl core covered by glass on each side of the core.
4. Glass made by molding glass around a special wire mesh.
6. Walls constructed on the line between two properties.
11. Describes how quickly a material will conduct heat.
12. The purpose for which a building, or part thereof, is used or intended to be used.
16. Horizontal roofs often found on commercial or industrial occupancies.
17. Built-up unit of construction or combination of materials such as brick, clay tiles, or stone set in mortar.

Down

1. Interior walls extending from the floor to the underside of the floor above.
2. The weight of the building contents.
3. A lightweight steel truss used as a horizontal beam.
5. The weight of a building.
7. Fire-rated assemblies used to enclose vertical openings such as stairwells, elevator shafts, and chases for building utilities.
8. Chipping or pitting of concrete or masonry surfaces.
9. The destructive distillation of organic compounds in an oxygen-free environment that converts the organic matter into gases, liquids, and char.
10. A roof with sloping or inclined surfaces.
13. A wall with a fire-resistive rating and structural stability that separates buildings to prevent the spread of fire.
14. Joists that are mounted in an inclined position to support a roof.
15. A naturally occurring material composed of calcium sulfate and water molecules.

Fire Alarms

The following real case scenarios will give you an opportunity to explore the concerns associated with building construction. Read each scenario, and then answer each question in detail.

1. It is 10:00 on Friday night when your engine is dispatched to a fire at a bowling alley. Upon arrival, you find a masonry building with a curved roof and the rear exterior wall leaning out. The fire has involved the office area in the rear of the building. How will you proceed?

   _____
   _____
   _____
   _____
   _____
   _____

2. Your ladder truck company is dispatched to a structure fire at a three-story Type III construction nursing home. The fire is reported in the kitchen. You start to think about the contents and the building construction. What are your concerns?

   _____
   _____
   _____
   _____
   _____
   _____
   _____

Portable Fire Extinguishers

Workbook Activities

The following activities have been designed to help you. Your instructor may require you to complete some or all of these activities as a regular part of your fire fighter training program. You are encouraged to complete any activity that your instructor does not assign as a way to enhance your learning in the classroom.

Chapter Review

The following exercises provide an opportunity to refresh your knowledge of this chapter.

Matching

Match each of the terms in the left column to the appropriate definition in the right column.

_____ 1. Hydrostatic testing   A. A retaining device that breaks when the locking mechanism is released

_____ 2. Saponification   B. An extinguishing agent used in dry chemical fire extinguishers that can be used on Class A, B, and C fires

_____ 3. Aqueous film-forming foam   C. The initial stage of a fire

_____ 4. Fire load   D. The process of converting the fatty acids in cooking oils or fats to soap or foam

_____ 5. Tamper seal   E. Periodic testing of an extinguisher to verify it has sufficient strength to withstand internal pressures

_____ 6. Incipient   F. An extinguishing agent used on Class B fires that forms a foam layer over the liquid and stops the production of flammable vapors

_____ 7. Extinguishing agent   G. The weight of combustibles in a fire area or on a floor in buildings and structures

_____ 8. Ammonium phosphate   H. A material used to stop the combustion process

_____ 9. Lob   I. A gauge on a pressurized portable fire extinguisher that indicates the internal pressure of the expellant

_____ 10. Pressure indicator   J. A method of discharging extinguishing agent in an arc to avoid splashing or spreading the burning fuel

CHAPTER 7

Multiple Choice
Read each item carefully, and then select the best response.

_____ 1. Class A fire extinguishers include a number. This number is related to the
   A. type of fuel the fire extinguisher can extinguish.
   B. size of the discharge field.
   C. approximate area of burning fuel the fire extinguisher can extinguish.
   D. amount of water the fire extinguisher holds.

_____ 2. The safest and surest way to extinguish a Class C fire is to turn off the power and
   A. treat it like a Class A or B fire.
   B. treat it like a Class D fire.
   C. treat it like a Class K fire.
   D. treat it like a Class A, B, or D fire.

_____ 3. All fires require
   A. fuel, heat, and oxygen.
   B. fuel and oxygen.
   C. an ignition source.
   D. fuel, heat, oxygen, and carelessness.

_____ 4. What is the only dry chemical extinguishing agent rated as suitable for Class A fires?
   A. Potassium chloride
   B. Potassium bicarbonate
   C. Ammonium phosphate
   D. Ammonium bicarbonate

_____ 5. Disposable dry chemical extinguishers are
   A. refillable.
   B. not refillable.
   C. effective on Class K fires.
   D. not usable in freezing temperatures.

_____ 6. Electrical rooms should have extinguishers that are approved for use on
   A. Class K fires.
   B. Class A fires.
   C. Class B fires.
   D. Class C fires.

_____ 7. Carbon dioxide extinguishers have relatively short discharge ranges of:
   A. 1 to 3 feet.
   B. 3 to 8 feet.
   C. 10 to 15 feet.
   D. 15 to 30 feet.

_____ 8. Which lever is used to discharge the agent from a portable fire extinguisher?
   A. Trigger
   B. Nozzle
   C. Cylinder
   D. Handle

9. The sodium chloride-based extinguishing agent that is used in portable fire extinguishers can be
   A. stored in liquid form.
   B. harmful to the environment.
   C. used in all portable fire extinguishers.
   D. applied by hand.

10. Self-expelling agents do not require
    A. regular maintenance.
    B. tamper seals on the cylinders.
    C. a separate gas cartridge.
    D. maintenance personnel to be specially trained in their use.

11. Carbon dioxide extinguishers are not recommended for
    A. Class B fires.
    B. Class C fires.
    C. outdoor use.
    D. use in kitchens or laboratories.

12. Vegetable oil fires are classified as
    A. Class A fires.
    B. Class B fires.
    C. Class C fires.
    D. Class K fires.

13. Fires that have not spread past their point of origin are
    A. called introductory fires.
    B. called incipient fires.
    C. easily suppressed.
    D. most often suppressed with an exterior attack.

14. Where is the extinguishing agent in a fire extinguisher stored?
    A. Trigger
    B. Nozzle
    C. Cylinder
    D. Handle

15. Class B fire extinguishers can be identified by the
    A. solid red square.
    B. solid blue square.
    C. solid red circle.
    D. solid yellow five-point star.

16. An extinguisher rated 40-B should be able to control a liquid pan fire
    A. with a surface area of 40 square feet.
    B. within 40 seconds.
    C. 40 times more effectively than a normal Class B extinguisher.
    D. 4 times more effectively than a normal Class B extinguisher.

17. Fire extinguishers weighing more than 40 pounds (18.14 kilograms) should be mounted so that the top of the extinguisher is not more than:
    A. 5 feet above the floor.
    B. 3 feet above the floor.
    C. 2 feet above the floor.
    D. 6 feet above the floor.

18. The three risk classifications according to the amount and type of combustibles that are present in an area are
    A. light, ordinary, and extra hazards.
    B. light, medium, and extra hazards.
    C. normal, light, and extra hazards.
    D. normal, average, and extra hazards.

_____ 19. Carbon dioxide is a gas that is 1.5 times heavier than
   A. water.
   B. most extinguishing agents.
   C. air.
   D. carbon monoxide.

_____ 20. Two factors to consider when determining the number and types of fire extinguishers that should be placed in each area of occupancy are the
   A. quality and quantities of the fuels.
   B. fuels and ignition sources.
   C. types of fuels and area traffic.
   D. types and quantities of the fuels.

_____ 21. Class D fires are most often encountered in
   A. kitchens or restaurants.
   B. offices or schools.
   C. machine or repair shops.
   D. hayfields or woodland areas.

_____ 22. The best method of transporting a hand-held portable fire extinguisher depends on the
   A. training level of the operator.
   B. size, weight, and design of the extinguisher.
   C. type of extinguishing agent.
   D. size and type of fire.

Labeling

Label the following diagram with the correct terms.

1. Basic parts of the portable fire extinguisher.

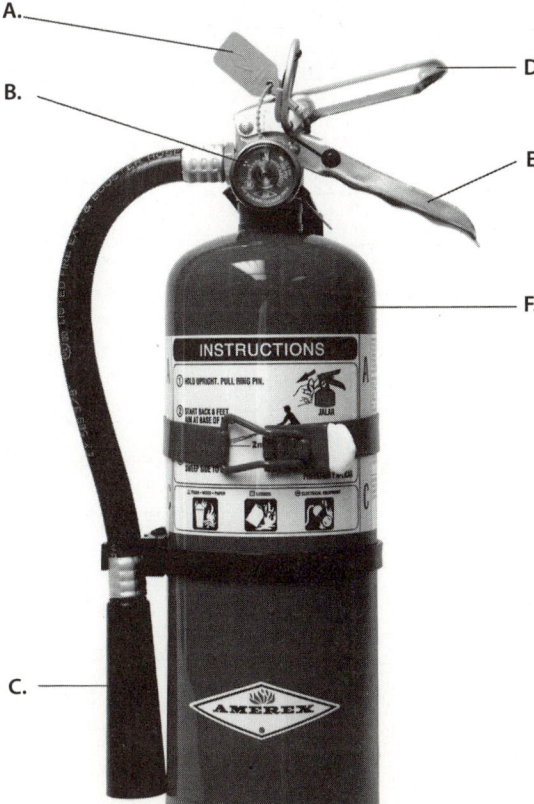

A. _____
B. _____
C. _____
D. _____
E. _____
F. _____

Figure 7-19

Vocabulary

Define the following terms using the space provided.

1. Cartridge/cylinder fire extinguisher:

2. Extra hazard locations:

3. Underwriters Laboratories, Inc. (UL):

4. Rapid oxidation:

5. Multipurpose dry chemical extinguisher:

6. Class K fires:

7. Extinguishing agent:

_____

_____

_____

8. Polar solvent:

_____

_____

_____

Fill-in

Read each item carefully, and then complete the statement by filling in the missing word(s).

1. The _____ is the discharge orifice of a portable fire extinguisher.
2. A(n) _____ is a volatile or gaseous fire extinguishing agent that does not leave a residue when it evaporates.
3. A Class _____ fire is one that involves wood, cloth, paper, grass, hay, or straw.
4. _____ is a colorless, odorless, electronically nonconductive, inert gas that puts out Class B and C fires by displacing oxygen and cooling the fuel.
5. The _____ temperature is the minimum temperature at which a substance will burn.
6. Class _____ labels are represented by a solid red square.
7. A fire extinguisher must be _____ after each and every use.
8. _____ extinguishers have a short discharge range.
9. _____ extinguishers are used primarily outdoors for fighting brush and grass fires.
10. An individual with _____ training should be able to use most fire extinguishers safely and effectively.

True/False

If you believe the statement to be more true than false, write the letter "T" in the space provided.
If you believe the statement to be more false than true, write the letter "F."

1. _____ Fire extinguishers can contain several hundred pounds of extinguishing agent.
2. _____ The primary disadvantage of fire extinguishers is their effectiveness.
3. _____ A Class B extinguisher with a 10-B rating indicates that it is capable of extinguishing the highest level of Class B fires.
4. _____ The bottom of an extinguisher should be mounted at least 4 inches (10.2 centimeters) above the floor.
5. _____ Halon 1211 should be used only when its clean properties are essential.
6. _____ Class K extinguishers are identified by a solid yellow five-point star.
7. _____ "Press the trigger" is the first step of PASS.
8. _____ Most offices or classrooms would be examples of light hazard areas.
9. _____ All fire fighters are trained to perform fire extinguisher maintenance.
10. _____ Time intervals for testing requirements for an extinguisher are based on construction material and vessel type.

Short Answer
Complete this section with short written answers using the space provided.

1. Identify the six basic parts of most hand-held portable fire extinguishers.

   _____
   _____
   _____
   _____
   _____
   _____

2. Identify and describe the seven types of fire extinguishers.

   _____
   _____
   _____
   _____
   _____
   _____
   _____
   _____

3. Identify the six basic steps in extinguishing a fire with a portable fire extinguisher.

   _____
   _____
   _____
   _____
   _____
   _____

4. Describe the PASS acronym used for fire extinguisher operations.

   _____
   _____
   _____

5. Identify four common indicators that a fire extinguisher needs maintenance.

   _____
   _____
   _____
   _____

Fundamentals of Fire Fighter Skills

CLUES

Across

1. Fires that involve energized electrical equipment, where the electrical conductivity of the extinguishing media is of importance.
6. The weight of combustibles in a fire area or on a floor in buildings or structures, including either contents or building parts, or both.
7. Fires involving combustible metals.
12. A chemical process that occurs when a fuel is combined with oxygen, resulting in the formation of ash or other waste products and the release of energy as heat and light.
13. A material used to stop the combustion process.
15. The grip used for holding and carrying a portable fire extinguisher.
17. A device that locks an extinguisher's trigger to prevent its accidental discharge.

Down

2. A fire extinguisher in which the agents have sufficient vapor pressure at normal operating temperatures to expel themselves.
3. Periodic testing of an extinguisher to verify it has sufficient strength to withstand internal pressures.
4. The process of converting the fatty acids in cooking oils or fats to soap or foam.
5. One of the water-soluble flammable liquids.
8. A retaining device that breaks when the locking mechanism is released.
9. A water-based fire extinguisher that uses an alkali metal salt as a freezing-point depressant.
10. The initial stage of a fire.
11. The tapered discharge nozzle of a carbon dioxide-type fire extinguisher.
14. The discharge orifice of a portable fire extinguisher.
16. Fires involving combustible cooking media such as vegetable oils, animal oils, or fats.

Word Fun

The following crossword puzzle is an activity provided to reinforce correct spelling and understanding of terminology associated with firefighting. Use the clues provided to complete the puzzle.

Fire Alarms

The following real case scenarios will give you an opportunity to explore the concerns associated with portable fire extinguishers. Read each scenario, and then answer each question in detail.

1. It is 8:00 on Saturday morning, and your Lieutenant is conducting the morning shift meeting. Saturday's duties are to do a detailed inspection of all equipment. The Lieutenant assigns you to inspect all of the extinguishers on the apparatus and report any that need maintenance. How should you proceed?

   _____
   _____
   _____
   _____
   _____
   _____
   _____

2. It is 7:00 on a Thursday evening when your engine is dispatched to a chimney fire. Upon arrival, you find a two-story, wood-frame residential structure with nothing showing. Upon further investigation, you confirm there is a fire in the chimney. Your Lieutenant tells you to extinguish the fire in the fireplace with an extinguisher. How should you proceed?

   _____
   _____
   _____
   _____
   _____
   _____
   _____

Skill Drills

Skill Drill 7-2: Operating a Carbon Dioxide Extinguisher
Test your knowledge of this skill drill by placing the photos below in the correct order. Number the first step with a "1," the second step with a "2," and so on.

_____ Aim the nozzle of the fire extinguisher at the base of the fire and squeeze the trigger.

_____ Back away from the fire. Overhaul the fire to ensure that it is completely extinguished.

_____ Give the trigger a quick squeeze to ensure that the extinguisher is operational and the agent discharges properly.

_____ Sweep the extinguishing agent from side to side, continuing to aim at the base of the flames. Continue to use the extinguisher until the fire is out or the extinguisher is empty.

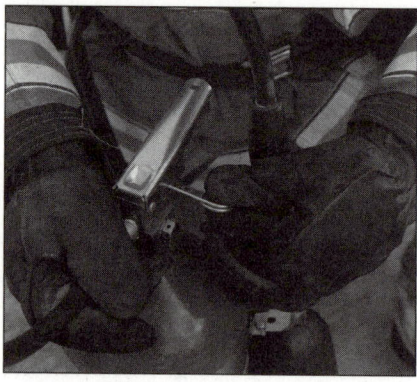

_____ Pull the pin on the handle.

_____ Remove the horn or nozzle from the secured position on the extinguisher and aim in the direction of your approach.

_____ Approach the fire with an exit to your back. Never let the fire get between you and the exit.

Skill Drill 7-3: Attacking a Class A Fire with a Stored-Pressure Water-Type Fire Extinguisher
Test your knowledge of this skill drill by filling in the correct words in the photo captions.

1. Begin to attack the fire from a(n) _____ distance.

2. Aim the stream _____ at the _____ of the flames. _____ the nozzle back and forth, moving _____ as the fire goes out.

3. After the flames are out, position your finger _____ _____ of the nozzle to create a spray and soak the fuel.

4. Break apart the fuel with a stick and apply the _____ _____ to any smoldering, smoking, or glowing surfaces.

5. Apply additional water spray to prevent the fire from _____.

Skill Drill 7-5: Attacking a Class B Flammable Liquid Fire with a Dry Chemical Fire Extinguisher
Test your knowledge of this skill drill by placing the photos below in the correct order.
Number the first step with a "1," the second step with a "2," and so on.

_____ Sweep the nozzle back and forth across the surface of the flammable liquid.

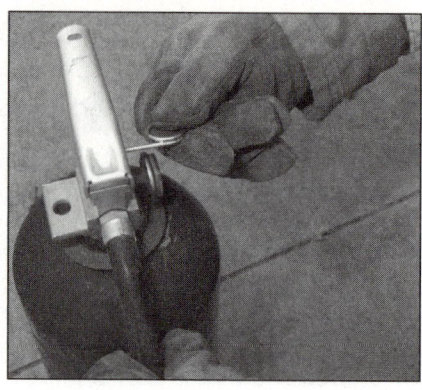

_____ Pull the pin on the handle.

_____ Discharge the stream at the base of the flame, starting at the near edge of the fire and working toward the back.

_____ Look for hot or smoldering objects that could reignite the liquid.

_____ Do not aim the initial discharge into the liquid at close range.

_____ Check the pressure gauge to ensure that the extinguisher is properly charged.

_____ Begin fighting the fire from a safe distance.

Skill Drill 7-7: Using a Wet Chemical Fire Extinguisher
Test your knowledge of this skill drill by filling in the correct words in the photo captions.

1. Begin to apply the agent to the deep fat fryer from a(n) _____ distance.

Do not direct the agent stream _____ into the burning liquid. _____ the extinguishing agent, which is expelled as a(n) _____, lightly onto the burning surface to create a(n) _____ blanket.

2. Continue to apply the agent until the _____ has extinguished all flames.

3. Do not _____ the foam blanket even after all flames have died down. If reignition occurs, _____ these steps.

Fire Fighter Tools and Equipment

Workbook Activities

The following activities have been designed to help you. Your instructor may require you to complete some or all of these activities as a regular part of your fire fighter training program. You are encouraged to complete any activity that your instructor does not assign as a way to enhance your learning in the classroom.

Chapter Review

The following exercises provide an opportunity to refresh your knowledge of this chapter.

Matching

Match each of the terms in the left column to the appropriate definition in the right column.

_____ 1. Vertical ventilation
_____ 2. Sledgehammer
_____ 3. Battering ram
_____ 4. Hydraulic spreader
_____ 5. Size-up
_____ 6. Box-end wrench
_____ 7. Thermal imaging device
_____ 8. Roofman's hook
_____ 9. Rapid intervention
_____ 10. Multipurpose hook

A. A long, heavy hammer that requires the use of both hands
B. A lightweight hand-operated tool that can produce up to 10,000 pounds of prying and spreading force
C. A tool made of hardened steel with handles on the sides used to force doors and to breach walls
D. The process of making openings so that the smoke, heat, and gases can escape vertically from a structure
E. A hand tool with a closed end used to tighten or loosen bolts
F. A long pole with a wooden or fiberglass handle and a metal hook on one end used for pulling
G. The ongoing observation and evaluation of factors that are used to develop objectives, strategy, and tactics for fire suppression
H. A minimum of two fully equipped personnel on site, in a ready state, for immediate rescue of injured or trapped fire fighters
I. Electronic devices that detect differences in temperature based on infrared energy and then generate images based on those data
J. A long pole with a solid metal hook used for pulling

CHAPTER 8

Multiple Choice

Read each item carefully, and then select the best response.

_____ 1. The best piece of equipment or tool for a fire fighter's duties is
   A. the newest equipment on the market.
   B. technologically advanced over regular tools.
   C. the one that is most effective and efficient for the job.
   D. usually more expensive than regular tools.

_____ 2. Which of the following is not considered part of a fire fighter's personal protective equipment (PPE)?
   A. Boots
   B. Approved firefighting gloves
   C. Approved firefighting prying tool
   D. Personal alert safety system

_____ 3. Which of the following is considered a tool for rotating?
   A. Claw bar
   B. Ceiling hook
   C. Axe
   D. Screwdriver

_____ 4. Which of the following tools is most often used in vehicular crashes to gain access to a victim who needs care?
   A. Spring-loaded center punch
   B. Sledgehammer
   C. Chainsaw
   D. Crowbar

_____ 5. Which of the following is not a mechanical saw?
   A. Chain saw
   B. Rotary saw
   C. Reciprocating saw
   D. Hacksaw

_____ 6. After use, all hand tools should be completely cleaned and
   A. scientifically tested.
   B. inspected.
   C. sharpened.
   D. placed in the tool cabinet.

_____ 7. All power equipment should be left in a ready state, which includes
   A. fuel tanks filled completely with fresh fuel.
   B. hydraulic hoses, if applicable, cleaned and inspected.
   C. the removal and replacement of any dull or damaged blades.
   D. all of the above.

_____ 8. Which of the following is considered a tool for striking?
   A. Crowbar
   B. Drywall hook
   C. Pick-head axe
   D. Hacksaw

78 FUNDAMENTALS OF FIRE FIGHTER SKILLS

_____ 9. Which of the following is a basic piece of equipment for interior firefighting?
   A. Hand light or portable light
   B. Thermal imaging device
   C. Chain saw
   D. Exhaust fan

_____ 10. Which of the following is considered a tool for cutting metal?
   A. Crowbar
   B. Drywall hook
   C. Pick-head axe
   D. Hacksaw

_____ 11. Which of the following is not a basic search and rescue hand tool?
   A. Halligan tool
   B. Axe
   C. Hand light
   D. K tool

_____ 12. Special equipment to be carried by an RIC should include
   A. a thermal imager.
   B. prying tools.
   C. striking tools.
   D. all of the above.

_____ 13. Tools used for vertical roof ventilation include
   A. shovels and brooms.
   B. axes and pike poles.
   C. negative-pressure fans.
   D. rakes and buckets.

_____ 14. Pike poles are commonly used for
   A. pulling ceilings and opening walls.
   B. opening floors.
   C. popping doors off hinges.
   D. car fires.

_____ 15. At a single-family residential house fire, tool staging should be located
   A. at the first-due firehouse.
   B. outside the building.
   C. inside the structure.
   D. on the second ladder company.

_____ 16. Wood handles on tools should be
   A. sanded and painted.
   B. sanded and varnished.
   C. sanded and linseed oil applied.
   D. left alone.

_____ 17. A pike pole
   A. is a lever used for prying.
   B. is a pushing tool.
   C. is a pulling tool.
   D. both B and C.

_____ 18. Which tool is used to cut fences or padlocks?
   A. Bolt cutter
   B. Battering ram
   C. Flat-head axe
   D. K tool

_____ 19. Which of these are prying tools?
   A. Sledgehammer
   B. Halligan bar and crowbar
   C. K tool and chisel
   D. Bucket and shovel

_____ 20. Which tool heats metal until it melts?
   A. Hydraulic spreader
   B. Power saw
   C. Air bag
   D. Cutting torch

Vocabulary
Define the following terms using the space provided.

1. Ceiling hook:

2. Crowbar:

3. Kelly tool:

4. Hydrant wrench:

5. Reciprocating saw:

6. Gripping pliers:

7. Cutting torch:

8. Seat belt cutter:

9. Pike pole:

10. Spanner wrench:

11. Claw bar:

12. Overhaul:

Fill-in
Read each item carefully, and then complete the statement by filling in the missing word(s).

1. _____ is a prime consideration when using tools and equipment.
2. A fire fighter must know how to use tools safely, effectively, and _____.
3. _____ are tools that use extremely high-temperature flames to cut through an object.
4. _____ or _____ tools allow a fire fighter to increase the power exerted upon an object and extend the fire fighter's reach.
5. To reduce the total number of tools needed to achieve a goal, a fire fighter may carry a tool that has a number of uses. This tool is categorized as a(n) _____ tool.
6. _____ is the phase during which fire fighters start thinking about the possible tools or equipment they may need during an incident.
7. During an incident, the tools or equipment that may be required are often placed in a designated area referred to as the _____ area.
8. The _____ manual provides recommendations and instructions on how to clean, care for, and inspect tools and equipment.
9. A(n) _____ has a pointed "pick" on one end of the head and an axe blade on the other end.
10. A(n) _____ is a specialized striking tool with an axe on one end of the head and a sledgehammer on the other end.

True/False
If you believe the statement to be more true than false, write the letter "T" in the space provided.
If you believe the statement to be more false than true, write the letter "F."

1. _____ Department guidelines or standard operating procedures usually guide the decision for which tool to use during an incident.
2. _____ Mechanically powered equipment is more effective than manually powered equipment.
3. _____ Air bags can be used to lift heavy objects.
4. _____ Horizontal ventilation may be achieved by opening a window.
5. _____ Vertical ventilation is always achieved by opening windows.
6. _____ All members of a fire department are personally responsible for understanding the purpose of each tool or piece of equipment that is carried on their apparatus.
7. _____ Striking tools should be assigned to crews only during the forcible entry phase of a response.
8. _____ Interior attack can be orchestrated by any response member at any time during an emergency response.
9. _____ The K tool is used to pull the lock cylinder out of a door.
10. _____ An RIC should carry SCBA and spare air cylinders.

Short Answer

Complete this section with short written answers using the space provided.

1. Why is it important to know which tools are needed for each phase of an incident?

2. Describe the general functions of the six categories of tools, and provide an example from each category.

3. Identify and describe the basic steps of fire suppression.

4. List the components of a full set of PPE.

5. Identify the basic set of tools used for interior firefighting. List some examples.

6. Identify the basic set of tools used for search and rescue. List some examples.

7. Identify the tools and special equipment the rapid intervention crew (RIC) should carry and have ready for immediate response.

8. List the special equipment needed for ventilation.

_____

_____

_____

9. Identify the tools used during overhaul operations.

_____

_____

_____

10. Explain the importance of tool and equipment maintenance for a fire fighter.

_____

_____

_____

Fundamentals of Fire Fighter Skills

CLUES

Word Fun

The following crossword puzzle is an activity provided to reinforce correct spelling and understanding of terminology associated with firefighting. Use the clues provided to complete the puzzle.

Across

1. One set or a pair of connection devices attached to a fire hose that allows the hose to be interconnected to additional lengths of hose.
8. A specialized striking tool with an axe on one end and a sledgehammer on the other end.
9. Hydraulic spreading tool designed to pry open doors that swing inward.
10. Activities that occur in preparation for an emergency and continue until the arrival of emergency apparatus at the scene.
11. A cutting tool used to cut through thick metal objects such as bolts, locks, and wire fences.
12. A cutting tool designed for use on metal.
13. Used to remove lock cylinders from structural doors so the locking mechanism can be unlocked.

Down

2. A combination tool, normally consisting of the Halligan tool and the flat-head axe.
3. A long pole with a pointed head and two retractable cutting blades on the side.
4. Techniques used by fire fighters to gain entry into buildings when normal means of entry are locked or blocked.
5. A multipurpose tool that can be used for several forcible entry and ventilation applications because of its unique head design.
6. A prying tool that incorporates a pick and a fork.
7. A multipurpose tool that can be used for several forcible entry and ventilation applications because of its unique design.

Fire Alarms

The following real case scenarios will give you an opportunity to explore the concerns associated with fire fighter tools and equipment. Read each scenario, and then answer each question in detail.

1. It is 6:00 on a Monday morning when your engine is dispatched to a commercial structure fire in a warehouse. You are the fifth engine to arrive at the scene. The IC assigns your engine to the RIC group. Your Lieutenant tells you to gather the RIC equipment and place it in the staging area. How will you proceed?

   _____
   _____
   _____
   _____
   _____

2. It is 7:30 on a Sunday morning and your Lieutenant is conducting a shift meeting. The Lieutenant tells you to ensure the power tools and equipment are clean and inspected. After the meeting, you start with the ladder truck by pulling off the chainsaws. How should you proceed?

   _____
   _____
   _____
   _____
   _____

Ropes and Knots

Workbook Activities

The following activities have been designed to help you. Your instructor may require you to complete some or all of these activities as a regular part of your fire fighter training program. You are encouraged to complete any activity that your instructor does not assign as a way to enhance your learning in the classroom.

Chapter Review

The following exercises provide an opportunity to refresh your knowledge of this chapter.

Matching

Match each of the terms in the left column to the appropriate definition in the right column.

_____ 1. Dynamic rope   A. Rope constructed of fibers twisted into strands, which are then twisted together

_____ 2. Hitches   B. Rope constructed without knots or splices in the yarns, ply yarns, strands, braids, or rope

_____ 3. Static rope   C. A U-shape created by bending a rope with the two sides parallel

_____ 4. Safety knot   D. A knot used to secure the leftover working end of the rope; also known as an overhand knot or keep knot

_____ 5. Bight   E. A knot that attaches to or wraps around an object

_____ 6. Block creel construction   F. A rope generally made out of synthetic materials that is designed to be elastic and stretch when loaded

_____ 7. Twisted rope   G. The part of a rope between the working end and the running end

_____ 8. Carabiner   H. An emergency use rope designed to carry the weight of only one person and to be used only once

_____ 9. Personal escape rope   I. An oval-shaped device with a spring-loaded clip that can be used for connecting together pieces of rope, webbing, or other hardware

_____ 10. Standing part   J. A rope generally made out of synthetic material that stretches very little under load

Multiple Choice

Read each item carefully, and then select the best response.

_____ 1. A two-person rope is designed to bear the weight of
   A. two rescuers' equipment.
   B. two victims.
   C. two people or 600 pounds.
   D. two people and 600 pounds.

_____ 2. Life safety rope is used solely for
   A. sectioning off safety areas.
   B. supporting people.
   C. medical response.
   D. hoisting equipment.

_____ 3. Life safety rope is made of
   A. continuous filament virgin fiber and woven of block creel construction.
   B. the strongest natural fiber available.
   C. lightweight water-resistant fibers.
   D. fibers tested by the NFPA.

_____ 4. The running end is the part of the rope used for
   A. forming the knot.
   B. life safety.
   C. securing the knot.
   D. lifting or hoisting.

_____ 5. A one-person life safety rope has a safety factor of
   A. 15:1.
   B. 1:600.
   C. 10:1.
   D. 5:1.

_____ 6. Polypropylene rope is often used in water rescue because
   A. it is light, does not absorb water, and floats.
   B. it is less expensive than natural rope.
   C. it is light and has built-in water repellents.
   D. it is easier to control.

_____ 7. What is the working end of the rope used for?
   A. Lifting or hoisting
   B. Securing the knot
   C. Carrying the rope
   D. Forming the knot

_____ 8. How is a loop formed?
   A. By using the standing part of the rope
   B. By rolling from the end of the rope
   C. By making a circle in the rope
   D. By using the standing part of the knot

_____ 9. The load-carrying capacity or strength of the rope will be
   A. increased by any knot.
   B. reduced by any knot.
   C. maintained by any knot.
   D. maintained by a bend or hitch.

_____ 10. When hoisting an axe, a good rule of thumb is to
   A. hoist the equipment in a vertical position.
   B. use one-person rope.
   C. use dynamic rope.
   D. execute the hoist as quickly as possible.

_____ 11. What are two primary types of rope used in the fire service?
   A. Polypropylene and synthetic
   B. Natural fiber and manila
   C. Life safety and utility
   D. Hauling and securing

_____ 12. Personal escape ropes
   A. Are for one fire fighter and one use only.
   B. Can safely support two victims.
   C. Have a 300-pound breaking weight.
   D. Are designed to bear 600 pounds.

_____ 13. Utility ropes
   A. May be used for a personal safety rope.
   B. Can bear a single person (300 pounds).
   C. Can bear two persons (600 pounds).
   D. Are not to be used as a life safety rope.

_____ 14. Natural fiber ropes
   A. Lose their load-carrying ability over time.
   B. Absorb water.
   C. Degrade quickly.
   D. All of the above.

_____ 15. Synthetic ropes
   A. Absorb more water than natural fiber ropes.
   B. Cannot be used for life safety rope.
   C. Can be damaged by ultraviolet light.
   D. Will not be damaged by acids.

_____ 16. Life safety ropes are rated under the minimum requirements set by
   A. NRA.
   B. NFPA.
   C. NIOSH.
   D. CDC.

_____ 17. Which device is used to connect one rope to another?
   A. Carabiner
   B. Kernmantle
   C. Harness
   D. Stokes

_____ 18. Fire department ropes may be cleaned by using
   A. Mild alkali solution.
   B. Mild acid solution.
   C. Mild soap and water.
   D. Window cleaner.

_____ 19. The knot used to secure the leftover working end of a rope is called a
   A. Granny knot and severe tangle.
   B. Becket bend.
   C. Slip knot.
   D. Safety knot.

_____ 20. A pike pole is hoisted using
   A. A figure eight and a clove hitch.
   B. Clove and half hitches.
   C. A figure eight on a bight.
   D. A safety knot and bowline.

Labeling
Label the following diagrams with the correct terms.

1. Twisted and braided rope.

Figure 9-9 (A and B)

A. _____

B. _____

Vocabulary

Define the following terms using the space provided.

1. Braided rope:

2. Kernmantle rope:

3. Running end:

4. Working end:

5. Knot:

6. Round turn:

7. Rope bag:

8. Harness:

_____

_____

_____

9. Shock load:

_____

_____

_____

10. Depressions:

_____

_____

_____

Fill-in
Read each item carefully, and then complete the statement by filling in the missing word(s).

1. A snap link, or _____, is commonly used to connect one rope to another rope, a harness, or itself.
2. A(n) _____ is a piece of rescue or safety equipment worn by a person, and used to secure that person to a rope or a solid object.
3. _____ rescue is often associated with environments that have poor ventilation and limited entry or exit areas.
4. Each piece of rope must have a(n) _____ that details its history, usage, type of use, and loads applied.
5. _____ are used to fasten rope or webbing to objects or to each other.
6. Any knot will _____ the load-carrying capacity or strength of the rope.
7. A(n) _____ is formed by reversing the direction of the rope to form a "U" bend with two parallel ends.
8. A(n) _____ is formed by making a loop and then bringing the two ends of the rope parallel to each other.
9. A(n) _____ is used to secure the leftover working end of the rope to the standing part of the rope.
10. _____ are knots that wrap around an object such as a pike pole or fencepost.

True/False

If you believe the statement to be more true than false, write the letter "T" in the space provided.
If you believe the statement to be more false than true, write the letter "F."

1. _____ Life safety ropes must be inspected after each use.
2. _____ Hitches are used to attach a rope around an object.
3. _____ A two-person rope is designed to bear the weight of two people or 800 pounds.
4. _____ Life safety ropes can be made of natural fibers if properly inspected, recorded, and maintained.
5. _____ Personal escape ropes can be used only once.
6. _____ A static rope is better suited for most fire rescue situations.
7. _____ A personal escape rope has a safety factor of 15:1.
8. _____ Kernmantle ropes are the only ropes subject to being shock-loaded.
9. _____ All knots reduce the load-carrying capacity of a rope.
10. _____ All knots require a safety knot for completion.

Short Answer

Complete this section with short written answers using the space provided.

1. What are some of the drawbacks of using natural fiber ropes?

   _____
   _____
   _____

2. What are some of the advantages of using synthetic fiber ropes?

   _____
   _____
   _____

3. List and describe the most common synthetic fiber ropes used for fire department operations.

   _____
   _____
   _____

4. List and describe the three types of rope construction.

   _____
   _____
   _____

5. Identify the four parts of the rope maintenance formula.

6. List the principles to preserve the strength and integrity of rope.

7. List three steps recommended for cleaning ropes.

8. List and describe eight simple knots and their usage in the fire service.

94 Fundamentals of Fire Fighter Skills

CLUES

Across

3 A piece of rope formed into a circle.
6 Rope constructed of fibers twisted into strands, which are then twisted together.
7 Rope constructed by intertwining strands in the same way that hair is braided.
11 The part of a rope between the working end and the running end.
12 Knots that attach to or wrap around an object.

Down

1 A fastening made by tying together lengths of rope or webbing in a prescribed way.
2 A bag used to protect and store rope so that the rope can be easily and rapidly deployed without kinking.
4 A piece of equipment worn by a rescuer that can be attached to a life safety rope.
5 Knots used to join two ropes together.
8 A rope generally made out of synthetic materials that is designed to be elastic and stretch when loaded.
9 A rope generally made out of synthetic material that stretches very little under load.
10 A U-shape created by bending a rope with the two sides parallel.

Word Fun

The following crossword puzzle is an activity provided to reinforce correct spelling and understanding of terminology associated with firefighting. Use the clues provided to complete the puzzle.

Fire Alarms

The following real case scenarios will give you an opportunity to explore the concerns associated with ropes and knots. Read each scenario, and then answer each question in detail.

1. You have just returned from an apartment fire. During the fire, you used some of the utility rope off your engine to raise and lower equipment from the roof. While using the rope, it became dirty and needs to be cleaned and inspected. How will you proceed?

   _____
   _____
   _____
   _____
   _____

2. It is 9:00 on a Thursday evening when your engine company is dispatched to a structure fire at a three-story townhouse. Upon arrival, you are assigned to staging. The fire has been extinguished on the third floor. The IC assigns you and your partner to retrieve a ventilation fan from your engine and secure it to the rope that is deployed from the third floor. How should you proceed?

   _____
   _____
   _____
   _____
   _____
   _____

Fundamentals of Fire Fighter Skills

Skill Drills

Skill Drill 9-5: Tying a Safety Knot

Test your knowledge of this skill drill by filling in the correct words in the photo captions.

1. Take the loose end of the rope, beyond the knot, and form a loop around the _____ of the rope.

2. Pass the _____ end of the rope through the loop.

3. Tighten the safety knot by _____ on both ends at the same time.

Skill Drill 9-6: Tying a Half Hitch

Test your knowledge of this skill drill by placing the photos below in the correct order. Number the first step with a "1," the second step with a "2," and so on.

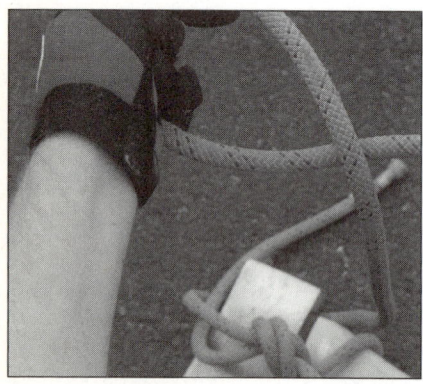

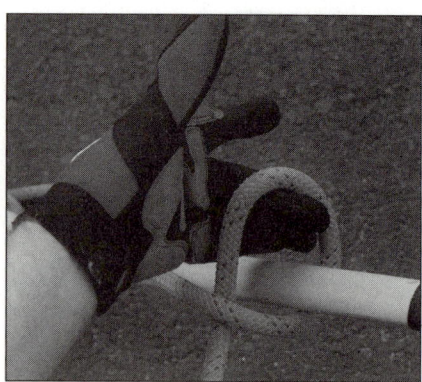

_____ Rotate your hand so your palm is facing toward you. This will make a loop in the rope.

_____ Grab the rope with your palm facing away from you.

_____ Pass the loop over the end of the object.

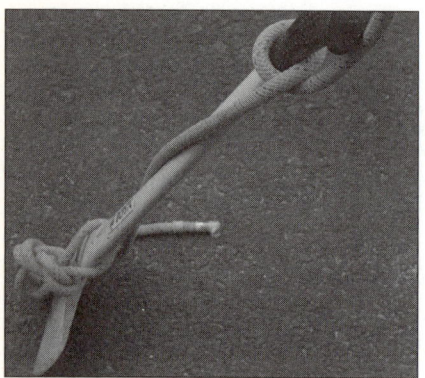

_____ Finish the half-hitch knot by positioning it and pulling tight.

Skill Drill 9-7: Tying a Clove Hitch in the Open
Test your knowledge of this skill drill by filling in the correct words in the photo captions.

1. Starting from _____ to _____ on the rope, grab the rope with crossed hands with the left positioned higher than the right.

2. Holding onto the rope, _____ your hands. This will create a loop in each hand.

3. Slide the _____ hand loop behind the _____ hand loop.

4. Slide both loops over the _____.

5. Pull in opposite directions to _____ the clove hitch. Tie a safety knot in the working end of the rope.

Skill Drill 9-9: Tying a Figure Eight Knot
Test your knowledge of this skill drill by filling in the correct words in the photo captions.

1. Form a(n) _____ in the rope.

2. Loop the _____ of the rope completely around the _____ of the rope.

3. Thread the _____ back through the bight.

4. Tighten the knot by pulling on both ends _____.

Skill Drill 9-10: Tying a Figure Eight on a Bight

Test your knowledge of this skill drill by placing the photos below in the correct order. Number the first step with a "1," the second step with a "2," and so on.

_____ Form a bight and identify the end of the bight as the working end.

_____ Feed the working end of the bight back through the loop.

_____ Secure the loose end of the rope with a safety knot.

_____ Pull the knot tight.

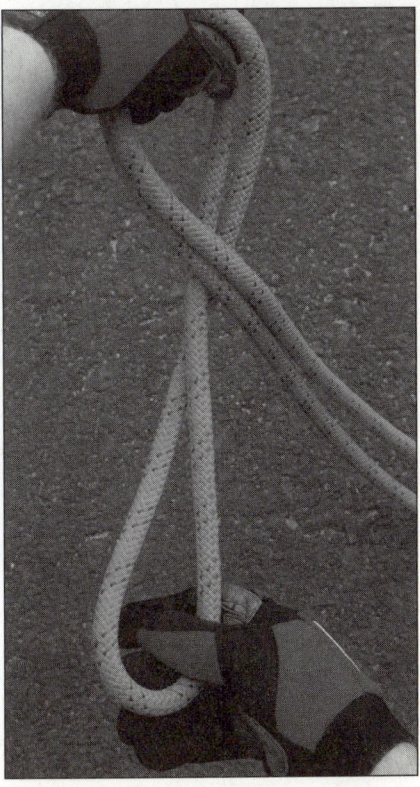

_____ Holding both sides of the bight together, form a loop.

Skill Drill 9-12: Tying a Bowline
Test your knowledge of this skill drill by filling in the correct words in the photo captions.

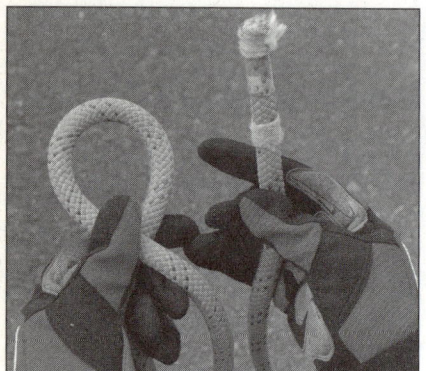
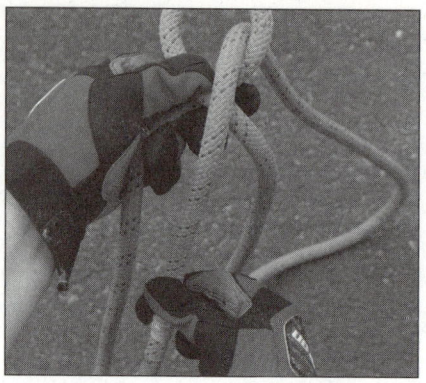

1. Make the desired sized _____ and bring the working end back to the standing part.

2. Form another small loop in the _____ of the rope with the section close to the _____ on top. Thread the working end up through this loop from the _____.

3. Pass the working end _____ the loop, around and under the standing part, and back down through the same opening.

4. Tighten the knot by holding the working end and pulling the standing part of the rope _____.

5. Tie a(n) _____ in the working end of the rope.

Skill Drill 9-14: Hoisting an Axe

Test your knowledge of this skill drill by placing the photos below in the correct order. Number the first step with a "1," the second step with a "2," and so on.

_____ Prepare to raise the axe.

_____ Tie one or two half hitches along the axe handle.

_____ The team that needs the axe should lower a rope with enough extra rope to tie the required knot around the axe.

_____ Pass the standing part of the rope around the head of the axe.

_____ Place the standing part of the rope parallel to the axe handle.

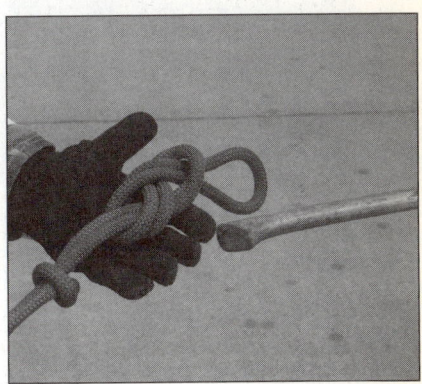

_____ Tie a figure eight on a bight to make a small loop.

_____ Place the loop over the axe handle near the head.

Skill Drill 9-15: Hoisting a Pike Pole
Test your knowledge of this skill drill by filling in the correct words in the photo captions.

1. The team that needs the pike pole should lower a rope with enough extra rope available for the required knot and the _____.

2. Tie a clove hitch in the open and slip it over the handle. Secure it near the _____.

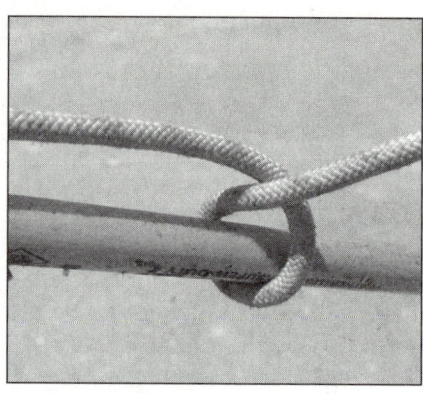

3. Place a(n) _____ around the handle below the clove hitch.

4. Place a second _____ around the handle near the bottom of the pike pole.

5. Prepare to raise the _____.

Skill Drill 9-16: Hoisting a Ladder

Test your knowledge of this skill drill by placing the photos below in the correct order. Number the first step with a "1," the second step with a "2," and so on.

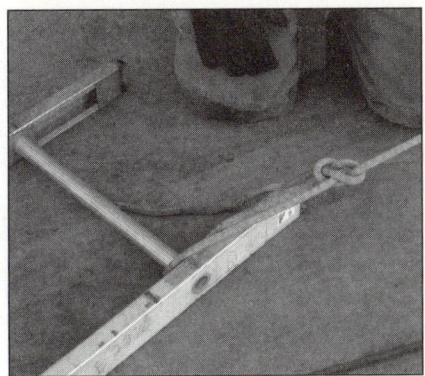

_____ Attach a tag line from below to control the ladder as it is hoisted.

_____ Prepare to raise the ladder.

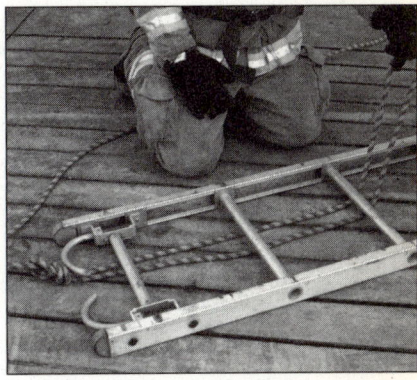

_____ Pass the rope between the rungs of the ladder, three or four rungs from the top. Pull the loop under the rungs toward the top of the ladder.

_____ The team that needs the ladder should lower a rope with enough extra rope to tie onto the ladder.

_____ Remove the slack from the rope and allow the loop to slide down the ladder.

_____ Tie a figure eight on a bight to make a loop 3 or 4 feet in diameter.

_____ Place the loop around the top of the ladder.

Response and Size-Up

Workbook Activities

The following activities have been designed to help you. Your instructor may require you to complete some or all of these activities as a regular part of your fire fighter training program. You are encouraged to complete any activity that your instructor does not assign as a way to enhance your learning in the classroom.

Chapter Review

The following exercises provide an opportunity to refresh your knowledge of this chapter.

Matching

Match each of the terms in the left column to the appropriate definition in the right column.

_____ 1. Salvage

_____ 2. Extension

_____ 3. Overhaul

_____ 4. Exposure

_____ 5. Size-up

_____ 6. Response

_____ 7. Freelancing

_____ 8. Personnel accountability system

_____ 9. Preincident plan

_____ 10. Resources

A. Any person or property that may be endangered by flames, smoke, gases, heat, or runoff from a fire

B. The ongoing observation and evaluation of factors that are used to develop objectives, strategy, and tactics for fire suppression

C. Removing or protecting property that could be damaged during firefighting or overhaul operations

D. Activities that occur in preparation for an emergency and continue until the arrival of emergency apparatus at the scene

E. The movement of fire into uninvolved areas of a structure

F. Examination of the building and contents involved in a fire to ensure that the fire is completely extinguished

G. All of the means available to fight a fire or conduct emergency operations

H. Provides details about a building's construction, layout, contents, special hazards, and fire protection systems

I. An updated list of fire fighters assigned to each vehicle or crew

J. Taking action on your own without regard to standard operating procedures, command structure, or the strategic plan

CHAPTER 10

Multiple Choice

Read each item carefully, and then select the best response.

_____ 1. Dispatch information will include
   A. the location of the incident and the type of attack required.
   B. the location of the incident and initial scene assessment.
   C. the location of the incident, the type of emergency, and the units due to respond.
   D. the type of emergency, scene assessment, and initial scene assessment.

_____ 2. An emergency vehicle must always be operated with
   A. due regard for the safety of everyone on the road.
   B. the assurance that all drivers will yield to the emergency vehicle's right of way.
   C. the skills learned in training.
   D. the intention to arrive on scene as soon as possible.

_____ 3. To track all fire fighters there should be a(n) _____ at every incident scene.
   A. incident commander
   B. personnel accountability system
   C. incident management system
   D. accountability officer

_____ 4. During an incident, shutting off electrical service eliminates potential
   A. damage to fire department equipment.
   B. interference with communications equipment.
   C. ignition sources.
   D. structural obstructions.

_____ 5. The initial size-up of an incident is conducted
   A. by the bystanders on scene.
   B. by the first officers on scene.
   C. when the first unit arrives on scene.
   D. by reviewing the preincident plan.

_____ 6. Events and outcomes that can be predicted based on facts, observations, common sense, and previous experience are called
   A. estimates.
   B. report items.
   C. part of the size-up.
   D. probabilities.

_____ 7. A preincident plan is helpful during size-up because it contains
   A. information about the structure.
   B. the potential number of units needed for response.
   C. the potential equipment requirements for a response.
   D. information about the weather.

_____ 8. Why is the age of the building often an important consideration in size-up?
   A. Older buildings burn faster.
   B. Building and fire safety codes change over time.
   C. Newer buildings have higher property values.
   D. Ventilation is often easier to perform on new homes.

_____ 9. A fire department's basic resources are
   A. its personnel and apparatus.
   B. its preincident plans and trained personnel.
   C. its specialized equipment and apparatus.
   D. its specially trained personnel.

_____ 10. If an incident requires more resources than the local community can provide, most departments have
   A. relief workers.
   B. agreements with state or provincial training institutions.
   C. mutual aid agreements.
   D. support response teams.

_____ 11. Who develops the incident action plan that outlines the steps needed to control the situation?
   A. The initial attack team
   B. The dispatcher
   C. The incident management system
   D. The incident commander

_____ 12. The gas supply to a building is usually controlled by
   A. a single valve at the entry point of the gas piping.
   B. an underground valve that requires the use of a special wrench.
   C. qualified technicians.
   D. the property owner.

_____ 13. What is the first consideration at any emergency incident?
   A. Protecting property
   B. Protecting lives
   C. Controlling traffic
   D. Completing a full size-up

_____ 14. When fire fighters advance into the fire building with hose lines to overpower the fire, they are part of a(n)
   A. defensive attack.
   B. defensive response.
   C. offensive attack.
   D. rapid intervention team.

_____ 15. What is the secondary objective at any emergency incident?
   A. Protecting property
   B. Protecting lives
   C. Protecting fire fighters
   D. Protecting bystanders

_____ 16. The removal or protection of property that could be damaged during firefighting is called
   A. postincident reporting.
   B. overhaul.
   C. recovery.
   D. salvage.

_____ 17. The main area of the fire is the
   A. hot spot.
   B. seat of the fire.
   C. attack area.
   D. target area.

_____ 18. The incident commander assembles, interprets, and bases decisions on information presented in the
   A. call-out.
   B. size-up.
   C. reconnaissance report.
   D. dispatch message.

_____ 19. During the response phase, the fire fighter should begin to
   A. consider any factors that could affect the situation.
   B. rest and prepare for the upcoming incident.
   C. have constant communications with the driver.
   D. organize equipment.

_____ 20. Fire fighters who respond to an incident on the fire apparatus deposit their personnel accountability tags
   A. in the command post.
   B. with the incident commander.
   C. with the rapid intervention team.
   D. on a designated location on the apparatus.

Vocabulary

Define the following terms using the space provided.

1. Thermal imaging devices:

2. Freelancing:

3. Personnel accountability tag (PAT):

4. Personal alert safety system (PASS):

5. Size-up:

Fill-in
Read each item carefully, and then complete the statement by filling in the missing word(s).

1. The _____ report is created for the incident commander based on the inspection and exploration of a specific area.
2. A(n) _____ attack occurs when fire fighters advance into the fire with extinguishing agents to overpower the fire.
3. An older style of building that contains channels in the wall extending from the basement to the attic is an example of a(n) _____-frame construction
4. _____ are events that can be predicted or anticipated, based on facts, observations, and previous experiences.
5. A(n) _____ attack occurs when fire fighters attempt to suppress a fire and protect exposures externally.
6. After a fire is under control, a(n) _____ is conducted to completely extinguish any remaining fires.
7. The _____ information is essential for determining the appropriate strategy and tactics for an incident.
8. All equipment should be properly mounted, stowed, or _____ on the fire apparatus to prevent injury.
9. The shut-off valve for a natural gas system is usually a(n) _____-turn valve with a locking device.
10. The size-up process requires a(n) _____ approach to managing information.

True/False
If you believe the statement to be more true than false, write the letter "T" in the space provided.
If you believe the statement to be more false than true, write the letter "F."

1. _____ Freelancing is dangerous.
2. _____ The driver of the fire apparatus is legally responsible for the safe operation of the apparatus at all times.
3. _____ During transport, it is important to have constant contact with the driver to make him or her aware of all response information.
4. _____ Fire fighters should not respond to an emergency incident unless they are dispatched.
5. _____ Defensive strategies are used to protect exposed properties because defensive strategies are ineffective in extinguishing fires.
6. _____ Never attempt to mount fire apparatus while it is moving.
7. _____ Visible flames provide all of the information needed for the incident commander to make decisions on where to set up the attack teams.
8. _____ Shutting off water service in a residential fire is often difficult because of the number of valves that need to be closed within the structure.
9. _____ Traffic is often a major concern for fire fighter safety at an emergency incident.
10. _____ A fire fighter can make a better size-up by standing up in the apparatus while approaching the scene of a working fire.

Short Answer

Complete this section with short written answers using the space provided.

1. Identify why the three main utilities need to be controlled at an emergency incident.

2. Why is it important for fire fighters to understand how size-up is completed?

3. Identify and describe the two basic categories of information utilized in size-up.

4. In order of priority, identify the five basic fire-ground objectives.

Word Fun

The following crossword puzzle is an activity provided to reinforce correct spelling and understanding of terminology associated with firefighting. Use the clues provided to complete the puzzle.

CLUES

Across

2 Activities that occur in preparation for an emergency and continue until the arrival of emergency apparatus at the scene.

3 Removing or protecting property that could be damaged during firefighting or overhaul operations.

9 Devices that detect differences in temperature based on infrared energy and then generate images based on that data.

10 A method of tracking the identity, assignment, and location of fire fighters operating at an incident scene.

12 A situation where a fire, which was thought to be completely extinguished, reignites.

Down

1 The ongoing observation and evaluation of factors that are used to develop objectives, strategy, and tactics for fire suppression.

2 The inspection and exploration of a specific area so as to gather information for the incident commander.

4 A written document resulting from the gathering of general and detailed information to be used by public emergency response agencies and private industry for determining the response to reasonable anticipated emergency incidents at a specific facility.

5 Exterior fire suppression operations directed at protecting exposures.

6 Examination of all areas of the building and contents involved in a fire to ensure that the fire is completely extinguished.

7 An advance into the fire building by fire fighters with hose lines or other extinguishing agents to overpower the fire.

8 The movement of fire into uninvolved areas of a structure.

11 Any person or property that may be endangered by flames, smoke, gases, heat, or runoff from a fire.

Fire Alarms

The following real case scenarios will give you an opportunity to explore the concerns associated with response and size-up. Read each scenario, and then answer each question in detail.

1. Your engine company is dispatched to a commercial structure fire in a large pizza restaurant. The building is a wood-frame construction with a lightweight truss. The building is 75 percent involved with fire in the attic area. You are the acting officer and need to decide the overall strategy. How will you proceed?

   _____
   _____
   _____
   _____
   _____

2. You are about to eat dinner when you are dispatched to a vehicle fire. It is dusk when you arrive, and it is rush hour. Your Lieutenant orders everyone to put on their high-visibility traffic vests and tells you to put out warning devices to help secure the scene. How should you proceed?

   _____
   _____
   _____
   _____
   _____

Skill Drills

Skill Drill 10-1: Mounting Apparatus
Test your knowledge of this skill drill by filling in the correct words in the photo captions.

1. When mounting (climbing aboard) fire apparatus, always have at least one hand firmly grasping a(n) _____ and at least one foot firmly placed on a foot surface. Maintain the one hand and one foot placement until you are seated.

2. Fasten your seat belt and then don any other required safety equipment for response, such as _____. Eye and face protection are required for seating areas that are not fully enclosed.

Forcible Entry

Workbook Activities

The following activities have been designed to help you. Your instructor may require you to complete some or all of these activities as a regular part of your fire fighter training program. You are encouraged to complete any activity that your instructor does not assign as a way to enhance your learning in the classroom.

Chapter Review

The following exercises provide an opportunity to refresh your knowledge of this chapter.

Matching

Match each of the terms in the left column to the appropriate definition in the right column.

_____ 1. Bite
_____ 2. Door
_____ 3. Interior wall
_____ 4. Latch
_____ 5. Solid-core
_____ 6. A tool
_____ 7. Prying tool
_____ 8. Shackles
_____ 9. Rabbet
_____ 10. Halligan

A. The part of the door lock that catches and holds the door frame
B. The U-shaped part of a padlock
C. A wall inside a building that divides a large space into smaller areas
D. A cutting tool with a pry bar built into the cutting part of the tool
E. A small opening made to enable better tool access in forcible entry
F. A door design that consists of wood core blocks inside the door
G. A tool used to pry doors and windows from their frames
H. A type of door frame that has the stop for the door cut into the frame
I. Usually the best point for forcing entry into a vehicle or structure
J. One part of the forcible entry tool "irons"

Multiple Choice

Read each item carefully, and then select the best response.

_____ 1. Gaining access to a structure when the normal means of entry are unable to be used is referred to as
   A. structure entry.
   B. forcible entry.
   C. operating access.
   D. forced access.

_____ 2. Company officers usually select both the point of entry and the
   A. rate of entry.
   B. equipment to be used.
   C. method to be used.
   D. point of exit.

_____ 3. Which types of tools are used to generate a force directly on an object or another tool?
   A. Striking tools
   B. Cutting tools
   C. Prying tools
   D. Through-the-lock tools

CHAPTER 11

_____ 4. A tool that has evolved from the use of a large log and is used to force doors or breach walls is a
   A. hammer.
   B. maul.
   C. lift.
   D. battering ram.

_____ 5. Which types of tools are used to spread doors and windows?
   A. Striking tools
   B. Cutting tools
   C. Prying tools
   D. Through-the-lock tools

_____ 6. The adz, the pick, and the claw are all incorporated into the
   A. hammer.
   B. Halligan tool.
   C. maul.
   D. pick axe.

_____ 7. Which type of circular saw blade is used to cut through metal doors, locks, or gates?
   A. Steel blade
   B. Masonry-cutting blade
   C. Carbide-tipped blade
   D. Metal-cutting blade

_____ 8. Which type of tool is designed to cut into a lock cylinder and has a pry bar built into the cutting part of the tool?
   A. A tool
   B. K tool
   C. J tool
   D. Adz

_____ 9. The slab door typically used for entrance doors because it is heavy and difficult to force open is a
   A. metal door.
   B. solid-core door.
   C. hollow-core door.
   D. slab door.

_____ 10. Doors that have a wood frame inset with solid wood panels are called
   A. tempered doors.
   B. ledge doors.
   C. panel doors.
   D. slab doors.

_____ 11. If you can see the hardware of a door, it is a(n)
   A. sliding door.
   B. inward-swinging door.
   C. outward-swinging door.
   D. overhead door.

_____ 12. Doors that have two sections and a double track where one side is fixed and the other slides are known as
   A. sliding doors.
   B. tempered doors.
   C. slab doors.
   D. honeycomb doors.

_____ 13. Larger pieces or panes of glass are called
   A. annealed glass.
   B. tempered glass.
   C. laminated glass.
   D. plate glass.

_____ 14. Vehicle windshields are most commonly made of
   A. tempered glass.
   B. laminated glass.
   C. plate glass.
   D. glazed glass.

_____ 15. The windows that are similar to sliding doors are
   A. jalousie windows.
   B. casement windows.
   C. horizontal-sliding windows.
   D. projected windows.

_____ 16. The part of the door lock that catches and holds the door frame is called the
   A. latch.
   B. operator lever.
   C. deadbolt.
   D. lock body.

_____ 17. Another term for the doorknob is the
   A. latch.
   B. operator lever.
   C. deadbolt.
   D. lock body.

_____ 18. The main part of a padlock that houses the locking mechanisms is the
   A. lock body.
   B. shackle.
   C. unlocking device.
   D. deadbolt.

_____ 19. The most common locks on the market today are
   A. mortise locks.
   B. cylindrical locks.
   C. padlocks.
   D. rim locks.

_____ 20. The two door locks that can be surface mounted on the interior of the door frame are rim locks and
   A. mortise locks.
   B. combination locks.
   C. key locks.
   D. deadbolts.

_____ 21. Walls that support the rafters and/or ceiling are
   A. nonbearing.
   B. exterior walls.
   C. load bearing.
   D. partitions.

Labeling
Label the following diagrams with the correct terms.

1. The parts of a door.

Figure 11-8

A. _____
B. _____
C. _____
D. _____

2. The parts of a lock.

Figure 11-22

A. _____
B. _____
C. _____
D. _____
E. _____
F. _____
G. _____

Vocabulary

Define the following terms using the space provided.

1. Projected windows:

2. Jalousie windows:

3. Mortise locks:

4. Cylindrical locks:

5. Casement windows:

6. K tool:

7. Jamb:

8. Tempered glass:

___

9. Rabbet:

___

Fill-in
Read each item carefully, and then complete the statement by filling in the missing word(s).

1. Forcible entry is usually required at emergency incidents where time is a(n) _____ factor.
2. All tools should be kept in a(n) _____ state.
3. _____ are used to cut metal components such as bolts, padlocks, and chains.
4. _____-powered tools are portable and can be placed into operation quickly, but have limited power and operating times.
5. The circular saw blade that can stay sharp for long periods of time and can cut through hard surfaces or wood is the _____-tipped blade.
6. A _____ is a small hydraulic spreader operated by a hand-powered pump.
7. _____ doors are usually made of four glass panels with metal frames.
8. The quickest way to force entry through a(n) _____ roll-up door is to cut the door with a saw or torch.
9. _____ provide air flow and light to the inside of the buildings, but can also provide emergency entrances.
10. _____ windows have two movable sashes that move freely up and down.
11. _____ windows are similar in operation to jalousie windows, except that they usually have one large or two medium-sized glass panels instead of many small ones.

True/False

If you believe the statement to be more true than false, write the letter "T" in the space provided.
If you believe the statement to be more false than true, write the letter "F."

1. _____ Fire fighters must consider, when making forcible entry, the need to secure the premises after operations are completed because they must never leave the premises in a state that would allow unauthorized entry.
2. _____ A pry axe should be used only for cutting.
3. _____ A rabbet tool is a small hydraulic spreader operated by a hand-powered pump.
4. _____ Duck-billed lock breakers are cutting tools used to snip off the shackles of a lock.
5. _____ The best point to attempt forcible entry to a vehicle or building is the door or window.
6. _____ Outward-opening doors are most often used in residential occupancies.
7. _____ The locking mechanisms on sliding doors are not strong and can be pried open.
8. _____ When breaking a window, always stand downwind.
9. _____ Double-pane glass is being used in many homes because it improves home insulation.
10. _____ Interior walls are usually constructed of wood or metal studs and covered by plaster, gypsum, or sheetrock.
11. _____ The two most popular floor materials found in residences and commercial buildings are tile and steel.

Short Answer

Complete this section with short written answers using the space provided.

1. Identify the four general safety tips for using tools.

2. Identify the four general carrying tips that apply to all tools.

3. List the four categories of forcible entry tools.

4. List the four basic components of a door.

Word Fun

The following crossword puzzle is an activity provided to reinforce correct spelling and understanding of terminology associated with firefighting. Use the clues provided to complete the puzzle.

CLUES

Across

1. A secondary locking device used to secure a door in its frame.
4. The part of a doorway that secures the door to the studs in a building.
5. The forked end of a tool.
7. A tool designed to remove the face plate of a lock cylinder.
8. Glass that is heat treated so that it will break into small pieces that are not as dangerous.
10. The parts of a door or window that enable it to be locked or opened.
13. A wall that supports structural members or upper floors of a building.
14. Windows in a steel or wood frame that open away from the building via a crank mechanism.
16. The process of forming standard glass.

Down

1. The primary choice for forcing entry into a structure.
2. Also known as safety glass.
3. The U-shaped part of a padlock that runs through a hasp and secures back into the lock body.
4. A tool that is designed to fit between double doors equipped with panic bars.
6. A small opening made to enable better tool access in forcible entry.
9. A wall that does not support a ceiling or structural member, but simply divides a space.
11. A door frame that has the stop for the door cut into the frame.
12. A combination tool, normally the Halligan tool and the flat-head axe.
15. The prying part of the Halligan tool.

Fire Alarms

The following real case scenarios will give you an opportunity to explore the concerns associated with forcible entry. Read each scenario, and then answer each question in detail.

1. It is 1:30 in the morning when your engine company is dispatched to an alarm activation at an elementary school. Upon arrival, you find light smoke in the hallway and the alarm annunciator panel tells you that water is flowing. Upon his walk-around, your Captain finds a single sprinkler head that has activated in a classroom and reports that the fire has been contained. He directs your crew to force entry through the classroom doors and overhaul the fire. Upon size-up, you determine the entrance doors are glass, outward-opening doors with a steel frame. How should you proceed?

   _____
   _____
   _____
   _____
   _____
   _____

2. It is 3:00 in the afternoon when your ladder truck company is dispatched to a commercial structure fire in a food storage warehouse. Upon arrival, your ladder truck is assigned to force entry on side C of the building. Your Lieutenant performs a size-up and determines that the best route of entry will be through an overhead-rolling door. She tells you and your partner to gather the appropriate tools and force entry through the door. How do you proceed?

   _____
   _____
   _____
   _____
   _____
   _____

Skill Drills

Skill Drill 11-1: Forcing Entry into an Inward-Opening Door

Test your knowledge of this skill drill by placing the photos below in the correct order. Number the first step with a "1," the second step with a "2," and so on.

_____ Once the Halligan tool is in position, have your partner, on your command, drive the tool farther into the gap between the rabbeted jamb or stop and the door. Make sure that the tool is not driven into the door jamb itself.

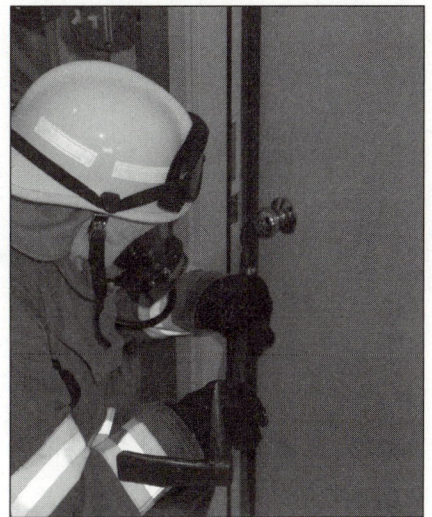

_____ Once the tool is past the stop, but between the door and the jamb, push the Halligan tool toward the door to force it open. If more leverage is needed, your partner can slide the axe head between the bevel of the Halligan tool and the door. It may be necessary to push in on the door. Secure the door to prevent it from closing behind you.

_____ Size up the door, looking for any safety hazards. Inspect the door for the location and number of locks and mechanisms.

_____ Place the forked end of the Halligan tool into the door frame between the door jamb and the door stop. Insert the tool near the lock, with the beveled end of the tool against the door.

Skill Drill 11-2: Forcing Entry into an Outward-Opening Door
Test your knowledge of this skill drill by filling in the correct words in the photo captions.

 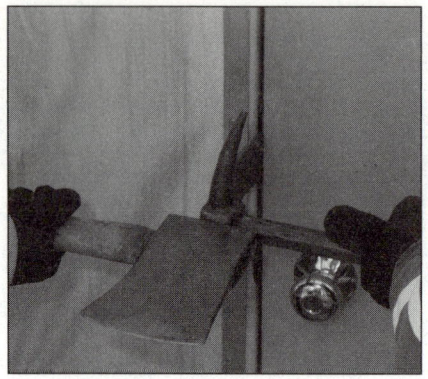 

1. _____ the door looking for any safety hazards. Place the _____ end of the Halligan tool between the door and the frame either near the locking mechanism, or between the mechanism and a secondary lock.

2. Once the _____ is in position, have your partner strike the _____ on your command and drive the _____ end farther into the gap.

3. Pry in a(n) _____ direction with the fork end of the tool and then force the door _____. Always secure the door to prevent it from closing behind you.

Skill Drill 11-7: Forcing Entry Using a K Tool
Test your knowledge of this skill drill by placing the photos below in the correct order. Number the first step with a "1," the second step with a "2," and so on.

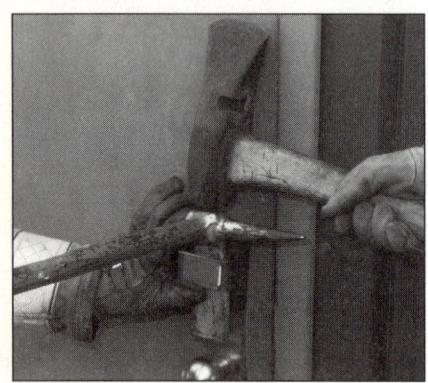  

_____ Place the adz end of the Halligan tool into the slot on the K tool, and strike the end of the Halligan tool with a flat-head axe to drive the K tool farther into the lock cylinder.

_____ Pry up on the Halligan tool to pull out the lock and expose the locking mechanism. Using the small tools that come with the K tool, turn the mechanism to open the lock.

_____ Place the K tool over the face of the cylinder, noting the location of the keyway. Using a Halligan or similar pry tool, tap the K tool down into the cylinder.

Skill Drill 11-8: Forcing Entry Using an A Tool
Test your knowledge of this skill drill by filling in the correct words in the photo captions.

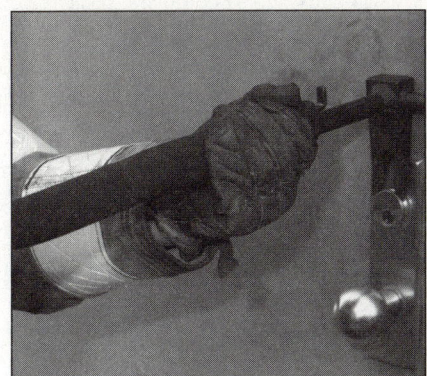 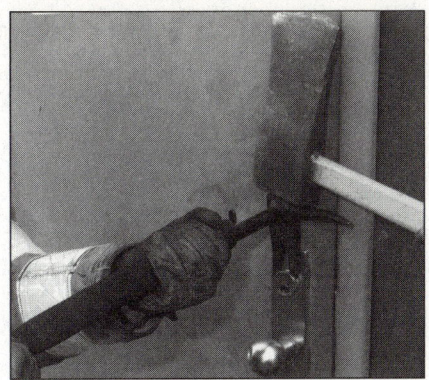

1. Place the _____ of the A tool at the top of the cylinder, between the cylinder and the door frame.

2. Using a flat-head axe or similar tool, drive the _____ into the _____. Pry _____ on the A tool to remove the _____ from the door. Insert a(n) _____ into the hole to manipulate the locking mechanism and open the door.

Ladders

Workbook Activities

The following activities have been designed to help you. Your instructor may require you to complete some or all of these activities as a regular part of your fire fighter training program. You are encouraged to complete any activity that your instructor does not assign as a way to enhance your learning in the classroom.

Chapter Review

The following exercises provide an opportunity to refresh your knowledge of this chapter.

Matching

Match each of the terms in the left column to the appropriate definition in the right column.

_____ 1. Dogs A. The end of the ladder that is placed against the ground when the ladder is raised

_____ 2. Beam B. The rope or cable used to extend or hoist the fly section(s) of an extension ladder

_____ 3. Egress C. The top and bottom surfaces of an I-beam ladder

_____ 4. Stop D. The ladder component that supports the rungs

_____ 5. Butt E. A mechanical locking device used to secure the fly section(s) of a ladder

_____ 6. Halyard F. A method of exiting from an area or a building

_____ 7. Rail G. A ladder crosspiece that provides a climbing step for the user

_____ 8. Rung H. The very top of the ladder

_____ 9. Tip I. A piece of material that prevents the fly section(s) of a ladder from overextending and collapsing the ladder

_____ 10. Staypoles J. Part of a Bangor ladder

Multiple Choice

Read each item carefully, and then select the best response.

_____ 1. Ladder gins are used to access
   A. below-grade sites.
   B. above-grade sites.
   C. grade-level positions.
   D. interior above-grade sites.

_____ 2. The rail is the
   A. top section of a solid beam.
   B. top section of a trussed beam.
   C. top or bottom section of a trussed beam.
   D. handrail at the tip of the ladder.

CHAPTER 12

_____ **3.** A truss block is a piece that connects the two
   A. rails of a trussed beam.
   B. rungs of a trussed beam.
   C. pulleys of an I-beam.
   D. rungs of an I-beam.

_____ **4.** Transferring the weight of the user to the beams is done through the
   A. halyards.
   B. rungs.
   C. butt spurs.
   D. tie rods.

_____ **5.** The metal bar that runs from one beam of the ladder to the other and keeps the beams from separating is the
   A. butt spur.
   B. rung.
   C. rail.
   D. tie rod.

_____ **6.** Butt spurs prevent the ladder from
   A. losing contact with the exterior of the structure.
   B. damaging the structure.
   C. chaffing other surfaces.
   D. slipping out of position.

_____ **7.** The part of an extension ladder that is raised or extended from the bed section is the
   A. fly section.
   B. elevating section.
   C. lift section.
   D. aerial section.

_____ **8.** The rope or cable used to extend the fly section is the
   A. dog.
   B. guide.
   C. halyard.
   D. lift.

_____ **9.** Pawls, ladder locks, or rung locks are also referred to as
   A. roof hooks.
   B. dogs.
   C. guides.
   D. truss locks.

_____ **10.** Ladders with staypoles or tormentors are typically referred to as
   A. aerial ladders.
   B. portable ladders.
   C. Bangor ladders.
   D. Fresno ladders.

_____ 11. Fire service portable ladders are limited to a maximum length of
   A. 25 feet.
   B. 50 feet.
   C. 75 feet.
   D. 100 feet.

_____ 12. A straight ladder equipped with retractable hooks to secure the tip of the ladder to a pitched roof is a
   A. roof ladder.
   B. Bangor ladder.
   C. combination ladder.
   D. Fresno ladder.

_____ 13. Staypoles are required on ladders of
   A. 10 to 20 feet.
   B. 20 to 30 feet.
   C. 30 to 40 feet.
   D. 40 feet or greater.

_____ 14. Ladders that are designed to allow access to attic scuttle holes and confined areas are
   A. folding ladders.
   B. pompier ladders.
   C. scaling ladders.
   D. combination ladders.

_____ 15. A ladder that has no halyard, is generally short, and designed for attic access is the
   A. pompier ladder.
   B. Bangor ladder.
   C. Fresno ladder.
   D. roof ladder.

_____ 16. The horizontal-bending test evaluates the
   A. manufacturer's specifications.
   B. structural strength of the ladder.
   C. service testing of the ladder.
   D. extension hardware on the ladder.

_____ 17. The most important safety check is confirming the
   A. proximity to direct flames.
   B. location of overhead utility lines.
   C. identification of stable, level surfaces.
   D. size of the structure.

_____ 18. The proper climbing angle for maximum load capacity and strength is
   A. 30 degrees.
   B. 45 degrees.
   C. 60 degrees.
   D. 75 degrees.

_____ 19. Most portable ladders are designed to support a weight of
   A. 500 pounds.
   B. 750 pounds.
   C. 1000 pounds.
   D. 1500 pounds.

_____ 20. Aerial equipment operators are trained to extend the bucket or tip of the ladder
   A. above a trapped individual.
   B. below a trapped individual.
   C. even with a trapped individual.
   D. in contact with a trapped individual.

_____ 21. What is the common rule of thumb when identifying the length of ladder to use on a structure?
   A. The tip of the ladder is in contact with the structure.
   B. The butt of the ladder is in contact with the structure.
   C. At least five ladder rungs show above the roofline.
   D. The ladder provides access to all possible incident outcomes.

_____ 22. In most cases, a single fire fighter can safely carry a straight or roof ladder
   A. less than 18 feet long.
   B. less than 24 feet long.
   C. less than 4 feet wide.
   D. less than 2 feet wide.

_____ 23. Most ladders are carried with the
   A. tip forward.
   B. fly-section forward.
   C. butt end forward.
   D. beam on top.

_____ 24. The number of fire fighters required to raise a ladder depends on the
   A. length and width of the ladder.
   B. weight and target surface of the ladder.
   C. width and clearance of the ladder.
   D. length and weight of the ladder.

_____ 25. The two common techniques for raising portable ladders are the
   A. team and self raises.
   B. beam and rung raises.
   C. beam and halyard raises.
   D. team and aerial raises.

_____ 26. When a fire fighter stands between the ladder and the structure, grasps the beams, and leans to pull the ladder into the structure, he or she is
   A. butting the ladder.
   B. guiding the ladder.
   C. heeling the ladder.
   D. dogging the ladder.

_____ 27. When climbing a ladder, the fire fighter's eyes should be looking
   A. forward, with an occasional glance upward
   B. upward.
   C. downward.
   D. at the ladder's tip.

_____ 28. Establishing verbal contact as quickly as possible is important when rescuing
   A. an unconscious patient.
   B. an infant.
   C. an elderly person.
   D. any conscious person.

Labeling

Label the following diagrams with the correct terms.

1. Basic components of a straight ladder.

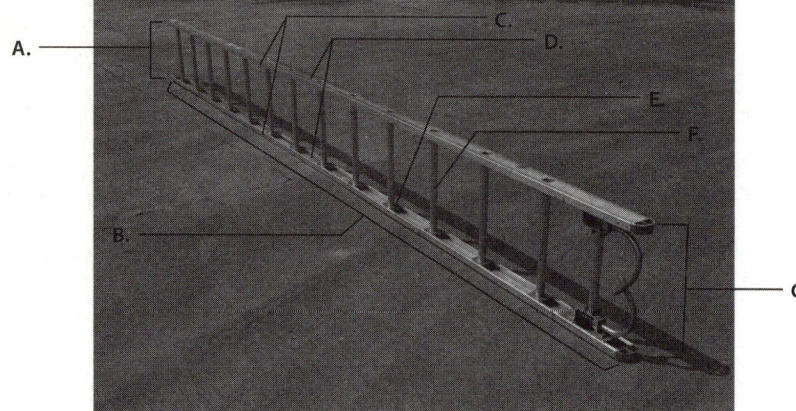

Figure 12-3

A. _____
B. _____
C. _____
D. _____
E. _____
F. _____
G. _____

2. Components of an extension ladder.

Figure 12-5

A. _____
B. _____
C. _____
D. _____
E. _____

Vocabulary

Define the following terms using the space provided.

1. Heat sensor label:

_____
_____
_____

2. Protection plates:

_____
_____
_____

3. Halyard:
___
___
___

4. Ladder belt:
___
___
___

5. Pulley:
___
___
___

6. Pawls:
___
___
___

7. Bed section:
___
___
___

8. Guides:
___
___
___

9. Roof hooks:
___
___
___

10. Tie rod:
___
___
___

Fill-in
Read each item carefully, and then complete the statement by filling in the missing word(s).

1. The _____ ladder is one of the most functional, versatile, and rapidly deployable tools used by fire fighters.
2. Ladders provide elevated platforms for _____ as well as for fire fighters.
3. The _____ serve as the steps of a ladder.
4. A(n) _____ ladder is an assembly of two or more ladder sections that can be extended or retracted to adjust the length.
5. _____ ladders are permanently mounted, power-operated ladders with a working length of at least 50 feet.
6. A(n) _____ ladder is a single-section, fixed-length ladder.
7. A(n) _____ ladder can be converted from a straight ladder to a stepladder configuration.
8. Ladders should always be inspected and maintained in accordance with the _____ recommendations.
9. _____ is a fire fighter's most important and unpredictable duty.
10. Fire fighters who are working from a ladder should use a safety belt or a(n) _____ to secure themselves to the ladder.
11. A rope, a rope-hose tool, or _____ can be used to secure a ladder in place.
12. When dismounting a ladder, the fire fighter should try to maintain contact with the ladder at _____ points.

True/False
If you believe the statement to be more true than false, write the letter "T" in the space provided.
If you believe the statement to be more false than true, write the letter "F."

1. _____ The ladder is one of the fire fighter's basic tools.
2. _____ A ladder can be used as a work platform.
3. _____ Ladders consist of two rungs connected by a series of parallel beams.
4. _____ There are two tips on a portable ladder.
5. _____ The butt and the heel of a ladder are at opposite ends of a portable ladder.
6. _____ When roof ladders are properly attached, they will not support the weight of the ladder and a fire fighter.
7. _____ Pulling on the halyard extends the bed sections of a combination ladder.
8. _____ SCBA is not required when working on the roof at a chimney fire.
9. _____ A fire fighter working from a ladder is in a less stable position than a fire fighter working on the ground.
10. _____ During a three-fire fighter shoulder carry, the middle fire fighter should be on the opposite side of the other two.
11. _____ When using an extension ladder, never wrap the halyard around your hand.
12. _____ In general, ladder manufacturers recommend that the fly sections be placed toward the structure.

Short Answer

Complete this section with short written answers using the space provided.

1. Describe the three basic types of beam construction.

   _____
   _____
   _____

2. Identify five fundamental ladder maintenance tasks.

   _____
   _____
   _____
   _____
   _____

3. Identify five basic safety issues when using ladders.

   _____
   _____
   _____
   _____
   _____

4. In the proper order, list the five steps to work from a ladder.

   _____
   _____
   _____
   _____
   _____

132 Fundamentals of Fire Fighter Skills

Word Fun

The following crossword puzzle is an activity provided to reinforce correct spelling and understanding of terminology associated with firefighting. Use the clues provided to complete the puzzle.

CLUES

Across

2. The rope or cable used to extend the fly section(s) of an extension ladder.
4. The level at which the ground intersects the foundation of a structure.
7. A metal rod that runs from one beam of the ladder to the other to keep the beams from separating.
8. The top or bottom piece of a trussed-beam assembly used in the construction of a trussed ladder.
9. The metal spikes attached to the butt of a ladder.
14. A power-operated ladder permanently mounted on a piece of apparatus.
15. Strips of metal or wood that serve to guide a fly section during extension.
17. A straight ladder equipped with retractable hooks so that the ladder can be secured to the peak of a pitched roof.
19. A piece of heat-sensitive material on each section of a ladder that identifies when the ladder has been exposed to high heat conditions.
22. An adjustable-length, multiple-section ladder.
23. A piece of material that prevents the fly section(s) of a ladder from overextending and collapsing the ladder.
24. A narrow, two-section extension ladder that has no halyard.

Down

1. The end of the ladder that is placed against the ground when the ladder is raised.
3. A belt specifically designed to secure a fire fighter to a ladder or elevated surface.
5. A method of exiting from an area or a building.
6. Reinforcing material placed on a ladder at chaffing and contact points to prevent damage from friction and contact with other surfaces.
9. These structures support the rungs.
10. A small, grooved wheel through which the halyard runs.
11. A ladder beam constructed of one continuous piece of I-shaped metal or fiberglass to which the rungs are attached.
12. A ladder beam constructed of a solid rectangular piece of material to which the ladder rungs are attached.
13. An A-shaped structure formed with two ladder sections.
16. A mechanical locking device used to secure the fly section(s) of a ladder after they have been extended.
18. A ladder that collapses by bringing the two beams together for portability.
19. The spring-loaded, retractable, curved metal pieces that allow the tip of a roof ladder to be secured to the peak of a pitched roof.
20. A ladder equipped with staypoles that stabilize the ladder during raising and lowering operations.
21. The very top of the ladder.

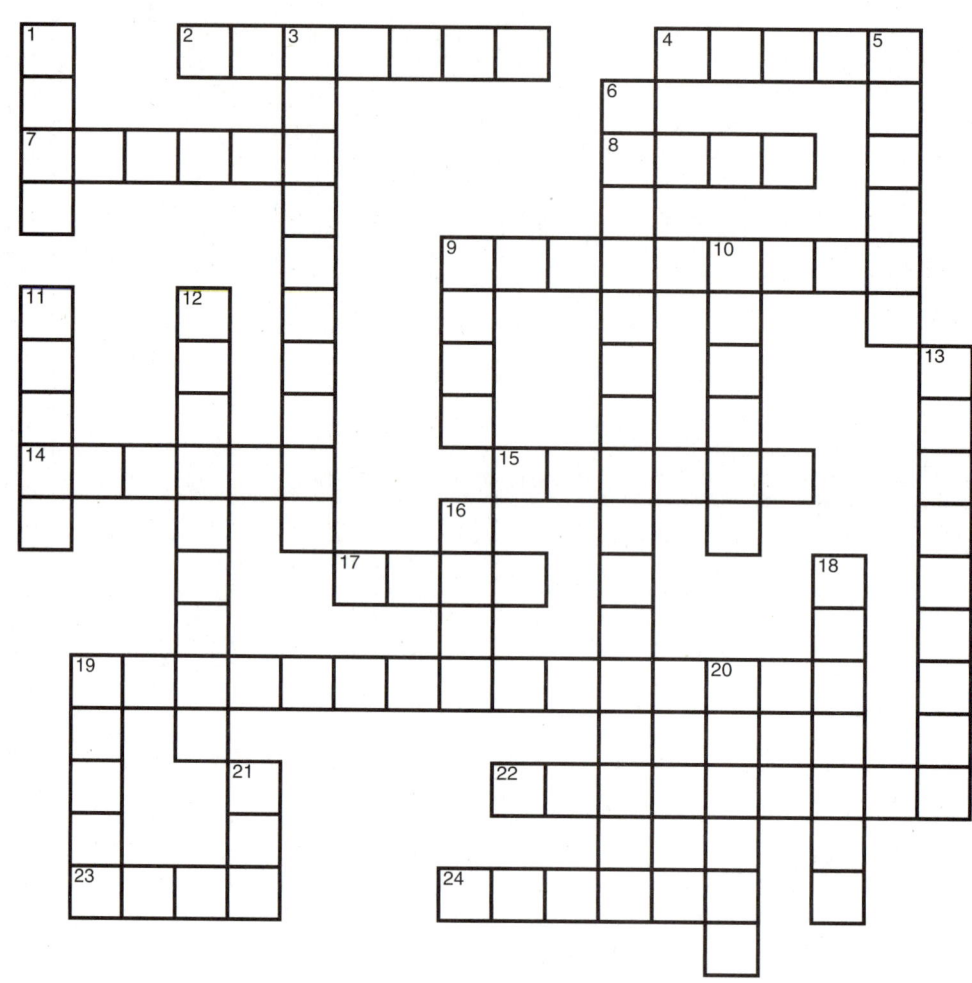

Chapter 12: Ladders

Fire Alarms

The following real case scenarios will give you an opportunity to explore the concerns associated with ladders. Read each scenario, and then answer each question in detail.

1. It is 11:30 in the evening when your engine company is dispatched to a two-story residential building for a structure fire with a life threat. As you exit your engine, you can hear cries for help. When you look toward the structure, you see a middle-aged woman straddling a second-floor window with smoke pouring out from behind her. The adjacent window is engulfed with fire. Your Lieutenant tells you to ladder the window for a rescue. How should you proceed?

   _____
   _____
   _____
   _____
   _____

2. It is 10:45 in the morning when your company is dispatched to an older two-story apartment complex. The fire is located in two units on the second floor. Upon arrival, the incident commander assigns you to Division C. Your Captain receives instructions to deploy a roof ladder utilizing the extension ladder already in place on side C of the structure. The Captain tells you and your partner to remove the roof ladder from the apparatus and place it on the roof. How should you proceed?

   _____
   _____
   _____
   _____
   _____

Skill Drills

Skill Drill 12-2: Performing a One-Fire Fighter Carry
Test your knowledge of this skill drill by filling in the correct words in the photo captions.

1. Locate the _____ of the ladder.

2. Place one arm through two rungs, just to one side of the _____ rung.

3. Bring the top _____ to rest on the fire fighter's shoulder.

Skill Drill 12-3: Performing a Two-Fire Fighter Shoulder Carry
Test your knowledge of this skill drill by placing the photos below in the correct order. Number the first step with a "1," the second step with a "2," and so on.

 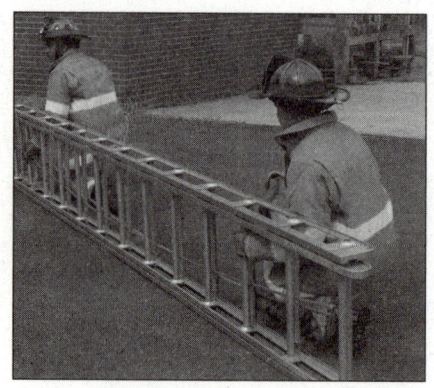 

_____ The butt spurs are covered with a gloved hand while the ladder is transported.

_____ Both fire fighters approach the ladder from the same side, facing the butt. One fire fighter stands near the butt, and the other stands near the tip.

_____ Each fire fighter places an arm between two rungs and lifts the ladder onto the shoulder.

Skill Drill 12-5: Performing a Two-Fire Fighter Suitcase Carry
Test your knowledge of this skill drill by filling in the correct words in the photo captions.

1. Both fire fighters face the _____ of the ladder, at opposite ends.

2. The fire fighters grasp the upper _____ of the ladder.

3. Pick up the ladder using good _____ techniques.

Skill Drill 12-11: Performing a One-Fire Fighter Rung Raise for Ladders Less Than 14 Feet Long
Test your knowledge of this skill drill by placing the photos below in the correct order. Number the first step with a "1," the second step with a "2," and so on.

_____ Place the ladder flat on the ground, with the heel positioned where it will be when the ladder is raised. By the tip, raise the ladder to hip level. Walk hand-over-hand down the rungs until the ladder is vertical.

_____ Carry the ladder to the structure. Check for overhead hazards before raising the ladder.

_____ Heel the ladder and lean it into place.

Skill Drill 12-13: Tying the Halyard
Test your knowledge of this skill drill by placing the photos below in the correct order. Number the first step with a "1," the second step with a "2," and so on.

_____ Pull the knot tight and add a safety knot.

_____ Wrap the excess halyard around two rungs and pull it tight over the upper rung.

_____ Tie a clove hitch around the upper rung.

Skill Drill 12-18: Climbing a Ladder while Carrying a Tool
Test your knowledge of this skill drill by filling in the correct words in the photo captions.

1. Hold the tool in one hand and place it against the _____.

Slide the tool up the beam, while sliding the opposite _____ up the other beam.

Skill Drill 12-19: Working from a Ladder
Test your knowledge of this skill drill by placing the photos below in the correct order. Number the first step with a "1," the second step with a "2," and so on.

_____ The fire fighter notes the side where the work will be performed. The opposite leg is extended between the rungs.

_____ The foot is secured around the lower rung or the beam. The fire fighter moves the other leg down one rung.

_____ The knee is bent around the rung and the foot is brought back under the rung.

_____ The fire fighter is now free to work with both hands.

_____ The fire fighter climbs the ladder to the desired work height and then one rung higher.

Search and Rescue

Workbook Activities

The following activities have been designed to help you. Your instructor may require you to complete some or all of these activities as a regular part of your fire fighter training program. You are encouraged to complete any activity that your instructor does not assign as a way to enhance your learning in the classroom.

Chapter Review

The following exercises provide an opportunity to refresh your knowledge of this chapter.

Matching

Match each of the terms in the left column to the appropriate definition in the right column.

_____ 1. Search
_____ 2. Rescue
_____ 3. Secondary search
_____ 4. Rekindle
_____ 5. Search rope
_____ 6. Thermal imaging device
_____ 7. Primary search
_____ 8. Left-hand search
_____ 9. Search and rescue
_____ 10. Rescue techniques

A. The removal of a person from danger
B. The process of looking for victims who are in danger
C. A more thorough search undertaken after the fire is under control
D. A situation where a fire, which was thought to be completely extinguished, reignites
E. An electronic device that detects differences in temperature based on infrared energy and then generates images based on those data
F. A guide rope that allows fire fighters to maintain contact with a fixed point
G. Assists, drags, and carries
H. Often the highest risk to fire fighters
I. A clockwise search pattern
J. Quick initial search for victims

Multiple Choice

Read each item carefully, and then select the best response.

_____ 1. The removal of a person from a confined space is classified as a
   A. search.
   B. rescue.
   C. primary rescue.
   D. secondary search.

_____ 2. When a building is occupied, fire fighters should first rescue the occupants who are
   A. in the most immediate danger.
   B. in the least danger.
   C. the most easily accessed.
   D. closest to the exits.

CHAPTER 13

_____ 3. A search begins with the areas where
   A. the greatest number of hazards exist.
   B. the building experiences the greatest traffic.
   C. occupants are expected.
   D. victims are at the greatest risk.

_____ 4. After the area immediately around the fire is searched in an apartment building, the next priority is to search the
   A. area directly above the fire.
   B. area directly below the fire.
   C. highest floors in the building.
   D. hallways and exits.

_____ 5. When conducting searches in high-rise buildings, it is important to work
   A. from the bottom floor up.
   B. from the middle floors out.
   C. from the walls to the middle of the rooms.
   D. as teams, coordinating searches.

_____ 6. The three most important senses during a search are
   A. sight, sound, and taste.
   B. touch, sight, and taste.
   C. sight, sound, and touch.
   D. sound, taste, and touch.

_____ 7. Information gathered from search operations needs to be communicated to the
   A. secondary search team.
   B. incident commander.
   C. safety officer.
   D. rapid intervention company/crew.

_____ 8. Search ropes should be used to
   A. help fire fighters exit the area.
   B. keep search teams connected.
   C. search wide open spaces.
   D. encourage search teams' progress.

_____ 9. After the fire is under control and the structural stability of the building is confirmed, fire fighters should begin a
   A. primary search.
   B. secondary search.
   C. rescue.
   D. safety search.

_____ 10. Searchers can use the hose line to guide them out of the building if they follow the
   A. female coupling.
   B. main attack line.
   C. male coupling.
   D. secondary attack line.

_____ 11. If the person is capable of walking, rescuers may only need to use the
   A. one-person walking assist.
   B. two-person seat carry.
   C. cradle-in-arms carry.
   D. three-person walking assist.

_____ 12. The two-person seat carry is used when the victim
   A. is very large.
   B. must be carried up or down stairs.
   C. is a child.
   D. is disabled or paralyzed.

_____ 13. The clothes drag is used to move a victim who is on the floor and
   A. is too heavy for one rescuer.
   B. must be carried up or down stairs.
   C. is disabled or paralyzed.
   D. is difficult to reach.

_____ 14. The fire fighter drag utilizes the victim's
   A. clothing as a handle.
   B. tied hands.
   C. weight to assist the movement.
   D. ability to assist moving.

_____ 15. To rescue a victim through a window, raise the ladder and secure it in the rescue position with the tip
   A. above the windowsill.
   B. in the open window.
   C. just below the windowsill.
   D. upwind from the window.

_____ 16. When removing an unresponsive victim on a ladder, the fire fighter on the ladder should
   A. maintain continuous eye contact with the victim.
   B. maintain eye contact with the second fire fighter.
   C. face the victim.
   D. remain on the ladder until a rescue team can assist the descent.

_____ 17. When a ladder rescue involves a conscious victim, the fire fighter should
   A. establish verbal contact.
   B. urge the victim to jump.
   C. have the victim climb down facing the fire fighter.
   D. have the victim exit head first.

_____ 18. When rescuing heavy adults using a ladder, the rescuer should
   A. get more help—three rescuers at a minimum.
   B. use two ladders.
   C. place two ladders side by side.
   D. all of the above.

_____ 19. Long backboard rescues are used to
   A. carry a conscious victim down a ladder.
   B. remove a victim from a vehicle.
   C. aid with a one-rescuer drag.
   D. aid with a fire fighter drag.

_____ 20. A webbing sling provides a secure grip around
   A. the victim's upper body.
   B. the victim's waist.
   C. the victim's arms.
   D. the victim's legs and feet.

Vocabulary

Define the following terms using the space provided.

1. Two-in/two-out rule:

   _____
   _____
   _____

2. Exit assist:

   _____
   _____
   _____

3. Primary search:

   _____
   _____
   _____

4. Shelter-in-place:

   _____
   _____
   _____

Fill-in

Read each item carefully, and then complete the statement by filling in the missing word(s).

1. A(n) _____ is done to look for victims who need assistance to leave a dangerous area.
2. Saving _____ has the highest priority of all operations.
3. _____ can reduce interior temperatures and improve visibility, enabling search teams to locate victims more rapidly.
4. The _____ process at every fire should include a specific evaluation of critical factors for search and rescue.
5. Occupants who are asleep are at a(n) _____ risk than occupants who are awake.
6. Upon completion of all searches, the priority can shift to _____ and _____ the fire.
7. Search team members may have to _____ to stay below layers of hot gases and smoke.
8. Search teams must have a(n) _____ escape route in case fire conditions change.
9. Searched rooms should be _____ so other searchers will know they have been searched.
10. The incident commander is responsible for managing the level of _____ during emergency operations.

True/False

If you believe the statement to be more true than false, write the letter "T" in the space provided.
If you believe the statement to be more false than true, write the letter "F."

1. _____ Fire fighters need to focus on fire suppression so they can prepare for search and rescue operations.
2. _____ Observations of the size and condition of the building can provide valuable information when trying to determine how many people may need rescue.
3. _____ Search and rescue operations should utilize only one search and rescue team.
4. _____ More fire fighters will be needed for search and rescue operations in a nursing home than in a similar-sized office building.
5. _____ It is often difficult to obtain accurate information from people who have just escaped from a burning building.
6. _____ Standard operating procedures emphasize various search patterns to ensure complete searches.
7. _____ It is justifiable to risk the safety of fire fighters if there is a potential to save lives.
8. _____ The two-in/two-out rule can be broken if there is a life-threatening situation.
9. _____ Most people who realize they are in danger will attempt to escape on their own.
10. _____ In some situations, the best option is to shelter occupants in place.

Short Answer

Complete this section with short written answers using the space provided.

1. Identify four considerations for search and rescue size-up.

   _____
   _____
   _____
   _____

2. Identify five pieces of valuable information a preincident plan can provide for search and rescue operations.

   _____
   _____
   _____
   _____
   _____

3. Describe three benefits of thermal imaging devices.

   _____
   _____
   _____

4. Identify the six tips fire fighters need to remember during search and rescue operations.

_____
_____
_____
_____
_____
_____

5. List six pieces of search and rescue equipment.

_____
_____
_____
_____
_____
_____

6. Identify the four simple carries that can be used to move a victim who is conscious and responsive, but incapable of standing or walking.

_____
_____
_____
_____

Fundamentals of Fire Fighter Skills

Fire Alarms

The following real case scenarios will give you an opportunity to explore the concerns associated with search and rescue. Read each scenario, and then answer each question in detail.

1. It is 1:00 in the morning when you are dispatched to an old three-story apartment complex. The main stairway is located in the center of the building. The fire is located on the first floor. A suppression crew has been sent to the seat of the fire. There are mixed reports on whether the building has been evacuated. Your engine company has been assigned to the third floor for search and rescue. How will you proceed?

   _____
   _____
   _____
   _____

2. You are the leader of a search and rescue team about to enter a carpet warehouse. Two employees are missing and thought to be located in the fire building. The warehouse has sprinklers, and the fire is confined. Your officer warns that the inside floor plan is complex owing to the numerous remodels. To make matters worse, the smoke is thick and black because of the burning carpet, and it is cold and hugging the floor because of the activation of the sprinkler system. Knowing that this is a very dangerous task, you want to ensure the safety of yourself and your crew. How should you proceed?

   _____
   _____
   _____
   _____
   _____

Skill Drills

Skill Drill 13-5: Performing a Two-Person Extremity Carry

Test your knowledge of this skill drill by placing the photos below in the correct order. Number the first step with a "1," the second step with a "2," and so on.

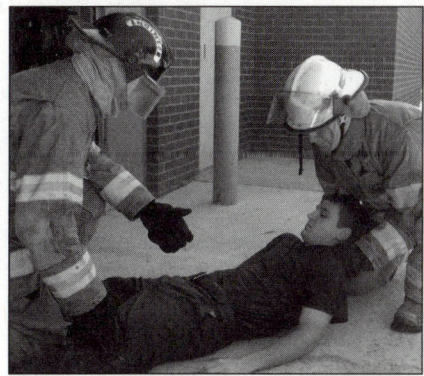

_____ Two fire fighters help the victim to sit up.

_____ The first fire fighter gives the command to stand and carry the victim away, walking straight ahead. Both fire fighters must coordinate their movements.

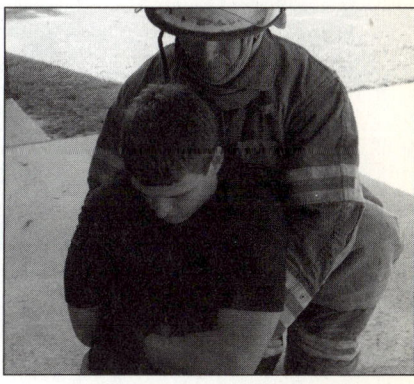

_____ The first fire fighter kneels behind the victim, reaches under the victim's arms, and grasps the victim's wrists.

_____ The second fire fighter backs in between the victim's legs, reaches around, and grasps the victim behind the knees.

Skill Drill 13-6: Performing a Two-Person Seat Carry
Test your knowledge of this skill drill by filling in the correct words in the photo captions.

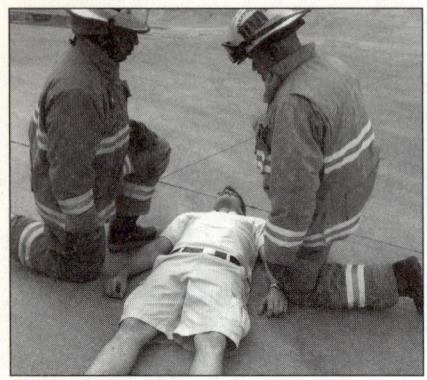

1. Kneel beside the victim near the victim's _____.

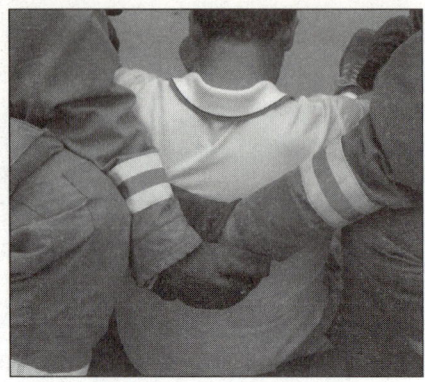

2. Raise the victim to a(n) _____ position and link arms behind the victim's _____.

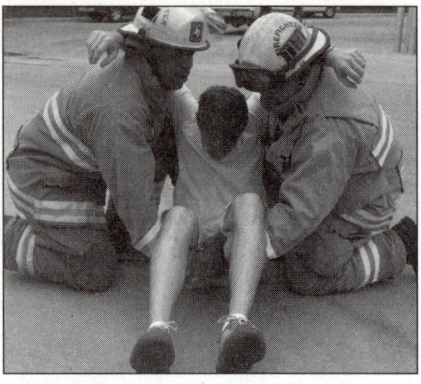

3. Place your free arms under the victim's _____ and link arms.

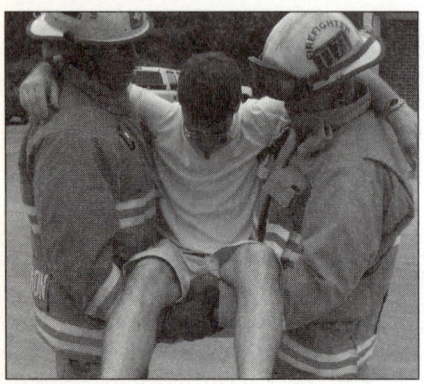

4. If possible, the victim puts his or her arms around your _____ for additional support.

Skill Drill 13-9: Performing a Clothes Drag
Test your knowledge of this skill drill by filling in the correct words in the photo captions.

 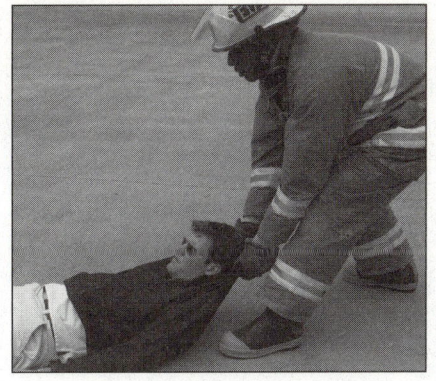

1. _____ behind the victim's _____, and grab the shirt or jacket around the collar and shoulder area.

2. Lift with your legs until you are fully upright. Walk _____, dragging the victim to safety.

Skill Drill 13-12: Performing a Webbing Sling Drag
Test your knowledge of this skill drill by placing the photos below in the correct order. Number the first step with a "1," the second step with a "2," and so on.

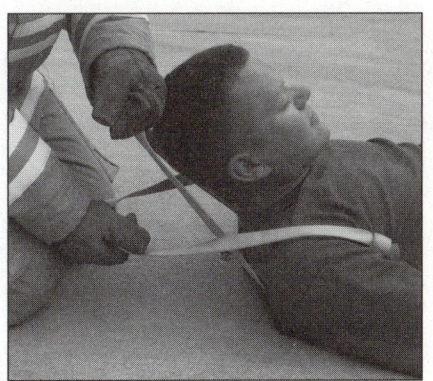 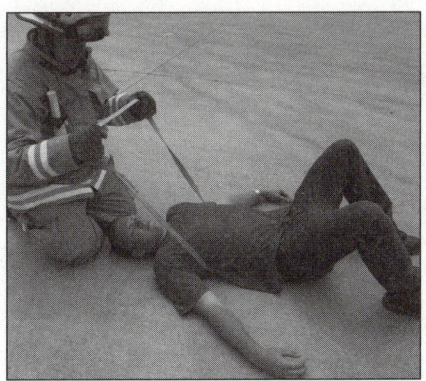 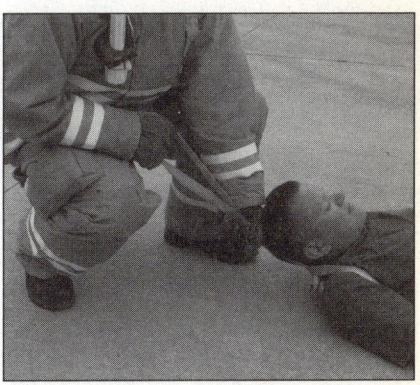

_____ Take the large loop over the victim and place it above the victim's head. Reach through, grab the webbing behind the victim's back, and pull through all the excess webbing. This creates a loop at the top of the victim's head and two loops around the victim's arms.

_____ Place the victim in the center of the loop so the webbing is behind the victim's back.

_____ Adjust hand placement to protect the victim's head while dragging.

Skill Drill 13-13: Performing a Fire Fighter Drag
Test your knowledge of this skill drill by filling in the correct words in the photo captions.

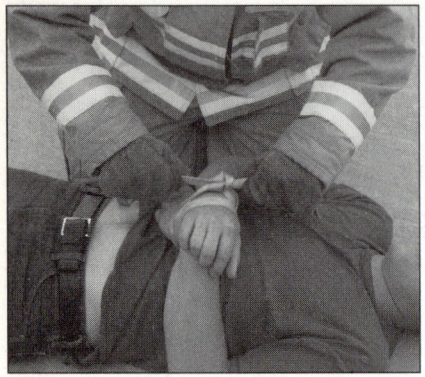  

1. _____ the victim's _____ together with anything that is handy.

2. Get down on your _____ and _____ and straddle the victim.

3. Pass the victim's tied hands around your _____, straighten your arms, and drag the victim across the floor by _____ on your _____ and _____.

Skill Drill 13-14: Performing a One-Person Emergency Drag from a Vehicle
Test your knowledge of this skill drill by placing the photos below in the correct order. Number the first step with a "1," the second step with a "2," and so on.

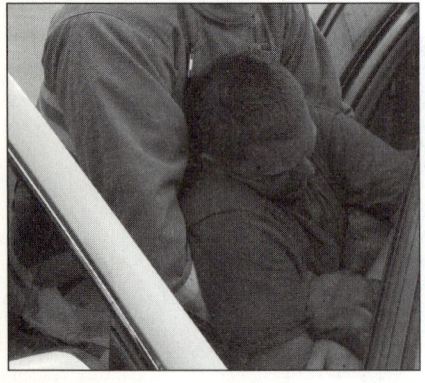 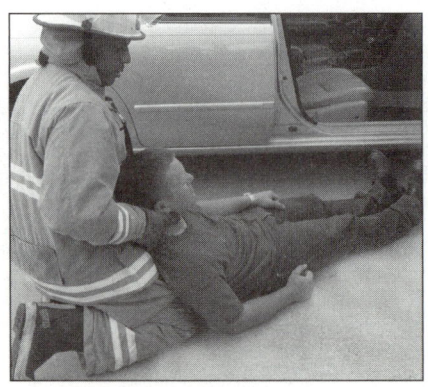 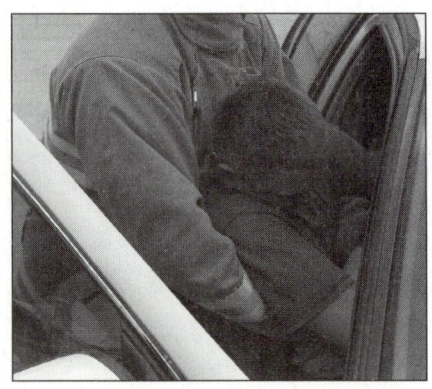

_____ Gently pull the victim out of the vehicle.

_____ Lower the victim down into a horizontal position in a safe place.

_____ Grasp the victim under the arms and cradle the head between your arms.

Skill Drill 13-17: Rescuing an Unconscious Victim from a Window

Test your knowledge of this skill drill by filling in the correct words in the photo captions.

1. The tip of the ladder is placed just _____ the windowsill.

2. One fire fighter _____ to assist the victim. The second fire fighter climbs to the window.

3. The fire fighter waiting on the ladder places both hands on the _____, with one leg _____ and the other leg horizontal to the ground with the knee at an angle of _____. The interior fire fighter will then pass the victim through the window and onto the ladder, keeping the victim's _____ toward the ladder.

4. The victim is lowered so that he or she _____ the fire fighter's leg. The fire fighter's _____ should be positioned under the victim's arms holding onto the _____.

5. Step down one rung at a time, _____ the victim's _____ from one leg to the other. The victim's _____ can also be secured around the fire fighter's neck.

Ventilation

Workbook Activities

The following activities have been designed to help you. Your instructor may require you to complete some or all of these activities as a regular part of your fire fighter training program. You are encouraged to complete any activity that your instructor does not assign as a way to enhance your learning in the classroom.

Chapter Review

The following exercises provide an opportunity to refresh your knowledge of this chapter.

Matching

Match each of the terms in the left column to the appropriate definition in the right column.

_____ 1. Chase
_____ 2. Parapet walls
_____ 3. Cockloft
_____ 4. Roof decking
_____ 5. Rafters
_____ 6. Convection
_____ 7. Ejectors
_____ 8. Laths
_____ 9. Pitched roofs
_____ 10. Flashover

A. Solid structural components that support a roof
B. Sloped roofs, often found on residences
C. Electrical fans used in negative-pressure ventilation
D. Open space within walls for wires and pipes
E. The concealed space between the top-floor ceiling and the roof of a building
F. Walls on a flat roof that extend above the roofline
G. The sudden ignition of all combustible objects in a room
H. Thin strips of wood used to make the supporting structure for roof tiles
I. The rigid component of a roof covering
J. The transfer of heat through a circulating medium of liquid or gas

Multiple Choice

Read each item carefully, and then select the best response.

_____ 1. The process of removing smoke, heat, and toxic gases from a burning building is called
   A. ventilation.
   B. convection.
   C. conduction.
   D. smoke inversion.

_____ 2. As a fire progresses and grows, it produces smoke, heat, and toxic gases, which are collectively known as the
   A. products of combustion.
   B. products of ventilation.
   C. ejectors.
   D. exhaust.

_____ 3. The transfer of heat through liquid or gas is called
   A. churning.
   B. conduction.
   C. leap-frogging.
   D. convection.

CHAPTER 14

_____ 4. A vertical space within a wall where heat and toxic gases can climb to a higher level is a
   A. gap.
   B. converter.
   C. chase.
   D. space.

_____ 5. When smoke, heat, and gases cannot find a vertical path, they
   A. roll.
   B. mushroom.
   C. flow.
   D. ventilate.

_____ 6. A backdraft occurs when smoke, heat, and gases accumulate with a rich supply of partially burned fuels and are suddenly introduced to
   A. a flame.
   B. increased temperature.
   C. clean air.
   D. open fuels.

_____ 7. Which type of ventilation occurs when fans are used to pull smoke through openings?
   A. Positive-pressure ventilation
   B. Natural ventilation
   C. Hydraulic ventilation
   D. Negative-pressure ventilation

_____ 8. Which type of ventilation occurs when fans are used to push clean air into a space to displace smoke?
   A. Positive-pressure ventilation
   B. Natural ventilation
   C. Hydraulic ventilation
   D. Negative-pressure ventilation

_____ 9. In a heated high-rise building, on a cold day there will probably be
   A. a strong downdraft.
   B. a strong updraft.
   C. an increase in air movement.
   D. greater resistance to air movement.

_____ 10. If it is safe to do so, ventilation operations should be conducted
   A. in front of the fire.
   B. beside the fire.
   C. as close to the fire as possible.
   D. as soon as possible.

_____ 11. Horizontal ventilation is most effective when the opening goes directly
   A. to another space within the structure.
   B. into a stairwell.
   C. into the space where the fire is located.
   D. past the attack team.

_____ 12. When fire fighters need quick or immediate ventilation, they often use
   A. natural ventilation.
   B. mechanical ventilation.
   C. hydraulic ventilation.
   D. vertical ventilation.

152   FUNDAMENTALS OF FIRE FIGHTER SKILLS

_____ 13. When breaking glass for ventilation purposes, the fire fighter should always use a(n)
  A. "All clear" call.
  B. hand tool.
  C. hose line.
  D. safety break before splintering.

_____ 14. Thermopane windows are more energy efficient and
  A. are more heat resistant.
  B. are less heat resistant.
  C. are easier to break.
  D. provide greater ventilation.

_____ 15. Negative-pressure ventilation fans are called
  A. ventilators.
  B. conductors.
  C. smoke ejectors.
  D. HVAC systems.

_____ 16. _____ fans are powered by internal combustion engines and can increase carbon monoxide levels if they run for significant periods of time after the fire is extinguished.
  A. Negative-pressure
  B. Horizontal
  C. Mechanical
  D. Positive-pressure

_____ 17. Hydraulic ventilation is most useful for clearing smoke and heat out of a room because it creates
  A. a low-pressure area behind the nozzle.
  B. a high-pressure area behind the nozzle.
  C. a mist that traps smoke particles and heat.
  D. water vapor.

_____ 18. Ventilation openings should never be
  A. opened directly into the atmosphere.
  B. between the fire fighters and the escape route.
  C. created without IC direction.
  D. opened before proper sounding.

_____ 19. All roofs have two major components:
  A. a support structure and a roof shoring.
  B. beams and rafters.
  C. a support structure and a roof covering.
  D. a platform and a support system.

_____ 20. Trusses are connected with heavy-duty staples or by
  A. triangular plates.
  B. gusset plates.
  C. plate locks.
  D. truss plates.

_____ 21. A bearing wall is used
  A. as an exterior wall.
  B. to support the weight of a floor or roof.
  C. as an interior wall.
  D. to extend the firewall.

_____ 22. Which type of roof has a visible slope for rain or snow runoff?
  A. Bowstring roof
  B. Arched roof
  C. Flat roof
  D. Pitched roof

_____ 23. A triangular examination hole created with three small cuts is a
   A. kerf cut.
   B. rectangular cut.
   C. louver cut.
   D. triangular cut.

_____ 24. A cut that can create a large opening quickly and is particularly suitable for flat or sloping roofs with plywood decking is a
   A. kerf cut.
   B. rectangular cut.
   C. louver cut.
   D. triangular cut.

_____ 25. A cut that works well on metal roofing because it prevents the decking from rolling away is a
   A. kerf cut.
   B. rectangular cut.
   C. louver cut.
   D. triangular cut.

Vocabulary

Define the following terms using the space provided.

1. Fire-resistive construction:

2. Ordinary construction:

3. Stack effect:

4. Ventilation:

5. Smoke inversion:

FUNDAMENTALS OF FIRE FIGHTER SKILLS

6. Sounding:
___
___
___

7. Vertical ventilation:
___
___
___

8. Horizontal ventilation:
___
___
___

9. Gusset plates:
___
___
___

Fill-in
Read each item carefully, and then complete the statement by filling in the missing word(s).

1. _____ ventilation takes advantage of the doors and windows on the same level as the fire, as well as any other horizontal openings that are available.

2. _____ is when the smoke or heat rises to reach a horizontal barrier and then begins to move out and back down.

3. Positive-pressure fans operate at a(n) _____ velocity to increase the air pressure in a space.

4. Fire fighters should be _____ from the ventilation openings so the wind will push the heat and smoke away.

5. The collapse of a(n) _____ truss roof is usually very sudden; for this reason, the presence of such a roof must be noted during preincident planning.

6. _____ ventilation creates a large opening ahead of the fire to reduce the fuel for the fire to spread and increase smoke and gas flow out of a building.

7. _____, _____ and _____ are powerful natural factors that can greatly affect ventilation.

8. _____ windows are used to increase the energy efficiency of a structure.

9. _____ ventilation is most useful after a fire is under control.

10. The most obvious risk to fire fighters performing vertical ventilation is _____.

True/False

If you believe the statement to be more true than false, write the letter "T" in the space provided.
If you believe the statement to be more false than true, write the letter "F."

1. _____ Vertical ventilation operations often involve opening or breaking a window.
2. _____ During ventilation, cutting several smaller holes is better than making one large hole.
3. _____ When carrying tools up a ladder, hold the beam of the ladder instead of the rungs.
4. _____ The type of roof construction is a major factor that is considered when fire fighters are determining which type of cut to use.
5. _____ A strong concern that arises when structures have metal roofs is the release of flammable vapors, which can be the result of leaking roof coverings.
6. _____ Basement fires are often easy to ventilate, because the smoke, heat, and gases can rise only within the structure.
7. _____ HVAC or other building systems can cause fire to spread through leap-frogging.
8. _____ Ventilation should occur as close to the fire as possible.
9. _____ Backdrafts are the results of negative ventilation.
10. _____ Smoke can be cooled with automatic sprinkler systems.

Short Answer

Complete this section with short written answers using the space provided.

1. What is the objective of any roof ventilation operation?

2. Describe the difference between a primary cut and a secondary cut.

3. Identify three factors that affect ventilation.

4. List five indicators that it is time for immediate retreat from the roof of a structure.

Word Fun

The following crossword puzzle is an activity provided to reinforce correct spelling and understanding of terminology associated with firefighting. Use the clues provided to complete the puzzle.

Clues

Across

4. The process of making openings so that smoke can escape horizontally from a building.
7. The sudden ignition of all combustible objects in a room.
8. The process of making openings so that smoke can escape vertically from a building.
9. Recirculation of exhausted air that is drawn back into a negative-pressure fan in a circular motion.
10. A wall that is designed to support the weight of a floor or roof.
14. The transfer of heat through a circulating medium of liquid or gas.
15. Electrical fans used in negative-pressure ventilation.
17. A cut that is used to inspect cockloft spaces from the roof.
18. Ventilation that relies on the natural movement of heated smoke and wind currents.
19. The sudden explosive ignition of fire gases when oxygen is introduced into a superheated space previously deprived of oxygen.

Down

1. The main ventilation opening made in a roof to allow gases to escape.
2. Ventilation that uses mechanical devices to move air.
3. Ventilation that relies on the movement of air caused by a fog stream.
5. A fire spread from one floor to the other through exterior windows.
6. The vertical movement of smoke and products of combustion within a high-rise building caused by temperature differentials between the interior and the exterior.
11. Solid structural components that support a roof.
12. A cut that is made using power saws and axes along and between roof supports so that the sections created can be tilted into the opening.
13. The process of striking a roof with a tool to determine if the roof is solid enough to support the weight of a fire fighter.
16. Open space within walls for wires and pipes.

Fire Alarms

The following real case scenarios will give you an opportunity to explore the concerns associated with ventilation. Read each scenario, and then answer each question in detail.

1. It is 2:00 in the morning when your engine is dispatched to a residential structure fire. You are on the second engine to arrive on scene. The interior attack team is met with smoke and high heat at the front door. The attack team notifies the IC that ventilation is needed prior to entering the structure. The IC assigns you and your partner to coordinate positive-pressure ventilation with the attack team. How will you proceed?

   _____
   _____
   _____
   _____
   _____

2. It is 12:30 in the afternoon when your engine is dispatched to a two-story apartment building. Dispatch reports that the fire is located on the second floor. The interior attack teams are fighting the fire as you arrive. The attack team reports that the fire is knocked down on the second floor but there is fire in the attic. The IC tells your Lieutenant to ladder the building and perform vertical ventilation. Your Lieutenant tells two members of your crew to ladder the building, and you and your partner to grab the chainsaw and complete a louver cut over the seat of the fire. How should you proceed?

   _____
   _____
   _____
   _____
   _____

Skill Drills

Skill Drill 14-1: Breaking Glass with a Hand Tool
Test your knowledge of this skill drill by filling in the correct words in the photo captions.

1. Position yourself to the _____ of the window.

2. With your back facing the wall, swing _____ forcefully with the tip of the tool striking the top one-third of the glass.

3. Clear remaining glass from the opening with the _____.

Skill Drill 14-5: Delivering Positive-Pressure Ventilation

Test your knowledge of this skill drill by placing the photos below in the correct order. Number the first step with a "1," the second step with a "2," and so on.

_____ Provide an exhaust opening at or near the fire.

_____ Place the fan in front of the opening to be used for the fire attack.

_____ Start the fan and allow the smoke to clear.

Skill Drill 14-6: Sounding a Roof

Test your knowledge of this skill drill by filling in the correct words in the photo captions.

1. Use a hand tool to check the roof before _____ onto it.

2. Use the tool to sound ahead and to both _____ as you walk. Locate support members by sound and rebound. Check conditions around your work area periodically.

3. Sound the roof along your _____ path.

Skill Drill 14-8: Making a Rectangular or Square Cut

Test your knowledge of this skill drill by placing the photos below in the correct order. Number the first step with a "1," the second step with a "2," and so on.

_____ Make two cuts perpendicular to the roof supports and then make the final cut parallel to another roof support.

_____ Pull out or push in the triangle cut.

_____ Punch out the ceiling below. Be wary of a sudden updraft of hot gases or flames.

_____ Make a triangle cut at the first corner.

_____ Locate the roof supports by sounding. Make the first cut parallel to the roof support.

Skill Drill 14-9: Making a Louver Cut
Test your knowledge of this skill drill by filling in the correct words in the photo captions.

1. Locate the roof supports by _____.

2. Make two parallel cuts _____ to the roof supports.

3. Cut parallel to the supports and between pairs of supports in a(n) _____ pattern.

4. Tilt the _____ to a vertical position.

Skill Drill 14-10: Making a Triangular Cut

Test your knowledge of this skill drill by placing the photos below in the correct order.
Number the first step with a "1," the second step with a "2," and so on.

_____ Make the first cut from just inside a support member in a diagonal direction toward the next support member.

_____ Make the final cut along the support member so as to connect the first two cuts. Cutting from this location allows fire fighters the full support of the member directly below them while performing ventilation.

_____ Begin the second cut at the same location as the first, and make it in the opposite diagonal direction, forming a "V" shape.

_____ Locate the roof supports.

Water Supply

Workbook Activities

The following activities have been designed to help you. Your instructor may require you to complete some or all of these activities as a regular part of your fire fighter training program. You are encouraged to complete any activity that your instructor does not assign as a way to enhance your learning in the classroom.

Chapter Review

The following exercises provide an opportunity to refresh your knowledge of this chapter.

Matching

Match each of the terms in the left column to the appropriate definition in the right column.

_____ 1. Flow pressure   A. The pressure in a water pipe when there is no water flowing
_____ 2. Dump valve   B. A gauge used to determine the flow of water from a hydrant
_____ 3. Pitot gauge   C. The pressure remaining while water is flowing
_____ 4. Static pressure   D. Any valve that can be used to shut down water flow
_____ 5. Shut-off valve   E. The amount of pressure created by moving water
_____ 6. Residual pressure   F. A large valve used for quick discharge
_____ 7. Reservoir   G. A small-diameter underground water pipe
_____ 8. Water main   H. The largest-diameter pipe in a water distribution system
_____ 9. Primary feeder   I. Any underground water pipe
_____ 10. Distributor   J. A water storage facility

Multiple Choice

Read each item carefully, and then select the best response.

_____ 1. The water source, treatment plant, and distribution system are parts of a
   A. municipal water system.
   B. private water system.
   C. static water source.
   D. reservoir.

_____ 2. The distribution system of underground pipes is known as
   A. reservoirs.
   B. piping.
   C. water mains.
   D. water traffic.

CHAPTER 15

_____ 3. What is the recommended minimum water pressure from a fire hydrant?
   A. 10 psi
   B. 20 psi
   C. 40 psi
   D. 60 psi

_____ 4. The smallest pipes in a water distribution system that carry the water to the users and hydrants are the
   A. primary feeders.
   B. secondary feeders.
   C. water mains.
   D. distributors.

_____ 5. The pipes that deliver large quantities of water to a section of a town or city are the
   A. primary feeders.
   B. secondary feeders.
   C. direct mains.
   D. distributors.

_____ 6. The size of water mains depends on the amount of water needed for both normal consumption and
   A. heavy consumption.
   B. extended delays.
   C. fire protection.
   D. business operations.

_____ 7. To assure that water flows to a fire hydrant from two or more directions, well-designed systems follow a
   A. mixed pattern.
   B. multiple-port pattern.
   C. grid pattern.
   D. center pattern.

_____ 8. Fire department hoses can be connected to a hydrant by the
   A. valves.
   B. outlets.
   C. ports.
   D. taps.

_____ 9. A large opening on a fire hydrant that is used to allow as much water as possible to flow directly into the pump is a(n)
   A. outlet.
   B. steamer port.
   C. valve.
   D. drain.

_____ 10. Most dry-barrel hydrants have _____ large valve(s) controlling the flow of water.
   A. one
   B. two
   C. three
   D. four

_____ 11. What are the first factors to check when inspecting hydrants?
   A. Stability and structural integrity
   B. Visibility and structural integrity
   C. Visibility and accessibility
   D. Stability and component location

_____ 12. Hydrants should be positioned so that the connections, and especially the large steamer connection,
   A. are parallel to the water system.
   B. will not be damaged by passing traffic.
   C. are near waste storage areas.
   D. face the street.

_____ 13. The flow or quantity of water moving through a pipe, hose, or nozzle is measured by
   A. volume.
   B. area.
   C. weight.
   D. mass.

_____ 14. Which unit is used to measure water pressure?
   A. Gallons
   B. Gallons per square inch
   C. Pounds per square inch
   D. Pounds

_____ 15. Water that is not moving has
   A. elevation pressure.
   B. static pressure.
   C. residual pressure.
   D. potential pressure.

_____ 16. What is the best indication of how much more water is available in the system?
   A. The elevation pressure
   B. The static pressure
   C. The residual pressure
   D. The potential pressure

_____ 17. The quantity of water flowing through an opening during a hydrant test is the
   A. residual pressure.
   B. flow pressure.
   C. normal operating pressure.
   D. elevation pressure.

_____ 18. Which type of hydrant includes a pipe with a strainer on one end and a cap threaded for a hard suction hose on the other end that can be used to access static water sources?
   A. Dry hydrant
   B. Wet hydrant
   C. Static hydrant
   D. Straining hydrant

_____ 19. If a large volume of water is needed for an extended period, tankers can be used to deliver water from a fill site to the scene, thereby creating:
   A. a mobile water system.
   B. portable tanks.
   C. a mobile water supply apparatus.
   D. a tanker shuttle.

_____ 20. Which type of system may not require pumps because the water source, the treatment plant, and storage facilities are located on ground higher than the end users?
   A. Gravity-feed system
   B. Wet hydrant system
   C. Tanker shuttle system
   D. Static system

Labeling
Label the following diagram with the correct terms.

1. Dry hydrant placed near a static water source.

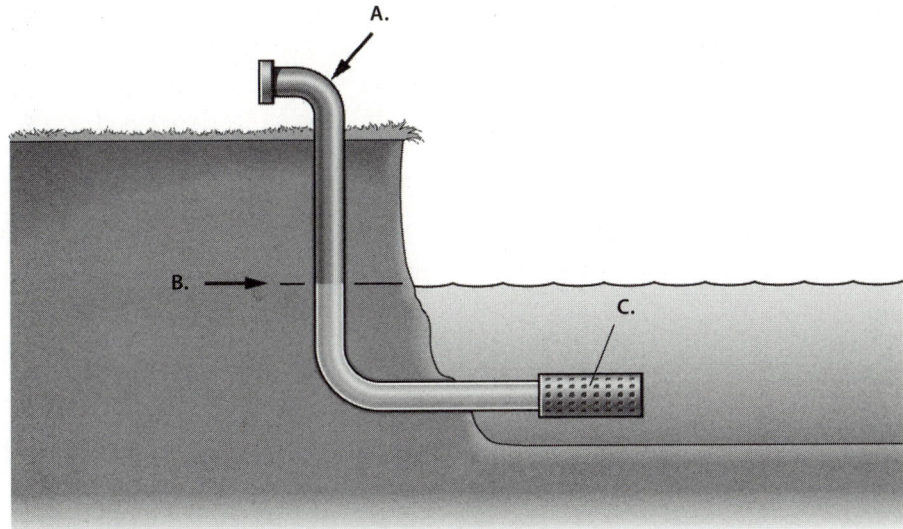

Figure 15-4

A. _____

B. _____

C. _____

Vocabulary
Define the following terms using the space provided.

1. Gravity-feed system:
   _____
   _____
   _____

2. Dry-barrel hydrant:
   _____
   _____
   _____

3. Tanker shuttle:

_____
_____
_____

4. Normal operating pressure:

_____
_____
_____

5. Municipal water system:

_____
_____
_____

Fill-in

Read each item carefully, and then complete the statement by filling in the missing word(s).

1. The importance of a dependable and adequate _____ for fire-suppression operations is self-evident.
2. Rural areas may depend on _____ such as lakes and streams.
3. Municipal water systems can draw water from human-made storage facilities called _____.
4. _____ valves allow different water main sections to be turned off or isolated.
5. Hydrants are equipped with one or more _____ to control the flow of water through the hydrant.
6. Most hydrants have an upright steel casing or _____ that is attached to the underground water distribution system.
7. Dry-barrel hydrants need to be either _____ opened or _____ closed.
8. In many communities, hydrants are painted in bright reflective colors for increased _____.
9. The flow or quantity of water moving through a pipe is measured by its _____, usually in gallons per minute.
10. Water that is not moving has _____ energy.
11. Elevation pressure can be created by _____.
12. Water can be transported through long hose lines, pumper relays, or _____ water supply tankers.

True/False

If you believe the statement to be more true than false, write the letter "T" in the space provided.
If you believe the statement to be more false than true, write the letter "F."

1. _____ The basic plan for fighting most fires depends on having an adequate supply of water.
2. _____ The backup water supply for some municipal systems can be large enough to store enough water for several months or years of municipal use.
3. _____ Generally, water pressure ranges from 40 psi to 60 psi at the delivery point.
4. _____ Elevated water storage towers are used to increase the efficiency of treatment facilities.
5. _____ The water department can often increase the flow of water within the system or to specific areas.
6. _____ When the hydrant valve is opened, the drain closes in a dry-barrel hydrant.
7. _____ All fire fighters must understand how to inspect and maintain a fire hydrant.
8. _____ Volume and water pressure are synonymous.
9. _____ Residual pressure is the amount of pressure that remains in the system when water is flowing.
10. _____ The Pitot gauge is used to measure the quantity of water flowing through an opening during a hydrant test.

Short Answer

Complete this section with short written answers using the space provided.

1. Identify the two water sources fire fighters rely upon.

   _____
   _____
   _____

2. Describe the differences between dry-barrel and wet-barrel hydrants.

   _____
   _____
   _____

3. Describe the procedure a fire fighter should follow to ensure there is no foreign matter in a dry-barrel hydrant.

   _____
   _____
   _____

4. List the duties that need to be included in a hydrant inspection.

   _____
   _____
   _____

Word Fun

The following crossword puzzle is an activity provided to reinforce correct spelling and understanding of terminology associated with firefighting. Use the clues provided to complete the puzzle.

CLUES

Across

1. A source of water.
7. The pressure remaining in a water distribution system while water is flowing.
9. A water source such as a pond that is not under pressure.
10. A water distribution system that depends on gravity to provide the required pressure.
12. The large-diameter port on a hydrant.
13. Folding or collapsible tanks that are used at the fire scene to hold water for drafting.

Down

2. The largest-diameter pipes in a water distribution system.
3. Relatively small-diameter underground pipes that deliver water to local users within a neighborhood.
4. The amount of pressure created by moving water.
5. A type of gauge that is used to measure the velocity pressure of water that is being discharged from an opening.
6. A large valve installed on a mobile water supply apparatus to allow the contents of the tank to be discharged quickly.
8. A hydrant where the barrel of the hydrant is normally filled with water.
9. The pressure in a water pipe when there is no water flowing.
11. A type of hydrant used in areas subject to freezing weather.

Fire Alarms

The following real case scenarios will give you an opportunity to explore the concerns associated with water supply. Read each scenario, and then answer each question in detail.

1. It is 3:00 on an August afternoon when your tanker is dispatched to a barn fire. The first engine on the scene reports a large barn that is 75 percent involved with multiple exposures. When you arrive on the scene, you are instructed to set up a portable tank at the top of the driveway for a tanker shuttle. The next-in engine will draft from the tank and hook into the supply line. Your department has metal frame portable tanks. How should you proceed?

   _____
   _____
   _____
   _____
   _____

2. You have just finished lunch. Your Lieutenant has scheduled you to inspect and maintain hydrants in a residential subdivision. Your engine arrives at the first hydrant. How should you proceed?

   _____
   _____
   _____
   _____
   _____

Skill Drills

Skill Drill 15-2: Operating a Fire Hydrant
Test your knowledge of this skill drill by filling in the correct words in the photo captions.

1. Remove the cap from the _____ you will be using.

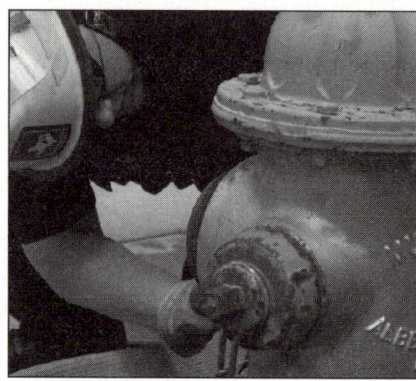

2. Quickly look inside the hydrant opening for _____ objects (dry-barrel hydrant only).

3. Check that the remaining _____ are snugly attached (dry-barrel hydrant only).

4. Attach the hydrant wrench to the _____ nut. Check for an arrow indicating the direction to turn to open.

5. Open the hydrant enough to verify the _____ and flush the hydrant (dry-barrel hydrant only).

6. Shut off the _____ of water (dry-barrel hydrant only).

7. Attach the hose or valve to the hydrant _____.

8. When instructed, turn the _____ to fully open the valve.

9. Open the hydrant slowly to avoid a pressure _____.

Skill Drill 15-3: Shutting Down a Hydrant

Test your knowledge of this skill drill by placing the photos below in the correct order. Number the first step with a "1," the second step with a "2," and so on.

_____ Turn the wrench to slowly close the hydrant valve.

_____ Drain the hose line. Slowly disconnect the hose from the hydrant outlet.

_____ Replace the hydrant cap.

_____ Leave one hydrant outlet open until the hydrant is fully drained (dry-barrel hydrant only).

Fire Hose, Nozzles, Streams, and Foam

Workbook Activities

The following activities have been designed to help you. Your instructor may require you to complete some or all of these activities as a regular part of your fire fighter training program. You are encouraged to complete any activity that your instructor does not assign as a way to enhance your learning in the classroom.

Chapter Review

The following exercises provide an opportunity to refresh your knowledge of this chapter.

Matching

Match each of the terms in the left column to the appropriate definition in the right column.

_____ 1. Forward lay
_____ 2. Reducer
_____ 3. Nozzles
_____ 4. Aeration
_____ 5. Siamese
_____ 6. Reverse lay
_____ 7. Wye
_____ 8. Supply hose
_____ 9. Attack engine
_____ 10. Adaptor
_____ 11. Gate valves
_____ 12. Fire hydraulics
_____ 13. Hose clamp
_____ 14. Protein foam
_____ 15. Rocker lug

A. A device used to split a single hose into two separate lines
B. A device that joins hose couplings of the same type
C. Attachments to the discharge end of attack hoses
D. Fittings on threaded couplings that aid in coupling the hoses
E. A device used to compress a fire hose so as to stop water flow
F. A device that allows two hoses to be connected together and flow into a single hose
G. Inducting air into the foam solution to expand and complete the foam
H. An organic foam that is made of animal by-products
I. The engine from which the attack lines have been pulled
J. The physical science of how water flows through a hose
K. The hose used to deliver water from a source to a fire pump
L. A method of laying a supply line where the line starts at the water source and ends at the attack engine
M. A method of laying a supply line where the line starts at the attack engine and ends at the water source
N. Valves found on hydrants and sprinkler systems
O. A device that can join two hoses of different sizes

CHAPTER 16

Multiple Choice

Read each item carefully, and then select the best response.

_____ 1. The properties of energy, pressure, and water flow are called
   A. fire dynamics.
   B. hose dynamics.
   C. fire hydraulics.
   D. hose hydraulics.

_____ 2. The flow of water is measured in
   A. gallons per minute.
   B. gallons per hour.
   C. gallons.
   D. gallons per second.

_____ 3. The amount of energy in a body of water is measured in
   A. flow.
   B. pressure.
   C. potential energy.
   D. kinetic energy.

_____ 4. Pressure is measured in
   A. gpm.
   B. csi.
   C. gsi.
   D. psi.

_____ 5. Friction loss is influenced by the diameter of the hose, the volume of water traveling through the hose, and
   A. the distance the water travels.
   B. the type of pump involved.
   C. the state of the water source.
   D. the temperature of the water.

_____ 6. To prevent water hammer, a fire fighter should always
   A. use the appropriate-size supply lines.
   B. select the most appropriate nozzles.
   C. open and close fire hydrant valves slowly.
   D. communicate with the driver/operator.

_____ 7. Attack hose must be tested annually at a pressure of at least
   A. 100 psi.
   B. 200 psi.
   C. 300 psi.
   D. 400 psi.

_____ 8. Supply hose line is intended to be used at a pressure up to
   A. 75 psi.
   B. 100 psi.
   C. 160 psi.
   D. 185 psi.

_____ 9. To connect and disconnect hose couplings, fire fighters use a(n)
   A. rocker wrench.
   B. spanner wrench.
   C. lug lock.
   D. adaptor.

_____ 10. Storz-type couplings are designed with
   A. quick-release lugs.
   B. dual swivel connectors.
   C. no male or female ends to the hose.
   D. no lugs or threads.

_____ 11. A 2½-inch handline hose is generally considered to flow
   A. 100 gallons of water per minute.
   B. 150 gallons of water per minute.
   C. 200 gallons of water per minute.
   D. 250 gallons of water per minute.

_____ 12. A short section of large-diameter hose used to connect a fire department engine directly to the large steamer outlet on a hydrant is a
   A. hard suction hose.
   B. soft suction hose.
   C. steamer sleeve hose.
   D. Siamese sleeve hose.

_____ 13. Any device that is used in conjunction with a fire hose for the purpose of delivering water is a
   A. hose appliance.
   B. nozzle.
   C. hydraulic appliance.
   D. delivering agent.

_____ 14. A gated wye with an additional 2½-inch outlet is a(n)
   A. water thief.
   B. open wye.
   C. Siamese.
   D. outlet wye.

_____ 15. A device used to stop a leaking section of hose is a
   A. sweep.
   B. water thief.
   C. hose cap.
   D. hose jacket.

_____ 16. Which type of valve is often found on the large pump intake connections where a hard suction or soft suction hose is connected and opened by rotating a handle one-quarter turn?
   A. Gate valve
   B. Ball valve
   C. Butterfly valve
   D. Gated wye valve

_____ 17. Which method for loading hose is the easiest and can be used for any size of hose?
   A. Split hose load
   B. Minuteman hose load
   C. Flat hose load
   D. Accordion hose load

_____ 18. Which hose load utilizes the perimeter of the hose bed and allows the hose to be laid in a U shape?
   A. Dutch hose load
   B. Horseshoe hose load
   C. Booster hose load
   D. Flat hose load

_____ 19. Which carry is used to transport full lengths of hose over a longer distance than it is practical to drag the hose?
   A. Shoulder carry
   B. Fireman's carry
   C. Hose carry
   D. Advancing carry

_____ 20. The three categories of nozzles are low volume, master stream, and
   A. high-volume nozzles.
   B. secondary stream nozzles.
   C. apparatus nozzles.
   D. handline nozzles.

_____ 21. Which nozzles separate the water into droplets?
   A. Smooth-bore nozzles
   B. Fog-stream nozzles
   C. Breakaway nozzles
   D. Aeration nozzles

_____ 22. Bresnan distributor nozzles are used to fight fires in
   A. warehouses.
   B. open spaces.
   C. inaccessible places.
   D. defensive attacks.

_____ 23. Which type of foam is used to fight fires involving ordinary combustible materials?
   A. Class A foam
   B. Class B foam
   C. Protein foam
   D. Fluoroprotein foam

_____ 24. Which device mixes the foam concentrate into the fire stream?
   A. Foam eductor
   B. Foam injector
   C. Foam regulator
   D. Foam proportioner

_____ 25. Which foam application method uses an object to deflect the foam stream down onto the fire?
   A. Overhead method
   B. Deflection method
   C. Bankshot method
   D. Rain-down method

Labeling

Label the following diagrams with the correct terms.

1. Higbee indicators and Higbee cuts.

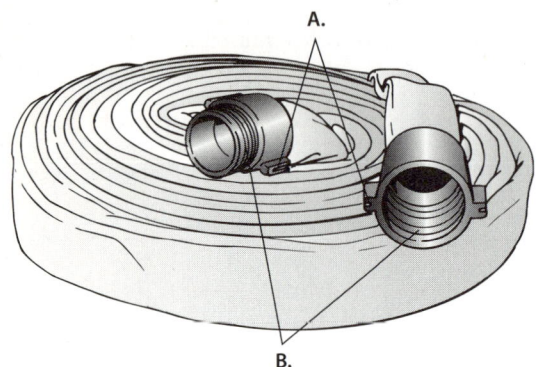

A. _____

B. _____

Figure 16-17

2. Ball valve, gate valve, and butterfly valve.

Figure 16-19

A. _____   B. _____   C. _____

3. Pin lugs, recessed lugs, and rocker lugs.

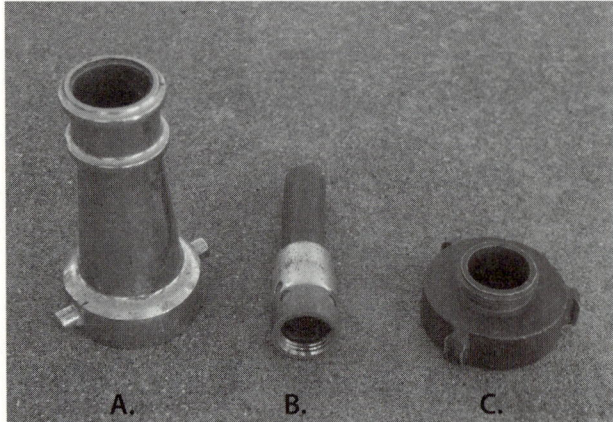

A. _____

B. _____

C. _____

Figure 16-6

Vocabulary

Define the following terms using the space provided.

1. Fixed-gallonage fog nozzle:

2. Higbee indicators:

3. Split hose lay:

4. Water hammer:

5. Hard suction hose:

Fill-in
Read each item carefully, and then complete the statement by filling in the missing word(s).

1. _____ refers to the volume of water that is being moved through a hose.
2. A pump is used to add energy to a water stream and causes a(n) _____ in pressure.
3. _____ is a surge in pressure caused by suddenly stopping the flow of a stream of water.
4. The hoses used to discharge water from an attack engine onto the fire are called _____.
5. _____ are used to make connections between hose lengths or between hose lengths and hydrants.
6. A(n) _____ is an important part of a threaded coupling, as it forms a seal to stop water from leaking.
7. Fire hose is likely to be damaged if run over by a vehicle; for this reason, hose _____ should be used in traffic.
8. _____ is a type of fungus that can grow on many fabrics and materials in warm, moist conditions.
9. A(n) _____ adapter is used to join two male hose couplings.
10. _____ are used to control the flow of water in a pipe or a hose line.
11. Fire hose evolutions are divided into _____ line operations and _____ line operations.
12. The _____ load attaches the female end of the hose to the preconnect discharge.
13. A hose line should be flaked out _____ it is charged with water.
14. If a hose line has to be advanced up a ladder, it should be charged _____ it is advanced.

True/False
If you believe the statement to be more true than false, write the letter "T" in the space provided.
If you believe the statement to be more false than true, write the letter "F."

1. _____ Elevation affects water pressure.
2. _____ Water hammer can rupture a hose, cause a coupling to separate, or damage underground piping systems.
3. _____ Larger-diameter hoses are used as attack lines, and smaller-diameter hoses are almost always used as supply lines.
4. _____ Booster lines are used for small outdoor fires.
5. _____ Double-jacket hose is constructed with two layers of hose liner.
6. _____ The male coupling has a swivel to reduce twisting of the hose.
7. _____ Charged hose lines should never be disconnected while the water inside the hose is under pressure.
8. _____ When fire fighters are working in below-freezing temperatures, water should be kept flowing to prevent freezing in hoses.
9. _____ A gated wye has two quarter-turn ball valves to allow for independent control of the water flow.
10. _____ A master stream device is used primarily during interior attacks.
11. _____ The attack hose is loaded so that it can be quickly and easily deployed.
12. _____ The best hose carry technique for a particular situation depends on the size of the hose, the distance it must be moved, and the number of fire fighters available to perform the task.
13. _____ A hose line should be charged before it is advanced up a stairway.
14. _____ Standpipe systems are used to provide a water supply for attack lines that will be operated inside the building.

Short Answer
Complete this section with short written answers using the space provided.

1. List the type of information contained in a hose record.
   _____
   _____
   _____

2. Identify four purposes of using a split hose bed.
   _____
   _____
   _____
   _____

3. List the steps of the procedure for unloading a hose bed.
   _____
   _____
   _____

Word Fun

The following crossword puzzle is an activity provided to reinforce correct spelling and understanding of terminology associated with firefighting. Use the clues provided to complete the puzzle.

Clues

Across
1. A firefighting foam for fires involving flammable liquids.
7. A device that is placed on the edge of a roof and is used to protect hose as it is hoisted up and over the roof edge.
10. Hose in the 1- to 2-inch range.
13. Hose that delivers water from a fire pump to the fire.
14. The inside portion of a hose that is in contact with the flowing water.

Down
2. A synthetic based foam suited for hydrocarbon fuels.
3. Foam that has been mixed with water.
4. Hose in the 4-, 5-, and 6-inch range
5. Device placed at the end of a fire hose that separates water into fine droplets to aid in heat absorption.
6. A nozzle that can be driven through sheet metal or other material to deliver a water stream to that area.
7. Any device used in conjunction with fire hose for the purpose of delivering water.
8. Nozzles that flow 40 gallons per minute or less.
9. Foam that is in its raw state.
11. Inducting air into the foam solution, which expands and finishes the foam.
12. A short fold placed in a hose when loading it into the bed to prevent the coupling from turning in the hose bed.

Fire Alarms

The following real case scenarios will give you an opportunity to explore the concerns associated with fire hoses, nozzles, streams, and foam. Read each scenario, and then answer each question in detail.

1. Your engine is dispatched to a house fire in a residential part of your running district. Your company is the first engine on scene and is assigned to be the attack engine. The closest hydrant is 200 feet down a side street, which your engine will pass as it responds to the fire.

    a. What would be an acceptable supply lay into your attack engine?
    b. Explain this evolution.

    _____
    _____
    _____
    _____
    _____
    _____

2. Your new engine company carries its attack lines preconnected in a minuteman hose load. Your company officer orders you to demonstrate loading a 150-foot, 1 ¾-inch minuteman hose load. How will you proceed?

    _____
    _____
    _____
    _____
    _____
    _____

3. Your company officer orders you to demonstrate advancing a 1 ¾-inch minuteman hose load. Explain this procedure In detail.

    _____
    _____
    _____
    _____
    _____
    _____

Skill Drills

Skill Drill 16-4: Performing the One-Fire Fighter Knee-Press Method of Uncoupling a Fire Hose
Test your knowledge of this skill drill by filling in the correct words in the photo captions.

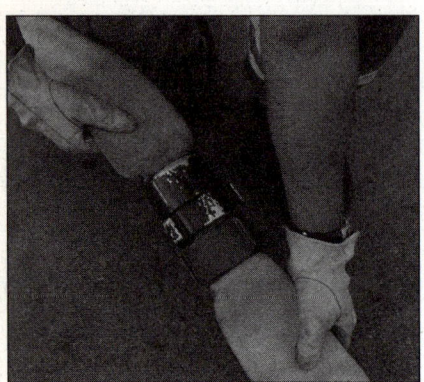

1. Pick up the connection by the _____ coupling end.

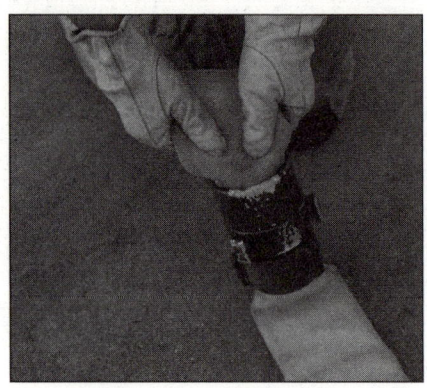

2. Turn the connection _____, resting the male coupling on a firm surface.

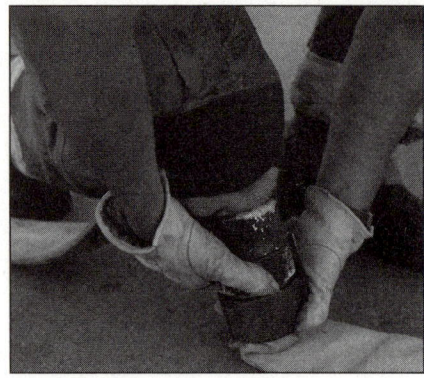

3. Place a knee on the _____ coupling and press down on it with your body weight. Turn the female swivel counterclockwise and loosen the coupling.

Skill Drill 16-6: Uncoupling a Hose with Spanners

Test your knowledge of this skill drill by placing the photos below in the correct order. Number the first step with a "1," the second step with a "2," and so on.

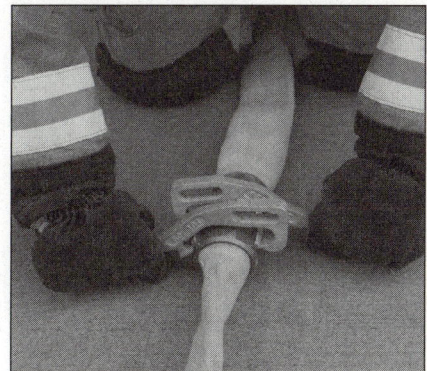

_____ Push both spanner handles down toward the ground, loosening the connection.

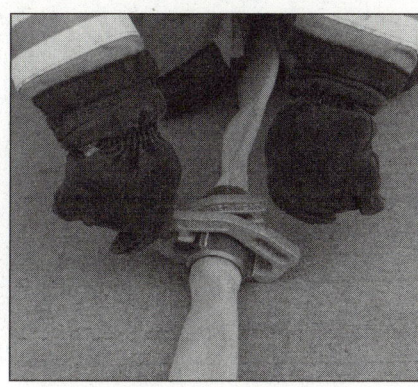

_____ Place the second spanner wrench on the male coupling with the handle of the wrench to the right.

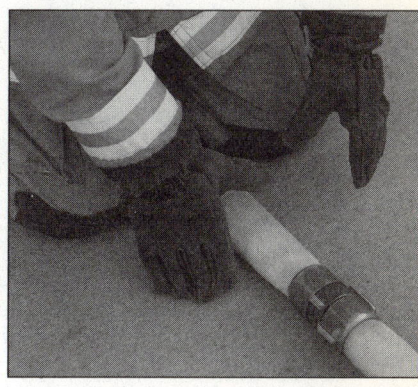

_____ With the connection on the ground, straddle the connection above the female coupling.

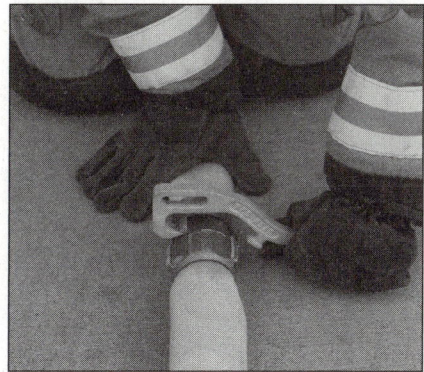

_____ Place one spanner wrench on the female coupling with the handle of the wrench to the left.

Skill Drill 16-12: Performing a Single-Doughnut Roll
Test your knowledge of this skill drill by filling in the correct words in the photo captions.

  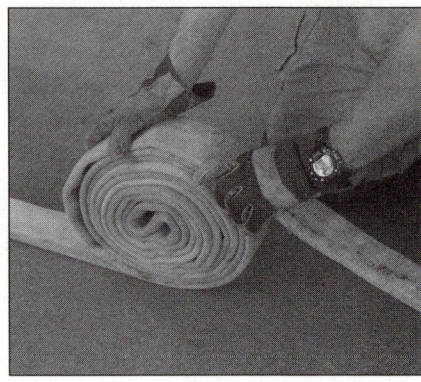

1. Place the hose _____ and in a straight line.

2. Locate the _____ of the hose.

3. From the midpoint, move _____ toward the male coupling end. Start rolling the hose toward the female coupling.

4. At the end of the roll, wrap the excess hose of the _____ end over the male coupling to protect the threads.

Skill Drill 16-19: Performing a Flat Hose Load

Test your knowledge of this skill drill by placing the photos below in the correct order. Number the first step with a "1," the second step with a "2," and so on.

_____ While laying the hose back to the rear of the hose bed, angle the hose to the side of the previous fold.

_____ Continue to lay the hose in neat folds until the whole hose bed is covered with a layer of hose. Continue to lead the layers of hose until the required amount of hose is loaded.

_____ Start the hose lay with the coupling at the front end of the hose compartment.

_____ Run the hose back to the front end on top of the previous length of hose. Fold the hose back on itself so the top of the hose is on the previous length.

_____ To set up the hose for a forward lay, place the male hose coupling in the hose bed first. To set up the hose for a reverse lay, place the female hose coupling in the hose bed first.

_____ Fold the hose back on itself at the rear of the hose bed.

Skill Drill 16-21: Performing an Accordion Hose Load

Test your knowledge of this skill drill by filling in the correct words in the photo captions.

1. Lay the first length of hose bed on its _____ against the side of the hose bed.

2. Double the hose back on itself at the rear of the hose bed. Leave the _____ end extended so that the two hose beds can be cross-connected.

3. Lay the hose next to the first length and bring it to the _____ of the hose bed. Fold the hose at the front of the hose bed so the bend is even to the edge of the hose bed. Continue to lay folds of hose across the hose bed.

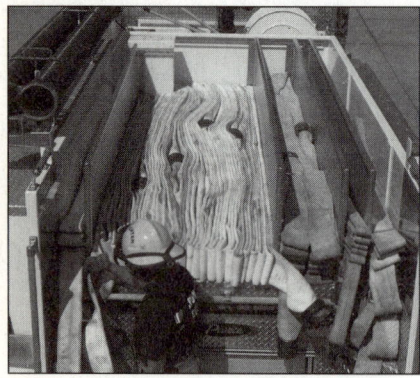

4. Alternate the length of the hose folds at each end to allow more room for the folded ends. When the bottom layer is completed, angle the hose _____ to begin the second tier. Continue the second layer by repeating the steps used to complete the first layer.

Skill Drill 16-22: Attaching a Soft Suction Hose to a Fire Hydrant

Test your knowledge of this skill drill by placing the photos below in the correct order. Number the first step with a "1," the second step with a "2," and so on.

_____ Remove the large hydrant cap.

_____ Open the hydrant slowly when so indicated by the pump operator. Check all connections for leaks. Tighten if necessary.

_____ Attach the soft suction hose to the hydrant.

_____ The driver/operator positions the apparatus so that its inlet is the correct distance from the hydrant.

_____ Unroll the hose.

_____ Remove the hose, any needed adaptors, and the hydrant wrench.

_____ Ensure that there are no kinks or sharp bends in the hose that will restrict the flow of water.

_____ Place chafing blocks under the hose where it contacts the ground to prevent mechanical abrasion.

_____ Attach the soft suction hose to the inlet of the engine.

Skill Drill 16-24: Performing a Minuteman Hose Load
Test your knowledge of this skill drill by filling in the correct words in the photo captions.

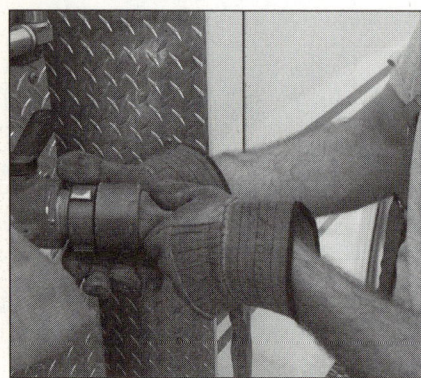

1. Connect the _____ of the first length of hose to the discharge outlet.

2. Flat load the hose _____ to the edges of the hose bed. At approximately the 30-feet mark, make a loop/ear (this will create a handle for advancing purposes). Continue the load until you reach the 60- to 70-feet mark. Make an additional loop/ear. Flat load the remainder of the 100-feet section, leaving extra hose to the side, in the opposite direction of deployment.

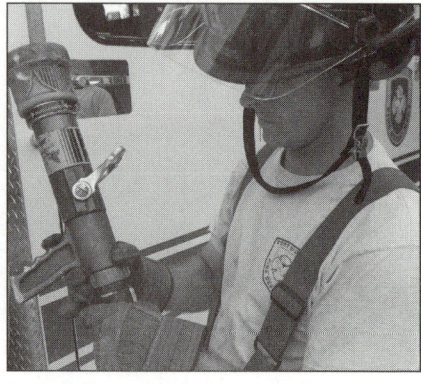

3. Assemble the remaining hose sections and attach the _____. Place the nozzle on the _____.

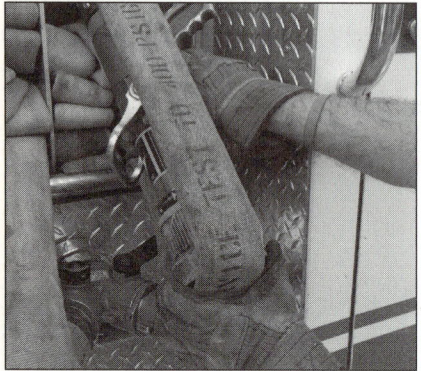

4. Load the remaining 100 feet of hose flat into the bed, _____ the folds from the left to the right sides of the bed.

5. _____ the last section loaded to the first section placed in the bed.

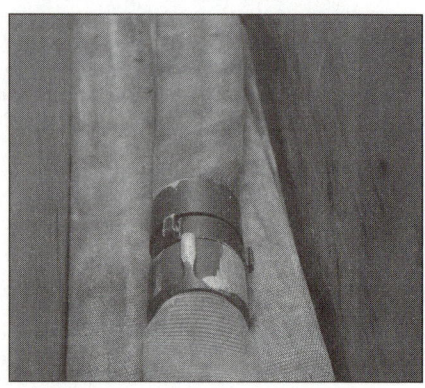

6. _____ the remaining loose hose on top of the load.

Skill Drill 16-25: Advancing a Minuteman Hose Load

Test your knowledge of this skill drill by placing the photos below in the correct order. Number the first step with a "1," the second step with a "2," and so on.

_____ Turn away from the hose bed and place the load on the shoulder. Walk away from the apparatus until all hose is clear from the hose bed.

_____ Grasp the nozzle and the folds next to it.

_____ Pull the load approximately one-third out of the bed.

_____ Continue walking away, pulling the remaining hose from the hose bed and then allowing the hose to deploy from the top of the load on the shoulder.

Skill Drill 16-26: Performing a Preconnected Flat Load
Test your knowledge of this skill drill by filling in the correct words in the photo captions.

1. Attach the _____ of the hose to the preconnect discharge.

2. Begin laying the hose _____ in the hose bed.

3. When about one-third of the hose is in the bed, make an 8-inch loop at the end of the hose bed. This loop will be used as a pulling handle. When two-thirds of the hose is loaded, make a second pulling loop that is about _____ the size of the first loop.

4. Finish loading the hose, attach the nozzle, and place it _____ of the hose bed. The preconnected flat load is now ready for use.

Skill Drill 16-27: Advancing a Preconnected Flat Hose Load

Test your knowledge of this skill drill by placing the photos below in the correct order. Number the first step with a "1," the second step with a "2," and so on.

_____ Walk away from the vehicle.

_____ Pull the load from the bed.

_____ Place the arm through the larger lower loop. Grasp the smaller loop with the same hand.

_____ As the load deploys, drop the small loop. Extend the remaining hose to length.

_____ Grasp the nozzle with the opposite hand.

Skill Drill 16-28: Performing a Triple-Layer Hose Load
Test your knowledge of this skill drill by filling in the correct words in the photo captions.

1. Attach the female end of the hose to the preconnect _____.

2. Connect the _____ of hose together.

3. Extend the hose directly from the hose bed. Pick up the hose _____ of the distance from the discharge to the hose nozzle.

4. Carry the hose back to the apparatus, forming a(n) _____ loop.

5. Pick up the entire length of _____ hose.

6. Lay the tripled folded hose in the hose bed in a(n) _____ with the nozzle on top.

Skill Drill 16-29: Advancing a Triple-Layer Hose Load

Test your knowledge of this skill drill by placing the photos below in the correct order. Number the first step with a "1," the second step with a "2," and so on.

_____ Turn away from the hose bed and place the hose on the shoulder.

_____ Extend the nozzle the remaining distance.

_____ Grasp the nozzle and the top fold.

_____ Walk away from the vehicle until the entire load is out of the bed.

_____ When the load is out of the bed, drop the fold.

Skill Drill 16-34: Advancing a Hose Line Up a Stairway
Test your knowledge of this skill drill by filling in the correct words in the photo captions.

1. Use a(n) _____ carry to advance up the stairs.

2. When ascending the stairway, lay the hose against the _____ of the stairs to reduce tripping hazards. Avoid sharp bends.

3. Arrange _____ hose so that it is available to fire fighters entering the fire floor.

Skill Drill 16-36: Advancing an Uncharged Hose Line Up a Ladder

Test your knowledge of this skill drill by placing the photos below in the correct order. Number the first step with a "1," the second step with a "2," and so on.

_____ Pick up the nozzle; place the hose across your chest with the nozzle draped over your shoulder.

_____ Advance the hose line to the ladder.

_____ Additional hose can be fed up the ladder until sufficient hose is in position. The hose can be secured to the ladder with a hose strap to support its weight and keep it from becoming dislodged.

_____ Climb up the ladder with the uncharged hose line. Once the first fire fighter reaches the first fly section of the ladder, a second fire fighter shoulders the hose to assist advancing the hose line up the ladder. To avoid overloading of the ladder, enforce a limit of one fire fighter per fly section.

_____ The nozzle is placed over the top rung of the ladder and advanced into the fire area.

Skill Drill 16-42: Operating a Smooth-Bore Nozzle
Test your knowledge of this skill drill by filling in the correct words in the photo captions.

1. Select the desired tip size and attach it to the nozzle _____ valve.

2. Attain a stable _____ (if standing).

3. Slowly open the _____, allowing water to flow.

4. Open the valve _____ to achieve maximum effectiveness.

5. Direct the _____ to the desired location.

Skill Drill 16-43: Operating a Fog-Stream Nozzle

Test your knowledge of this skill drill by placing the photos below in the correct order. Number the first step with a "1," the second step with a "2," and so on.

_____ Open the valve completely.

_____ Attain a stable stance (if standing).

_____ Select the desired water pattern, by rotating the bezel of the nozzle. Apply water where needed.

_____ Slowly open the valve, allowing water to flow.

Fire Fighter Survival

Workbook Activities

The following activities have been designed to help you. Your instructor may require you to complete some or all of these activities as a regular part of your fire fighter training program. You are encouraged to complete any activity that your instructor does not assign as a way to enhance your learning in the classroom.

Chapter Review

The following exercises provide an opportunity to refresh your knowledge of this chapter.

Matching
Match each of the terms in the left column to the appropriate definition in the right column.

_____ 1. Guideline
_____ 2. Safe haven
_____ 3. Rehabilitation
_____ 4. Self-rescue
_____ 5. Risk-benefit analysis
_____ 6. Mayday
_____ 7. RIC
_____ 8. Hazardous conditions
_____ 9. Hazard recognition
_____ 10. SOPs

A. A systematic process to provide periods of rest and recovery for emergency workers during an incident
B. The process of weighing predicted risks against potential benefits and making decisions based on the outcome of that analysis
C. A temporary place of refuge in which to await rescue
D. A rope used for orientation when inside a structure when there is low or no visibility
E. The activity of a fire fighter using techniques and tools to escape from a hazardous situation
F. Defines the manner in which a fire department conducts operations at an emergency incident
G. May not be evident by simple observation
H. Becomes easier through study and experience
I. An extension of the two-in/two-out rule
J. Indicates a fire fighter is in trouble

Multiple Choice
Read each item carefully, and then select the best response.

_____ 1. Comparing potential positive results to potential negative consequences is called
  A. causative factors.
  B. management factors.
  C. risk-benefit analysis.
  D. standard operating procedures.

_____ 2. During an incident, if fire fighters observe an increase in risk of their operations, they must report it to the
  A. company officer.
  B. sector officer.
  C. safety officer.
  D. incident commander.

CHAPTER 17

_____ 3. The manner in which a fire department conducts operations at an emergency incident is defined by
   A. general operating guidelines.
   B. the incident commander.
   C. department policies.
   D. standard operating procedures.

_____ 4. A systematic way to keep track of the location and function of all personnel operating at the scene of an incident is
   A. a team inventory.
   B. a personnel accountability system.
   C. the chain of command.
   D. the two-in/two-out rule.

_____ 5. A roll call taken by each supervisor at an emergency incident is known as a(n)
   A. team roll call.
   B. incident report.
   C. personnel accountability report.
   D. incident roll.

_____ 6. The standard radio terminology used to report a hazardous condition or situation is
   A. "Emergency traffic."
   B. "Mayday."
   C. "Halt operations."
   D. "Retreat."

_____ 7. A crew that is assigned to stand by fully dressed, equipped for action, and ready to deploy at an incident scene is called a(n)
   A. technical rescue crew.
   B. EMS team.
   C. special recovery team.
   D. rapid intervention crew.

_____ 8. To stay oriented when inside a burning structure, the fire fighter should use a hose line or a
   A. team member.
   B. radio.
   C. guideline.
   D. structure wall.

_____ 9. What is the first step of self-rescue?
   A. Manually set off your PASS alarm.
   B. Call for assistance.
   C. Exit the structure.
   D. Orient yourself within the structure.

_____ 10. While awaiting rescue, a fire fighter may find a temporary location that provides refuge. What is this location called?
   A. Safety point
   B. Safe haven
   C. Rescue point
   D. Landmark

11. Upon reaching a downed fire fighter, what is the most critical decision for the rescuers?
    A. How much time and effort will be needed to remove the fire fighter
    B. The treatment of the fire fighter's injuries
    C. The location of the fire fighter and rescuers
    D. How to exit the structure

12. What is the most common form of critical incident stress management (CISM)?
    A. Group mediation
    B. Professional counseling
    C. Peer defusing
    D. Self-reconciliation

13. CISM debriefings are usually held
    A. within 24 to 72 hours after a major incident.
    B. within hours after a major incident.
    C. in special operating facilities.
    D. in one-on-one interviews.

14. What is the primary desired outcome in any fire department operation?
    A. Fire fighter survival
    B. No rekindle
    C. Effective ICS
    D. Reduced water damage

15. Fire fighters will accept a higher level of risk in exchange for
    A. the possibility of saving lives.
    B. the possibility of saving property.
    C. property that is lost.
    D. persons who are already lost.

16. Observable factors that might indicate a hazard include
    A. building construction.
    B. weather conditions.
    C. occupancy.
    D. all of the above.

17. Which of the following must be learned and practiced before they can be implemented?
    A. CISMs
    B. SOPs
    C. GOPs
    D. DUIs

18. What does the NFPA 704 Diamond indicate?
    A. PARs are present.
    B. RICs are present.
    C. Ventilation is necessary.
    D. Hazardous materials are present.

19. When initiating a mayday,
    A. give a weather report.
    B. give a LUNAR report.
    C. give a CISD report.
    D. give a PASS report.

20. When a firefighter needs immediate assistance, the incident commander should immediately deploy
    A. the SOP.
    B. the PASS.
    C. the RIC.
    D. the radio.

Vocabulary

Define the following terms using the space provided.

1. Air management:
   _____
   _____
   _____

2. Rapid intervention company/crew (RIC):
   _____
   _____
   _____

3. Critical incident stress management (CISM):
   _____
   _____
   _____

4. Safe haven:
   _____
   _____
   _____

5. Self-rescue:
   _____
   _____
   _____

Fill-in

Read each item carefully, and then complete the statement by filling in the missing word(s).

1. The assessment of the risks and benefits and the decision to commit crews to the interior of a burning structure is the responsibility of the _____.

2. During an incident, company officers, sector officers, and safety officers are involved in risk analysis on a(n) _____ basis.

3. Hazardous conditions may or may not be evident by _____.

4. The only way to become proficient at a skill is through _____.

5. Team _____ means that a company arrives at a fire together, works together, and leaves together.

6. The word _____ is used to indicate that a fire fighter is in trouble and requires immediate assistance.

7. Safety and survival inside a fire building can be directly related to remaining _____ within the building.

8. Air _____ is important to all fire fighters and relates to the basic fact that air equals time.

9. The purpose of _____ is to reduce the effects of fatigue during an emergency operation.

True/False

If you believe the statement to be more true than false, write the letter "T" in the space provided.
If you believe the statement to be more false than true, write the letter "F."

1. _____ It is permissible to risk the life of a fire fighter only in a situation where there is a reasonable and realistic possibility of saving a life.

2. _____ Fire fighters must be capable of working in environments that include a wide range of hazards.

3. _____ Teamwork and communication are critical parts of all emergency operations.

4. _____ Rapid intervention crews/companies should be in place at any incident where fire fighters are in operation.

5. _____ The best method to remain oriented within an involved structure is to stay in contact with a team member.

6. _____ A room with a door and a window could be used as a safe haven.

7. _____ The SCBA cylinder's time rating is based on the minimum amount of time the cylinder will last during operations.

8. _____ Sometimes fire fighters react to critical incidents in ways that are not positive.

9. _____ Critical incident stress is created by an event that is interpreted by the fire fighter as traumatic.

10. _____ The stages of emotional reaction after a critical incident can occur within minutes or months of the incident.

Short Answer

Complete this section with short written answers using the space provided.

1. Identify a simply stated risk-benefit philosophy for a fire department.

   _____
   _____
   _____

2. Describe the procedure and steps to follow to initiate a mayday.

   _____
   _____
   _____

3. List the recognized stages of emotional reaction experienced by fire fighters and other rescue personnel to critical incidents.

   _____
   _____
   _____

Word Fun

The following crossword puzzle is an activity provided to reinforce correct spelling and understanding of terminology associated with firefighting. Use the clues provided to complete the puzzle.

CLUES

Across
2. A temporary place of refuge in which to await rescue.
5. A method of tracking the identity, assignment, and location of fire fighters operating at an incident scene.
6. A rope used for orientation when inside a structure when there is low or no visibility.

Down
1. A report confirming that all members of a company are present.
2. The activity of a fire fighter using techniques and tools to remove himself or herself from a hazardous situation.
3. The way in which an individual utilizes a limited air supply to be sure it will last long enough to enter a hazardous area, accomplish needed tasks, and return safely.
4. A systematic process to provide periods of rest and recovery for emergency workers during an incident.

Fire Alarms

The following real case scenarios will give you an opportunity to explore the concerns associated with fire fighter survival. Read each scenario, then answer each question in detail.

1. You have become confused and disoriented during a primary search of a second-floor apartment. You decide to attempt to locate a window for calling for assistance or exit. How will you proceed?

   _____
   _____
   _____
   _____
   _____
   _____

2. While performing a search in a commercial structure, you crawl into a portion of fallen suspended ceiling and become entangled in electric wire and communication cables. What steps will you take to remove yourself from the wires and cables?

   _____
   _____
   _____
   _____
   _____
   _____

Skill Drills

Skill Drill 17-1: Initiating a Mayday Call
Test your knowledge of this skill drill by filling in the correct words in the photo captions.

1. Use your radio to call, "Mayday, mayday, mayday." State your name and _____. State your location. State the type of problem.

2. Activate your _____. Attempt self-rescue. If you are able to move, identify a safe haven where you can await rescue.

3. Lie on your side in a(n) _____ position with your PASS device pointing out so that it can be heard.

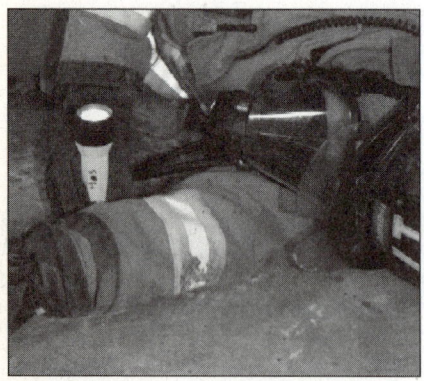

4. Point your flashlight toward the _____. Slow your breathing as much as possible to conserve your air supply.

Skill Drill 17-2: Performing Self-Rescue

Test your knowledge of this skill drill by placing the photos below in the correct order. Number the first step with a "1," the second step with a "2," and so on.

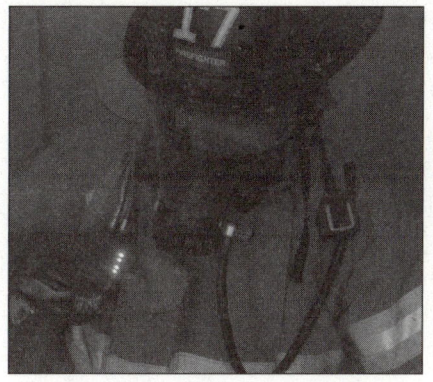

_____ Systematically search the room to locate a hose line. Follow the hose line to a hose coupling. Identify the male and female ends of the coupling. Move from the female coupling to the male coupling.

_____ Initiate a mayday over the portable radio. Manually activate the PASS device. Stay calm and control your breathing.

_____ Follow the hose out. Exit the hazard area. Notify command of your location.

Skill Drill 17-3: Locating a Door or Window for Emergency Exit

Test your knowledge of this skill drill by filling in the correct words in the photo captions.

1. Initiate a mayday over the portable radio. Manually activate the PASS device. Stay calm and control your _____.

2. Systematically locate a(n) _____.

3. Use a(n) _____ motion on the outside wall to locate an alternative exit. Identify the opening as a window, interior door, or external door.

(continues)

Skill Drill 17-3: Locating a Door or Window for Emergency Exit (continued)

4. If the first opening identified is not adequate for an exit, continue to search. Maintain your _____ and stay low. Exit the room safely if possible. If unable to exit, assume the downed fire fighter position in a safe haven or find refuge. Keep command informed of your situation.

Skill Drill 17-4: Opening a Wall to Escape
Test your knowledge of this skill drill by placing the photos below in the correct order. Number the first step with a "1," the second step with a "2," and so on.

_____ Identify deteriorating conditions that require exiting through a wall. Use the portable radio to initiate a mayday. Use a hand tool or your feet to open a hole in the wall between two studs. If using a hand tool, drive the hand tool completely through the wall to check for obstacles on the other side.

_____ Assist others through the opening in the wall. Report your status to command.

_____ Loosen the SCBA waist strap and remove one shoulder strap. Maintain control of the regulator side of the SCBA, which will assist in donning of the air pack later. Sling the SCBA to one side to reduce your profile.

(continues)

Skill Drill 17-4: Opening a Wall to Escape (continued)

_____ Enlarge the hole. Enter the hole head first to check the floor and fire conditions on the other side of the wall.

_____ Escape through the opening in the wall. Adjust the SCBA straps to their normal position.

Skill Drill 17-5: Escaping from an Entanglement
Test your knowledge of this skill drill by filling in the correct words in the photo captions.

 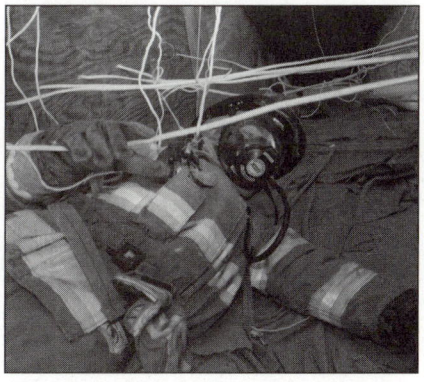 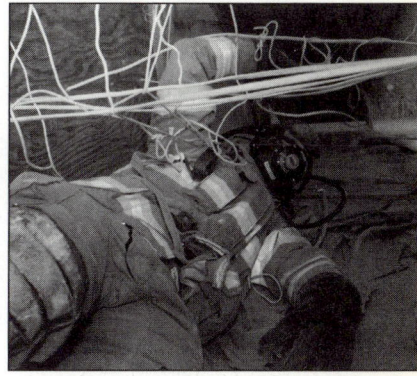

1. Initiate a(n) _____ over the portable radio. Activate the PASS device. Stay calm and control your breathing.

2. Change your _____ —back up or turn on your side to try to free yourself.

3. Use the swimmer _____ to try to free yourself.

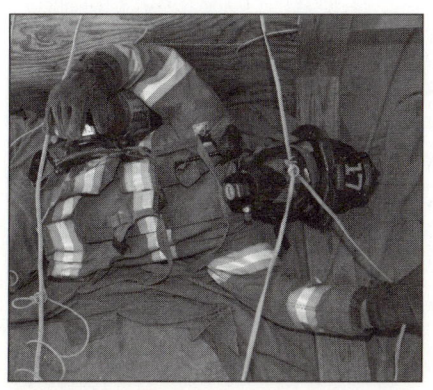 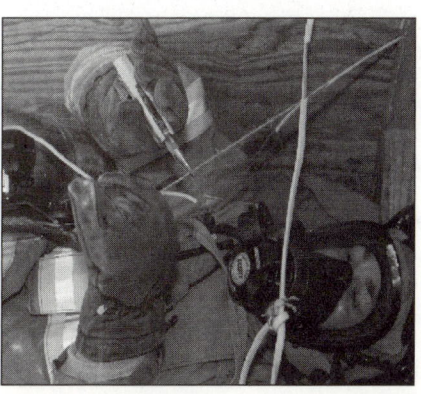

4. Loosen the _____ straps, remove one arm, and slide the air pack to the front of your body to try to free the SCBA.

5. Cut the wires or cables causing the entanglement. Be aware of any possible _____ risk. If you are unable to disentangle yourself, notify command of your situation. If you are able to exit, notify command that you are out of danger.

Salvage and Overhaul

Workbook Activities

The following activities have been designed to help you. Your instructor may require you to complete some or all of these activities as a regular part of your fire fighter training program. You are encouraged to complete any activity that your instructor does not assign as a way to enhance your learning in the classroom.

Chapter Review

The following exercises provide an opportunity to refresh your knowledge of this chapter.

Matching

Match each of the terms in the left column to the appropriate definition in the right column.

_____ 1. Spotlight
_____ 2. Inverter
_____ 3. Rekindle
_____ 4. Scupper
_____ 5. Salvage
_____ 6. Floodlight
_____ 7. Generator
_____ 8. Water catch-all
_____ 9. Water chute
_____ 10. Carryall

A. A situation where a fire thought to be out reignites
B. A light that can illuminate a broad area
C. Removing or protecting property that could be damaged during firefighting or overhaul operations
D. A device that converts the direct current from an apparatus electrical system into alternating current
E. An engine-powered device that provides electricity
F. An opening through which water can be removed from a building
G. A light designed to project a narrow, concentrated beam of light
H. A salvage cover that is folded to direct water
I. Heavy canvas with handles used to carry debris
J. A salvage cover that is folded to form a container to hold water

Multiple Choice

Read each item carefully, and then select the best response.

_____ 1. Efforts to protect property and belongings from damage are called
   A. overhaul.
   B. salvage.
   C. rescue.
   D. recovery.

_____ 2. Fire fighters' efforts at residential fires often focus on protecting
   A. expensive items or property.
   B. isolated property.
   C. personal property.
   D. market items.

_____ 3. Before a fire fighter can work without SCBA, the atmosphere must be tested and determined safe by the
   A. safety officer.
   B. incident commander.
   C. RIC leader.
   D. department captain.

_____ 4. Damage to the building's structural components during salvage and overhaul operations may potentially lead to
   A. atmospheric contamination.
   B. reignition.
   C. rekindling.
   D. structural collapse.

_____ 5. How much does a gallon of water weigh?
   A. 2.24 pounds
   B. 4.5 pounds
   C. 6 pounds
   D. 8.3 pounds

_____ 6. The most common method of protecting building contents is to cover them with
   A. heat-reflective blankets.
   B. salvage covers.
   C. overhaul tarps.
   D. salvage tarps.

_____ 7. The most efficient way to protect a room's contents is to move all the furniture to
   A. the center of the room.
   B. the wall farthest from the flames.
   C. the walls nearest the window(s) used for ventilation and fire suppression.
   D. the front of the room.

_____ 8. During salvage operations, smaller pictures and valuable objects should be placed in
   A. the pockets of the fire fighter's PPE.
   B. smaller tarps.
   C. drawers.
   D. the corner.

_____ 9. A long section of protective material used to cover a section of carpet is called a
   A. floor cover.
   B. floor runner.
   C. carpet tarp.
   D. drop tarp.

_____ 10. What is the best way to prevent water damage at a fire scene?
   A. Increase the number of ventilation sites
   B. Use higher flow rates
   C. Use the building sprinkler system
   D. Limit the amount of water used

_____ 11. Sprinklers should be shut down as soon as the IC declares the fire
    A. is under control.
    B. is out.
    C. has spread to other rooms of the structure.
    D. is being ventilated.

_____ 12. To stop the flow from a sprinkler, insert a
    A. hand tool.
    B. sprinkler wedge.
    C. sprinkler key.
    D. cloth.

_____ 13. The main control for a sprinkler system is usually a(n)
    A. sprinkler box.
    B. scupper.
    C. OS&Y valve or PIV.
    D. water valve.

_____ 14. Which type of opening drains water from an above-ground floor through an exterior wall hole?
    A. Water chute
    B. Water catch-all
    C. Drain
    D. Scupper

_____ 15. The building's construction, its contents, and the size of the fire are factors in determining
    A. salvage operations.
    B. the area that needs to be overhauled.
    C. who is the incident commander.
    D. the placement of the rapid intervention company/crew.

_____ 16. Fire can extend directly from the basement to the attic, without obvious signs of fire, in a
    A. balloon-frame building.
    B. platform-frame building.
    C. Type I building.
    D. remodeled building.

_____ 17. Mobile power outlets, which are placed in convenient locations for cords to be attached, are also known as
    A. junction boxes.
    B. generators.
    C. inverters.
    D. extension outlets.

_____ 18. Which type of lights are most often used during the first few critical minutes of an incident?
    A. 1500-watt portable lights
    B. Battery-powered lights
    C. 300-watt quartz lights
    D. 1000-watt halogen lights

_____ 19. To convert 12-volt DC current to 110-volt AC current, you would use a(n)
    A. power inverter.
    B. junction box.
    C. power outlet.
    D. extension.

_____ 20. How long should gasoline-powered generators be run to reduce deposit build-ups?
    A. 2 hours
    B. 1 hour
    C. 15–30 minutes
    D. 5–10 minutes

Vocabulary

Define the following terms using the space provided.

1. Floor runner:

   _____
   _____
   _____

2. Overhaul:

   _____
   _____
   _____

3. Salvage cover:

   _____
   _____
   _____

4. Balloon-frame construction:

   _____
   _____
   _____

5. Sprinkler wedge:

   _____
   _____
   _____

6. Sprinkler stop:

   _____
   _____
   _____

Fill-in
Read each item carefully, and then complete the statement by filling in the missing word(s).

1. _____ ensures that a fire is completely extinguished.
2. Salvage and overhaul have a(n) _____ priority than search and rescue.
3. During salvage and overhaul efforts, fire fighters must attempt to preserve _____ related to the cause of the fire, particularly when arson is expected.
4. Salvage efforts are usually aimed at preventing or limiting _____ that result from smoke and water damage.
5. Fire fighters should remember to _____ a door after a room is searched.
6. Sprinkler heads that have been activated must be _____ before they can be restored to normal operations.
7. A(n) _____ vacuum is a special piece of equipment used to suck up water during salvage operations.
8. A(n) _____ should always be present during overhaul operations to note any hazards and ensure that operations are conducted safely.
9. _____ teams remain at the fire scene and watch for signs of rekindling.
10. Look, listen, and _____ to detect signs of potential burning.
11. A(n) _____ can distinguish between objects or areas with different temperatures.
12. Buckets, tubs, wheelbarrows, and _____ can be used to remove debris from a building.

True/False
If you believe the statement to be more true than false, write the letter "T" in the space provided.
If you believe the statement to be more false than true, write the letter "F."

1. _____ Salvage efforts can be done during fire suppression.
2. _____ Often the damage caused to property by smoke and water can be more extensive and costly to repair or replace than the property that is burned.
3. _____ Water used in fire suppression can create potential hazards for fire fighters.
4. _____ Salvage crews begin on the floor of the fire to prevent water damage to room contents.
5. _____ When entering an area for salvage operations, fire fighters should roll a floor runner ahead of themselves.
6. _____ Sprinkler control valves should always be locked in the closed position.
7. _____ The IC may order hydraulic or standard overhaul procedures.
8. _____ The entire area around a fire building should be illuminated.
9. _____ Pike poles are used to pull down sections of ceiling.
10. _____ The cause of the fire can also indicate the extent of overhaul necessary.

Chapter 18: Salvage and Overhaul

Short Answer
Complete this section with short written answers using the space provided.

1. Identify five tools used in salvage operations.

   _____
   _____
   _____
   _____
   _____

2. List four potential hazards present during overhaul operations.

   _____
   _____
   _____
   _____

3. List five indicators of possible structural collapse.

   _____
   _____
   _____
   _____
   _____

4. List five tools used in overhaul operations.

   _____
   _____
   _____
   _____
   _____

216 Fundamentals of Fire Fighter Skills

Clues

Word Fun

The following crossword puzzle is an activity provided to reinforce correct spelling and understanding of terminology associated with firefighting. Use the clues provided to complete the puzzle.

Across

3. A piece of canvas or plastic material used to protect expensive flooring from dropped debris and/or dirt from shoes and boots.
6. A light that can illuminate a broad area.
7. An engine-powered device that provides electricity.
8. A piece of heavy canvas with handles, which can be used to tote debris, embers, and burning materials out of a structure.
9. A light designed to project a narrow, concentrated beam of light.
10. Examination of all areas of the building and contents involved in a fire to ensure that the fire is completely extinguished.
11. An opening through which water can be removed from a building.
12. A salvage cover folded to form a container to hold water until it can be removed.
13. A device that converts the direct current from an apparatus electrical system into alternating current.

Down

1. Property damage that occurs due to smoke, water, or other measures taken to extinguish the fire.
2. A sprinkler control valve with an indicator that reads either open or shut depending on its position.
4. A device that attaches to an electrical cord to provide additional outlets.
5. A salvage cover folded to direct water flow out of a building or away from sensitive items or areas.
9. Removing or protecting property that could be damaged during firefighting or overhaul operations.

Fire Alarms

The following real case scenarios will give you an opportunity to explore the concerns associated with salvage and overhaul. Read each scenario, and then answer each question in detail.

1. You are assigned to the second ladder company arriving at a third-floor apartment fire in a multistory residential structure. The IC has ordered your company to the floor below the fire for salvage operations. Water is running from the ceiling into a second-floor apartment, causing much water damage. Your company officer orders you to protect the contents of the family room and then build a water chute to divert water to an outside window.

    a. Which actions will you take to protect the family room contents?

    _____
    _____
    _____
    _____
    _____
    _____

    b. How will you build a water chute to divert the water from the ceiling?

    _____
    _____
    _____
    _____
    _____
    _____

2. After you return from the fire described in question 1, your company officer instructs you to clean and check the salvage covers for wear. What is the proper method for cleaning and inspecting salvage covers?

    _____
    _____
    _____
    _____
    _____
    _____

3. After the salvage covers have dried, you are to fold them for future use. How will you fold a salvage cover so it may be deployed by one fire fighter?

    _____
    _____
    _____
    _____
    _____
    _____

Skill Drills

Skill Drill 18-3: Using a Sprinkler Stop

Test your knowledge of this skill drill by filling in the correct words in the photo captions.

1. Have a _____ in hand.

2. Place the _____ part of the sprinkler stop over the sprinkler head orifice and between the frame of the sprinkler head.

3. Push the _____ to expand the sprinkler stop until it snaps into position.

Skill Drill 18-8: Constructing a Water Catch-All
Test your knowledge of this skill drill by placing the photos below in the correct order. Number the first step with a "1," the second step with a "2," and so on.

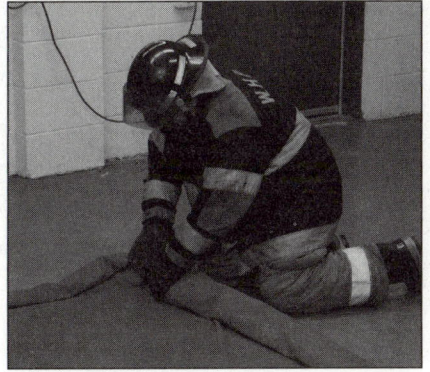

_____ Lift the rolled edge over the corner flaps, and tuck it in under the flaps, to lock the corners in place.

_____ Fold each of the four corners at a 90-degree angle, starting each fold approximately 3 feet in from the edge.

_____ Fully open a large salvage cover flat on the ground.

_____ Roll the remaining two edges inward approximately 2 feet.

_____ Roll two edges inward from the opposite sides, approximately 3 feet on each side.

Skill Drill 18-15: Pulling a Ceiling Using a Pike Pole
Test your knowledge of this skill drill by filling in the correct words in the photo captions.

1. Select the appropriate-length pike pole based on the height of the ceiling. For most residential applications, a(n) _____-feet-long pole is sufficient; longer poles are needed for higher ceilings.

2. Determine which area of the ceiling will be opened. Typically, the most heavily damaged areas are opened _____, followed by the surrounding areas. Position yourself to begin work with your back toward a door, so the debris you pull down will not block your access to the exit.

3. Using a strong, upward-thrusting motion, penetrate the ceiling with the _____ of the pike pole. Face the hook side of the tip away from you.

4. Pull down and away from your _____, so the ceiling material falls away from you.

5. Continue pulling down sections of the ceiling until the desired area is opened. Pull down any insulation, such as _____, found in the ceiling.

Skill Drill 18-16: Opening an Interior Wall

Test your knowledge of this skill drill by placing the photos below in the correct order. Number the first step with a "1," the second step with a "2," and so on.

_____ If necessary, remove items such as baseboards or window and door trim with a Halligan tool or axe. Continue opening sections of the wall until the desired area is open. Pull out any insulation, such as rolled fiberglass, found behind the wall.

_____ Determine which area of the wall will be opened up. The officer in charge usually makes this determination. Typically, the areas most heavily damaged by the fire are opened first, followed by the surrounding areas, working outward.

_____ After making two vertical cuts, use the pick end of the axe to pull the wall material away from the studs and open the wall. Work from top to bottom.

_____ Use the axe blade to begin cutting near the top of the wall. Cut downward between wall studs. Survey the wall for electrical switches or receptacles, as they are evidence of electrical wires behind the wall.

Fire Fighter Rehabilitation

Workbook Activities

The following activities have been designed to help you. Your instructor may require you to complete some or all of these activities as a regular part of your fire fighter training program. You are encouraged to complete any activity that your instructor does not assign as a way to enhance your learning in the classroom.

Chapter Review

The following exercises provide an opportunity to refresh your knowledge of this chapter.

Matching

Match each of the terms in the left column to the appropriate definition in the right column.

_____ 1. Rehabilitate

_____ 2. Dehydration

_____ 3. Electrolytes

_____ 4. Personal protective equipment (PPE)

_____ 5. Fully encapsulated suits

_____ 6. Glucose

_____ 7. Hypothermia

_____ 8. Revitalization

_____ 9. Frostbite

_____ 10. CISM

A. Gear worn by fire fighters that includes helmet, gloves, hood, coat, pants, SCBA, and boots

B. The source of energy for the body

C. A protective suit that completely covers the fire fighter, including the breathing apparatus

D. A state in which fluid losses are greater than fluid intake into the body

E. Certain salts and other chemicals that are dissolved in body fluids and cells

F. To restore to a condition of health or to a state of useful and constructive activity

G. A process that confronts and defuses the responses to critical incidents

H. Damage to tissue resulting from exposure to cold

I. Rest, fluid replacement, nutrition, and temperature stabilization

J. A condition in which the body temperature falls below 95 °F

Multiple Choice

Read each item carefully, and then select the best response.

_____ 1. Conditioning plays a significant role in
   A. endurance.
   B. glucose.
   C. intensity.
   D. reassignment.

_____ 2. Rehabilitation enables fire fighters to
   A. perform more safely and effectively.
   B. review personal protective equipment items.
   C. discuss emergency response strategies.
   D. observe and assess response tactics.

_____ 3. During a high-rise fire, a department may assign three companies to do the work that is normally done by one company
   A. because there is a larger structure to protect.
   B. because the higher the structure, the more distance to the ground.
   C. because there is more equipment to carry.
   D. to enable the companies to rotate duties.

_____ 4. Which of the following is *not* a function of an emergency incident rehabilitation center?
   A. Revitalization
   B. Medical monitoring
   C. Critical incident stress management
   D. Staging

_____ 5. What are the four components of revitalization?
   A. Rest, fluid replacement, nutrition, and temperature stabilization
   B. Vital signs, pulse, respiration, and blood pressure
   C. Assessment, pain tolerance, treatment, and recovery
   D. Knowing your limits, listening to your body, rehabilitating, and treatment

_____ 6. Which of the following factors can affect the need for rehabilitation?
   A. The type of incident
   B. The duration of the incident
   C. The environment of the incident
   D. All of the above

_____ 7. Which of the following is the best rehabilitation food source during a short incident?
   A. Water and a high-protein sports bar
   B. Coffee or tea and a high-protein sports bar
   C. Fruit juice and pasta
   D. A complete meal that includes complex carbohydrates

_____ 8. Signs and symptoms of fatigue can be affected by
   A. poor nutrition.
   B. physical exhaustion.
   C. emotional exhaustion.
   D. all of the above.

_____ 9. A fire fighter must stay at the rehabilitation center until his or her body temperature returns to
   A. a range consistent with the environment.
   B. a temperature above pre-entry levels.
   C. a normal range.
   D. below pre-entry levels.

_____ 10. Which of the following is the only fuel that the body can readily use during high-intensity physical activity?
   A. Carbohydrates
   B. Fats
   C. Proteins
   D. Electrolytes

Vocabulary

Define the following terms using the space provided.

1. Turnout gear:

2. Frostbite:

3. Critical incident stress management (CISM):

4. Emergency incident rehabilitation:

5. Hypothermia:

Fill-in
Read each item carefully, and then complete the statement by filling in the missing word(s).

1. _____ is a critical factor in maintaining health and well-being.
2. The amount of rest needed to recover from physical exertion is directly related to the _____ of the work performed.
3. The stomach can absorb only _____ quarts of fluid per hour, but the body can lose up to _____ quarts of fluid per hour.
4. Blood sugar, or _____, is the fuel the body uses for energy.
5. _____ are a major source of fuel for the body and can be found in grains, vegetables, and fruits.
6. _____ can actually cause a decrease in energy levels because it stimulates the production of insulin.
7. _____ occurs when body tissues are damaged due to prolonged exposure to the cold.
8. _____ happens only when a fire fighter is rested, rehydrated, refueled, and fit to return to active duty.
9. _____ is caused by drinking too much too quickly.
10. When a fire fighter has abnormal vital signs, is suffering pain, or is injured, he or she needs to have further _____ and _____.

True/False
If you believe the statement to be more true than false, write the letter "T" in the space provided.
If you believe the statement to be more false than true, write the letter "F."

1. _____ Rehabilitation enables fire fighters to perform more safely and effectively at an emergency scene.
2. _____ On a cold day, coffee or hot chocolate would be an appropriate beverage to rehabilitate fire fighters.
3. _____ The concept of rehabilitation needs to be addressed at all types of incidents.
4. _____ Rehabilitation helps improve the quality of decision making.
5. _____ The first stop at emergency incident rehabilitation is physical assessment.
6. _____ Thirst is a reliable indicator of dehydration.
7. _____ Carbohydrates are used by the body to grow and repair tissues and are used as a primary fuel source only in extreme conditions such as starvation.
8. _____ Fats are used for energy and breaking down some vitamins.
9. _____ The goal of the critical incident stress management (CISM) team is to save lives first and property second.
10. _____ Regular rehabilitation enables fire fighters to accomplish more work during a major incident.

Short Answer
Complete this section with short written answers using the space provided.

1. Which physical and mental symptoms may develop if the fire fighter does not get the opportunity to rest and recover?

2. Identify some of the factors that cause firefighting to be a stressful work environment.

3. What are the possible effects of dehydration on the fire fighter?

4. In addition to structure fires, what are some other incidents where full rehabilitation stations may be required?

5. Identify and describe the six functions of rehabilitation.

6. Why are caffeinated and sugar-rich beverages not recommended for rehabilitation?

7. Why is thirst not a good indicator of dehydration?
_____
_____
_____

8. Describe the types of meals a fire fighter should eat during rehabilitation for short and extended incidents.
_____
_____
_____

9. Identify the personal responsibilities of each fire fighter during rehabilitation.
_____
_____
_____

Clues

Across

1. A state in which fluid losses are greater than fluid intake into the body, leading to shock and even death if untreated.
5. Certain salts and other chemicals that are dissolved in body fluids and cells.
6. A condition in which the internal body temperature falls below 95°F.

Down

2. To restore to a condition of health or to a state of useful and constructive activity.
3. The source of energy for the body.
4. Damage to tissues as the result of exposure to cold.

Word Fun

The following crossword puzzle is an activity provided to reinforce correct spelling and understanding of terminology associated with firefighting. Use the clues provided to complete the puzzle.

Fire Alarms

The following real case scenarios will give you an opportunity to explore the concerns associated with fire fighter rehabilitation. Read each scenario, and then answer each question in detail.

1. You have just finished clearing brush and shrubs from the path of a wildland fire. It is a warm, 85°F summer day with no breeze. You begin to feel warm, are exhausted, are sweating profusely, and feel lightheaded. What should you do?

   _____
   _____
   _____
   _____
   _____
   _____

2. After fighting a winter-time defensive fire (temperature 15°F) for 2 hours, your partner comments that he has been wet for the last hour, and his feet and hands are hurting and causing a great deal of pain. How should you help him?

   _____
   _____
   _____
   _____
   _____
   _____

Wildland and Ground Fires

Workbook Activities

The following activities have been designed to help you. Your instructor may require you to complete some or all of these activities as a regular part of your fire fighter training program. You are encouraged to complete any activity that your instructor does not assign as a way to enhance your learning in the classroom.

Chapter Review

The following exercises provide an opportunity to refresh your knowledge of this chapter.

Matching

Match each of the terms in the left column to the appropriate definition in the right column.

_____ 1. Backfiring
_____ 2. Fuel volume
_____ 3. Fine fuel
_____ 4. Wildland
_____ 5. Slash
_____ 6. Pocket
_____ 7. Spot fire
_____ 8. Heel of the fire
_____ 9. Island
_____ 10. Black
_____ 11. Head of the fire

A. A new fire that starts outside areas of the main fire
B. Land in an uncultivated natural state that is covered by timber, woodland, brush, or grass
C. A planned operation to remove fuel by burning out large selected areas
D. The leftovers of a logging operation
E. Fuel that ignites and burns easily
F. The amount of fuel present in a given area
G. A deep indentation of unburned fuel along the fire's perimeter
H. The main or running edge of a fire
I. An area that has already been burned
J. An unburned area surrounded by burned land
K. The side opposite the head of the fire

Multiple Choice

Read each item carefully, and then select the best response.

_____ 1. Unplanned and uncontrolled fires burning in vegetative fuels that sometimes include structures are called
   A. ground cover fires.
   B. aerial fires.
   C. wildland fires.
   D. urban fires.

_____ 2. The partly decomposed organic material on a forest floor is called
   A. ground duff.
   B. slash.
   C. medium fuel.
   D. heavy fuel.

CHAPTER 20

_____ 3. Fuels that are located close to the surface of the ground are considered
   A. aerial fuels.
   B. subsurface fuels.
   C. supersurface fuels.
   D. surface fuels.

_____ 4. Which term describes the relative closeness of wildland fuels?
   A. Fuel compactness
   B. Fuel continuity
   C. Fuel volume
   D. Fuel moisture

_____ 5. The three causes of wildland fires are natural, accidental, and
   A. intentional fires.
   B. occupational fires.
   C. combustion fires.
   D. mechanical fires.

_____ 6. The study of elevation and the position of both natural and human-made features is known as
   A. geography.
   B. geology.
   C. physiology.
   D. topography.

_____ 7. As wildland and ground fires grow and reach into areas with new fuel, the traveling edge of the fire is called the
   A. heel of the fire.
   B. head of the fire.
   C. rear of the fire.
   D. arm of the fire.

_____ 8. An unburned area between a finger and the traveling edge of the fire is called a(n)
   A. island.
   B. lapse.
   C. pocket.
   D. spot fire.

_____ 9. Which combination tool is used to create a fire line?
   A. McLeod fire tool
   B. Adze
   C. Pulaski axe
   D. Halligan tool

_____ 10. A firefighting attack that involves building a fire line along natural fuel breaks, favorable breaks in topography, or at considerable distance from the fire and burning out the intervening fuel is called a(n)
   A. mounted attack.
   B. indirect attack.
   C. direct attack.
   D. counterattack.

_____ 11. A firefighting attack that requires only one team of fire fighters is called a(n)
   A. pincer attack.
   B. backfiring attack.
   C. flanking attack.
   D. indirect attack.

_____ 12. The firefighting attack most often used for large wildland and ground fires that are too dangerous for a direct attack is the
   A. pincer attack.
   B. backfiring attack.
   C. flanking attack.
   D. indirect attack.

_____ 13. A firefighting attack that requires two teams of fire fighters attacking both flanks of a wildland fire is called a(n)
   A. pincer attack.
   B. backfiring attack.
   C. flanking attack.
   D. counter attack.

_____ 14. How much water do small apparatus used for fighting wildland fires typically carry?
   A. 800 gallons
   B. 200–300 gallons
   C. 2000 gallons
   D. 50–100 gallons

_____ 15. What is the top priority in a wildland fire attack?
   A. Containment
   B. Extinguishment
   C. Minimization of damage
   D. Safety

_____ 16. The technique used to remove fuel by burning is called
   A. adze.
   B. backfiring.
   C. direct attack.
   D. flanking.

Vocabulary

Define the following terms using the space provided.

1. Fuel continuity:
   _____
   _____
   _____

2. Backpack pump extinguisher:
   _____
   _____

3. Heavy fuels:
   _____
   _____
   _____

4. Aerial fuels:

_____
_____
_____

5. Topography:

_____
_____
_____

Fill-in
Read each item carefully, and then complete the statement by filling in the missing word(s).

1. _____ conditions have a major impact on the behavior of wildland fires.
2. Vegetative fuels can be located _____, _____, or _____ the ground.
3. Fires spread more _____ in fine fuels than in heavy timber and brush.
4. The relative _____ is the ratio of the amount of water vapor present in the air compared to the maximum amount the air can hold at a given temperature.
5. _____-wing aircraft can take on a load of water from a lake and apply it to the fire.
6. _____ and _____ fires can advance quickly and can change directions quickly.
7. For small fires with a light fuel load, _____ may be an effective firefighting tactic.
8. The location where a wildland or ground fire begins is called the _____.
9. A(n) _____ fire is a new fire that starts outside the perimeter of the main fire.
10. The second side of the fire triangle is _____.

True/False
If you believe the statement to be more true than false, write the letter "T" in the space provided.
If you believe the statement to be more false than true, write the letter "F."

1. _____ The fire triangle consists of the elements fuel, oxygen, and heat.
2. _____ Fine fuels have a small surface area relative to their volume.
3. _____ The amount of moisture in a fuel is related to the season of the year.
4. _____ When relative humidity is high, the moisture from the air is absorbed by vegetative fuels, making them less susceptible to ignition.
5. _____ Fire shelters can be carried in a protective pouch on a fire fighter's belt.
6. _____ Wildland fires are unplanned and uncontrolled fires burning in vegetative fuel that sometimes includes structures.
7. _____ Roots, moss, duff, and decomposed stumps are examples of heavy fuels.
8. _____ The two most critical weather conditions that influence a wildland fire are moisture and wind.
9. _____ Rising of heated air in a wildland fire will preheat the fuels above the main body of the fire.
10. _____ A direct attack on a wildland fire is made by attacking the left flank of the main body.

Short Answer

Complete this section with short written answers using the space provided.

1. List three hazards of wildland fires.

_____

_____

_____

Word Fun

The following crossword puzzle is an activity provided to reinforce correct spelling and understanding of terminology associated with firefighting. Use the clues provided to complete the puzzle.

CLUES

Across

1. The room or area where a fire began.
5. The part of the fire that spreads with the greatest speed.
6. An area where undeveloped land with vegetative fuels is mixed with human-made structures.
8. The amount of fuel present in a given area.
9. A narrow point of fire caused by a shift in wind or a change in topography.
12. A new fire that starts outside areas of the main fire.
13. A hand tool used to chop brush for clearing a fire line or to mop up a wildland fire.
14. An area that has already been burned.

Down

2. Partly decomposed organic material on a forest floor; a type of fine fuel.
3. Fuels more than 6 feet off the ground.
4. An unburned area surrounded by fire.
7. A method of fire attack mounted by containing and extinguishing the fire at its burning edge.
10. The features of the earth's surface.
11. An area of unburned fuels.

Fire Alarms

The following real case scenarios will give you an opportunity to explore the concerns associated with wildland and ground fires. Read each scenario, and then answer each question in detail.

1. You are assigned to drive the brush truck for the first time. It is a larger-sized truck with a water capacity of 1000 gallons. You are dispatched to a wildland fire in a hilly, undeveloped part of your response district. As you approach the fire, you are instructed to drive down a very steep, unimproved road. How will you proceed?

   _____
   _____
   _____
   _____
   _____

2. You are clearing brush and extinguishing small fires in your assigned area when suddenly the wind picks up and shifts, building the fire and driving it toward you. Your escape route has been cut off and the flames are approaching. How will you protect yourself?

   _____
   _____
   _____
   _____
   _____

Fire Suppression

Workbook Activities

The following activities have been designed to help you. Your instructor may require you to complete some or all of these activities as a regular part of your fire fighter training program. You are encouraged to complete any activity that your instructor does not assign as a way to enhance your learning in the classroom.

Chapter Review

The following exercises provide an opportunity to refresh your knowledge of this chapter.

Matching

Match each of the terms in the left column to the appropriate definition in the right column.

_____ 1. Portable monitor
_____ 2. Deck gun
_____ 3. Elevated master stream device
_____ 4. Ladder pipe
_____ 5. Master stream device
_____ 6. Three team members
_____ 7. Combination attack
_____ 8. Direct attack
_____ 9. Indirect attack
_____ 10. Solid stream

A. Nozzle mounted on the end of an aerial device
B. Appliance designed to be set up and then left to operate unattended
C. Nozzle attached to the end of a straight ladder truck
D. Apparatus-mounted device intended to flow large amounts of water directly onto a fire or exposed building
E. Produced by a smooth-bore nozzle
F. Used to remove as much heat as possible
G. A straight or solid hose stream directed at the base of the fire
H. Employs both direct and indirect attack methods sequentially
I. Needed to advance a 2 ½-inch hose line inside a building
J. Used to produce high-volume water streams

Multiple Choice

Read each item carefully, and then select the best response.

_____ 1. Removing oxygen, fuel, or the heat from the combustion process is known to extinguish a fire by interrupting the fire
   A. square.
   B. triangle.
   C. cycle.
   D. tetrahedron.

_____ 2. When fire fighters advance hose lines into a building to attack a fire, which type of strategy are they using?
   A. Offensive
   B. Defensive
   C. Advancing
   D. Internal

CHAPTER 21

_____ 3. What is the primary objective in a defensive operation?
   A. To ensure the least amount of property damage
   B. To provide a safe environment for the fire fighter
   C. To prevent the fire from spreading
   D. To prepare the fire fighter for offensive attacks

_____ 4. The decision to conduct offensive or defensive operations is made by the
   A. fire fighter.
   B. captain.
   C. incident commander.
   D. fire chief.

_____ 5. Large handlines are defined as hoses that have at least a diameter of
   A. 2 ½ inches.
   B. 3 inches.
   C. 3 ½ inches.
   D. 5 inches.

_____ 6. Master streams are typically used for
   A. structure fires.
   B. hazardous materials incidents.
   C. offensive operations.
   D. defensive operations.

_____ 7. The pattern and form of the water that is discharged onto the fire is defined by the
   A. nozzle.
   B. hose line.
   C. fire size.
   D. strategy involved.

_____ 8. When fire fighters begin with an indirect attack and then continue with a direct attack, which type of attack are they utilizing?
   A. Aggressive
   B. Combination
   C. Multiple
   D. Progressive

_____ 9. In situations where the temperature is increasing and it appears that the room or space is ready to experience flashover, fire fighters should use a(n)
   A. indirect attack.
   B. exterior attack.
   C. indirect application of water.
   D. direct application of water.

_____ 10. When water is converted to steam, it expands to occupy a volume that is
   A. 10,000 times greater than the volume of water.
   B. 7000 times greater than the volume of water.
   C. 1700 times greater than the volume of water.
   D. 1000 times greater than the volume of water.

_____ 11. Master stream devices are used
   A. to produce high-volume water streams for large fires.
   B. to produce low-volume water streams for small fires.
   C. to produce increased air movement and ventilation.
   D. because they give the greatest control over the water flow.

_____ 12. The device that is permanently mounted on a vehicle and equipped with a piping system is a(n)
   A. master stream device.
   B. deck gun.
   C. ladder pipe.
   D. open monitor.

_____ 13. Actions that are taken to prevent the spread of a fire to areas that are not already burning are referred to as
   A. hosing down.
   B. maximum coverage.
   C. master stream spray.
   D. protecting exposures.

_____ 14. To decrease the need for manual overhaul during trashcan fires, it is very useful to apply
   A. Class A foam.
   B. Class B foam.
   C. Class C foam.
   D. Class D foam.

_____ 15. During vehicle fires under the hood or engine area, fire fighters should approach from the
   A. downhill and downwind side.
   B. downhill and upwind side.
   C. uphill and downwind side.
   D. uphill and upwind side.

_____ 16. After the main body of a vehicle fire has been extinguished, it is important to
   A. continue to apply a master stream from a safe distance.
   B. remove leftover flammable liquids.
   C. overhaul the vehicle.
   D. contain water overflow.

_____ 17. Vehicles that use compressed natural gas are powered by cylinders located
   A. under the hood of the vehicle.
   B. in the trunk of the vehicle.
   C. under the chassis of the vehicle.
   D. in front of the driver-side engine compartment.

_____ 18. Flammable liquid fires can be extinguished using
   A. Class A foam.
   B. Class B foam.
   C. Class C foam.
   D. Class D foam.

_____ 19. Storing propane as a liquid is very efficient because it has an expansion ratio of
   A. 27:1.
   B. 170:1.
   C. 270:1.
   D. 370:1.

_____ **20.** To prevent explosions in possible overheating situations, propane tanks are equipped with
   A. relief valves.
   B. release valves.
   C. connection valves.
   D. vapor space.

_____ **21.** When controlling and extinguishing electrical or electrical equipment fires, the fire fighter uses
   A. Class A agents.
   B. Class B agents.
   C. Class C agents.
   D. Class D agents.

Vocabulary

Define the following terms using the space provided.

1. Indirect application of water:

2. Master stream device:

3. Boiling-liquid, expanding-vapor explosion (BLEVE):

4. Portable monitor:

5. Straight stream:

Fundamentals of Fire Fighter Skills

Fill-in
Read each item carefully, and then complete the statement by filling in the missing word(s).

1. Fire suppression can be accomplished through a variety of methods that will stop the _____ process.
2. Directing water onto a fire from a safe distance is a(n) _____ operation.
3. Large handlines and master streams are more often used in _____ operations.
4. A(n) _____ stream divides water into droplets, which have a very large surface area and can absorb heat efficiently.
5. A fog stream moves a large quantity of _____ along with the mass of water droplets.
6. A fire that occurs inside a building is referred to as a(n) _____ structure fire.
7. A(n) _____ attack uses a single or solid hose stream to deliver water onto the base of the fire.
8. It is more difficult for fire fighters to advance and maneuver a(n) _____ handline inside a building.
9. Wetting neighboring exposures will keep the fuel from reaching its _____ temperature.
10. Fires in _____ present several challenges because they are below grade level and have limited routes of egress.
11. One of the main hazards of confined-space fires and incidents is low _____ levels.

True/False
If you believe the statement to be more true than false, write the letter "T" in the space provided.
If you believe the statement to be more false than true, write the letter "F."

1. _____ Successful offensive attacks often result in the least amount of property damage.
2. _____ If the risk factors are too great, the only acceptable option is an offensive strategy.
3. _____ Master stream devices are operated from a fixed position.
4. _____ A straight stream has more reach and penetrating power than a fog or solid stream.
5. _____ The air movement created by a fog stream can be used for ventilation.
6. _____ Large handlines are often used in offensive situations to direct a heavy stream of water onto a fire from an exterior position.
7. _____ A master stream device can be directed by remote control.
8. _____ A portable monitor is attached to a handline to create a fog stream.
9. _____ Before any interior attack is initiated, it is important that the structure be ventilated.
10. _____ Fires in stacked or piled materials can be approached aggressively, because they often burn evenly.
11. _____ During vehicle fires in hybrid automobiles, the orange cables that connect the batteries to the electric motors must be cut as quickly as possible to prevent sparking.
12. _____ The best method to prevent a BLEVE is to direct heavy water streams onto the tank from a safe distance.

Short Answer
Complete this section with short written answers using the space provided.

1. Identify five factors to be evaluated when considering whether to enter an involved structure or to mount an attack.

   _____
   _____
   _____
   _____
   _____

Fire Alarms
The following real case scenarios will give you an opportunity to explore the concerns associated with fire suppression. Read each scenario, and then answer each question in detail.

1. It is 6:30 in the evening when a thunder and windstorm hits your community. Winds are gusting up to 60 miles per hour. Your engine is dispatched to an electrical pole that is on fire. Upon arrival, you see that a transformer is on fire. How should you proceed?

   _____
   _____
   _____
   _____
   _____

2. It is 9:00 in the morning when you are dispatched to a residential structure fire. While you are en route, dispatch tells you that all of the occupants have evacuated. Upon arrival, your Captain gives a size-up: a one-story, wood-frame residential structure with fire and smoke coming out of one room. He tells you and your partner to complete an offensive direct attack on the fire room. How should you proceed?

   _____
   _____
   _____
   _____
   _____

Skill Drills

Skill Drill 21-2: Performing a Direct Attack
Test your knowledge of this skill drill by filling in the correct words.

1. Exit the fire apparatus wearing full PPE, including _____.
2. Select the proper hose line to fight the fire based on the fire's size, location, and _____.
3. Advance the hose line from the apparatus to the entry point of the _____.
4. Don a face piece and activate the SCBA and _____ device prior to entering the building.
5. Signal the operator/driver that you are ready for _____.
6. Open the nozzle to _____ air from the system and make sure water is flowing.
7. Make sure that _____ is completed or in progress.
8. Enter into the structure and locate the _____ of the fire.
9. Apply water in either a straight or solid stream onto the base of the fire until all visible _____ has been extinguished.
10. Watch for _____ in fire conditions.
11. Shut down the nozzle and _____.
12. Locate and _____ hot spots.

Skill Drill 21-3: Performing an Indirect Attack
Test your knowledge of this skill drill by placing the steps below in the correct order.
Number the first step with a "1," the second step with a "2," and so on.

_____ Don a face piece, and activate the SCBA and PASS device.

_____ Select the correct hose line to be used to attack the fire depending on the type of fire, its location, and its size.

_____ Advance the hose line from the apparatus to the opening in the structure where the indirect attack will be made.

_____ Notify the operator/driver that you are ready for water.

_____ Advance with a charged hose line to the location where you will apply water.

_____ Exit the fire apparatus wearing full PPE, including SCBA.

_____ Watch for changes and a reduction in the amount of fire. Once the fire is reduced, shut down the nozzle.

_____ Direct the water stream toward the upper levels of the room and ceiling into the heated area overhead, and move the stream back and forth. Flow water until the room begins to darken. Shut the nozzle off, and reassess the fire conditions.

_____ Attack any remaining fire and hot spots until the fire is completely extinguished.

_____ Open the nozzle and make sure that air is purged from the hose line and that water is flowing. If using a fog nozzle, ensure that it is set to the proper nozzle pattern for entry. Shut down the nozzle until you are in a position to apply water.

_____ Confirm that ventilation has been completed.

Skill Drill 21-4: Performing a Combination Attack
Test your knowledge of this skill drill by filling in the correct words.

1. Don full _____ and SCBA. Select the correct hose line to accomplish the suppression task at hand.

2. Stretch the hose line to the _____ of the structure, and signal the operator/driver that you are ready to receive water.

3. Open the nozzle to get the air out and make sure that water is _____.

4. Enter the structure, and locate the room or area where the fire _____.

5. Aim the nozzle at the _____ corner of the fire and make either a "T," "O," or "Z" pattern with the nozzle. Start high and then work the pattern down to the fire level.

6. Use only enough water to _____ down the fire without upsetting the thermal layering.

7. Once the fire has been reduced, find the remaining hot spots and complete fire extinguishment using a(n) _____ attack.

Skill Drill 21-5: Performing the One-Fire Fighter Method for Operating a Large Handline

Test your knowledge of this skill drill by placing the photos below in the correct order. Number the first step with a "1," the second step with a "2," and so on.

_____ Select the correct size of fire hose. Advance the hose into position. Signal that you are ready for water and open the nozzle to allow air to escape and to ensure water is flowing. Close the nozzle and then make a loop with the hose, ensuring that the nozzle is *under* the hose line that is coming from the fire apparatus.

_____ Allow enough hose to extend past the section where the line crosses itself for maneuverability.

_____ Open the nozzle and direct water onto the designated area.

_____ Lash the hose sections together where they cross, or use your body weight to kneel or sit on the hose line at the point where the hose crosses itself.

Skill Drill 21-6: Performing the Two-Fire Fighter Method for Operating a Large Handline
Test your knowledge of this skill drill by filling in the correct words in the photo captions.

1. _____ the hose line from the fire apparatus into position.

2. _____ that you are ready for water and open the nozzle to allow air to escape and to ensure water is flowing. Advance the hose line as needed.

3. Before attacking the fire, the fire fighter on the nozzle should cradle the hose on his or her hip while grasping the nozzle with one hand and supporting the hose with the other hand. The second fire fighter should stay approximately _____ feet behind the fire fighter who is on the nozzle. The second fire fighter should grasp the hose with two hands and may use a knee to stabilize the hose against the ground if necessary.

4. Open the _____ in a controlled fashion and direct water onto the fire or designated exposure.

Preincident Planning

Workbook Activities

The following activities have been designed to help you. Your instructor may require you to complete some or all of these activities as a regular part of your fire fighter training program. You are encouraged to complete any activity that your instructor does not assign as a way to enhance your learning in the classroom.

Chapter Review

The following exercises provide an opportunity to refresh your knowledge of this chapter. All questions in this chapter are Fire Fighter II level.

Matching

Match each of the terms in the left column to the appropriate definition in the right column.

_____ 1. Standpipe system **A.** A collection of lightweight structural components joined in a triangular configuration that can be used to support either floors or roofs

_____ 2. Horizontal evacuation **B.** A large fire, often involving multiple structures

_____ 3. Vertical ventilation **C.** A strategy in which victims are protected from the fire without relocation

_____ 4. Truss **D.** Where a pumper can draft water directly from a static source

_____ 5. Drafting sites **E.** Property that may be endangered by flames, smoke, gases, heat, or runoff from a fire

_____ 6. Defend-in-place **F.** Amount of combustibles within a fire area

_____ 7. Static water supply **G.** Properties that have an increased frequency of fires

_____ 8. Size-up **H.** Process of making openings so that the smoke, heat, and gases can escape vertically from a structure

_____ 9. Exposure **I.** Process used to gather information to develop a preincident plan

_____ 10. Preincident survey **J.** A strategy of moving occupants from a dangerous area to a safe area on the same floor level

_____ 11. Sprinkler system **K.** Ongoing observation and evaluation of factors that are used to develop objectives, strategy, and tactics for fire suppression

_____ 12. Conflagration **L.** Water supply that is not under pressure

_____ 13. Target hazard **M.** Arrangement of piping, valves, and hose connections installed in a structure to deliver water for fire hoses

_____ 14. Tanker shuttle **N.** System of tankers that transports water from a water source to the fire scene

_____ 15. Fire load **O.** Automatic fire protection system designed to turn on sprinklers if a fire occurs

CHAPTER 22

Multiple Choice

Read each item carefully, and then select the best response.

_____ 1. A preincident plan should include information about
   A. a building's floor plan.
   B. entrance and exit locations.
   C. hazardous materials stored in the building.
   D. all of the above.

_____ 2. Preincident planning assumes that a fire will occur and tries to
   A. correct the potential problem that could cause a fire.
   B. minimize the potential for a fire to occur.
   C. compile information that responders could use to be more effective.
   D. prevent a fire from occurring.

_____ 3. The preincident survey should be conducted in
   A. a systematic, uniform format.
   B. normal duty uniform.
   C. scheduled annual visits.
   D. teams of four.

_____ 4. The five types of building construction in descending order of fire resistance are
   A. fire resistive, noncombustible, ordinary, heavy timber, and wood frame.
   B. ordinary, noncombustible, fire resistive, wood frame, and heavy timber.
   C. fire resistive, noncombustible, ordinary, wood timber, and heavy frame.
   D. ordinary, fire resistive, noncombustible, wood frame, and heavy timber.

_____ 5. The classifications of buildings by major use group are
   A. lightweight, tested development, heavyweight, and open.
   B. public assembly, institutional, commercial, and industrial.
   C. renovated private, renovated public, commercial development, and industrial.
   D. public assembly, institutional, and commercial.

_____ 6. The primary role of a fire alarm system is to
   A. alert the occupants of a building when an incident occurs.
   B. alert the fire department of an incident.
   C. meet safety standards of the building code.
   D. all of the above.

_____ 7. Drafting sites should be included in a preincident plan because they identify
   A. which adjoining structures are most susceptible to fire spread.
   B. open areas that can trap fire fighters.
   C. the best routes for ventilation.
   D. locations where an engine can draft water directly from a static source.

_____ 8. Horizontal ventilation can be accessed through
   A. windows and doors.
   B. windows and chimneys.
   C. ceiling and pressure fans.
   D. windows and skylights.

_____ 9. The most challenging problem during an emergency incident at a healthcare facility
   A. is the limited access to patients.
   B. is the limitations of the floor plans.
   C. is protecting nonambulatory patients.
   D. is negotiating traffic en route to the facility.

_____ 10. Security and personnel safety concerns are a major concern at
   A. schools and daycare centers.
   B. hospitals and nursing homes.
   C. residential occupancies.
   D. detention and correctional facilities.

_____ 11. Lightweight construction
   A. can be found only in newer buildings.
   B. utilizes trusses as structural support materials.
   C. is the sturdiest of the newer construction types.
   D. is the most cost-effective type of construction.

Vocabulary

Define the following terms using the space provided.

1. Horizontal ventilation:

2. Dry hydrant:

3. Fire alarm annunciator panel:

4. HVAC system:

5. Ordinary construction:

6. Conflagration:

_____
_____
_____

7. Preincident plan:

_____
_____
_____

Fill-in
Read each item carefully, and then complete the statement by filling in the missing word(s).

1. The _____ can use the preincident information to direct the emergency operations more effectively.
2. The process of obtaining information about a building and storing the information in a system so that it can be retrieved quickly for future reference is often referred to as _____.
3. The use of _____ has greatly increased the ability of fire departments to capture, store, organize, update, and quickly retrieve preincident planning information.
4. Properties that pose unusual risks to fire fighters during an emergency response are identified as _____.
5. A preincident _____ is used by a team to gather information about a property to develop a preincident plan.
6. The _____ is the part of the fire alarm system that indicates the location of an alarm within a building.
7. Building layout and access information is particularly important during the _____ phase of an emergency incident.
8. Preincident plans should include the most efficient route to a property and a(n) _____ route in case of traffic interruptions.
9. _____ is the ongoing observation and evaluation of factors that are used to develop objectives, strategy, and tactics for fire suppression.
10. A Type I building can also be referred to as fire resistive and includes materials such as _____, _____, and _____.
11. A Type II building can also be referred to as _____ and is made of structure members that are noncombustible materials, but may not have _____.
12. Wood-frame buildings are classified as Type _____.
13. When the dimensions of the interior materials are greater than the dimensions of ordinary construction, the building construction is considered _____ or _____.
14. Schools and hospitals are in the major use classification _____.
15. Commercial use classification structures include the occupancy subcategories of _____, _____, _____, parking garages, and warehouses.
16. A(n) _____ is any other building or item that may be in danger if an incident occurs in another building or area.

17. A properly designed and maintained automatic _____ can help control or extinguish a fire before the arrival of the fire department.

18. _____ are installed in high-rise buildings to eliminate the need to extend hose lines from a pumper at the street level up to the fire level.

19. If the water is obtained from a lake or a stream it is considered a(n) _____.

20. During an emergency situation, it may be necessary to shut off the utilities such as _____ or _____ as a safety measure.

21. In preparation for possible search and rescue operations, it will greatly assist the fire fighters' efforts if they know where the _____, _____, and _____ are located.

22. The preincident survey should consider both _____ and _____ access problems.

23. A high-rise building is generally defined as a structure that is more than _____ feet high.

24. The _____ philosophy presumes that patients or occupants of some facilities will not be able to escape from a fire without assistance and, therefore, the facility itself is designed to protect patients from the fire.

25. Moving patients from a dangerous area to a safer area is known as _____ evacuation.

26. Buildings that have hazardous materials should have posted _____ documents to assist in the approach on the facility.

True/False

If you believe the statement to be more true than false, write the letter "T" in the space provided.
If you believe the statement to be more false than true, write the letter "F."

1. _____ The main purpose of the preincident plan is to provide information for more effective operations during emergency incidents.

2. _____ Residential property owners must post and forward documentation to the fire department if they have hazardous materials on-site.

3. _____ Vertical ventilation is a more valuable technique than horizontal ventilation.

4. _____ All new and renovated private residences are required to have a private water supply system.

5. _____ During an emergency incident, the property owner should be consulted prior to disconnecting any utilities.

6. _____ The fire load is the amount of combustible material and the rate of heat release a property may include.

7. _____ Lightweight construction uses materials too light to cause injury to a fire fighter in proper personal protective equipment.

8. _____ Preincident surveys should be conducted by contracted professionals.

9. _____ Preincident surveys of commercial and industrial properties should be conducted by independent contractors through the fire department.

10. _____ Wood-frame building construction has floors and walls made of combustible wood material.

11. _____ Buildings with unprotected steel beams are Type I: Fire Resistive, according to NFPA 220, *Standard on Types of Building Construction*.

12. _____ Potential natural barricades should be included in the preincident plan.

13. _____ All properties have the potential to create a conflagration.

Chapter 22: Preincident Planning 251

Short Answer
Complete this section with short written answers using the space provided.

1. Describe some of the benefits of preincident planning.
   _____
   _____
   _____

2. What is the objective of a preincident plan?
   _____
   _____
   _____

3. List five types of information that could be included in a preincident plan.
   _____
   _____
   _____
   _____
   _____

4. When developing a preincident plan, which questions should be asked when considering access to the exterior of a building?
   _____
   _____
   _____

5. When developing a preincident plan, which questions should be asked when considering access to the interior of a building?
   _____
   _____
   _____

6. List some of the possible target hazards in a community.
   _____
   _____
   _____

7. Identify some of the properties that have an increased life-safety hazard in a community.
   _____
   _____
   _____

8. Explain the difference between preincident planning and fire prevention.
   _____
   _____
   _____

9. Describe the general format of conducting a preincident survey.
   _____
   _____
   _____

10. Describe the characteristics of the following types of building construction:
    a. Fire resistive:
    _____
    _____
    _____

    b. Noncombustible:
    _____
    _____
    _____

    c. Ordinary:
    _____
    _____
    _____

    d. Heavy timber:
    _____
    _____
    _____

    e. Wood frame:
    _____
    _____
    _____

11. Identify examples of the following occupancy subcategories found within classifications of buildings:
    a. Public assembly:
    _____
    _____
    _____

b. Institutional:
_____
_____
_____

c. Commercial:
_____
_____
_____

12. Identify some of the questions that should be considered when addressing the water supply for a particular property in the preincident plan.
_____
_____
_____

13. If the preincident survey indicates that there are no hydrants available, what are some of the water supply considerations that you must address in the preincident plan?
_____
_____
_____

14. Identify the additional information you should gather when developing a preincident plan for a private residence.
_____
_____
_____

15. List several of the types of occupancies where fire fighters can expect to encounter hazardous materials.
_____
_____
_____

Word Fun

The following crossword puzzle is an activity provided to reinforce correct spelling and understanding of terminology associated with firefighting. Use the clues provided to complete the puzzle.

CLUES

Across

2. A water supply that is not under pressure, such as a pond, lake, or stream.
5. The process used to gather information to develop a preincident plan.
8. Moving occupants from a dangerous area to a safe area on the same floor level.
9. The ongoing observation and evaluation of factors that are used to develop objectives, strategy, and tactics for fire suppression.
10. The amount of combustibles within a fire area.

Down

1. A strategy where the victims are protected from the fire without relocation.
3. A large fire, often involving multiple structures.
4. Properties that have the potential to create large fires or present unusual situations for responding fire fighters.
6. Any property that may be endangered by flames or smoke from a fire.
7. The type of building construction where the structural members are of noncombustible materials that have a specified fire resistance.
8. A building that is more than 75 feet high.

Fire Alarms

The following real case scenarios will give you an opportunity to explore the concerns associated with preincident planning. Read each scenario, then answer each question in detail.

1. Your company has been assigned a list of buildings that must have a preincident survey conducted. As the rookie, you have been assigned the task of developing a plan for conducting each survey. Which steps are necessary for conducting a good preincident survey?

   _____
   _____
   _____
   _____
   _____
   _____

2. Your monthly probationary assignment is to list the target hazard properties in your first-response district.

   a. What are typical target hazard properties that could be included in your assignment?

   _____
   _____
   _____
   _____
   _____
   _____

   b. Include a sublist of properties with increased life-safety hazards for extra credit.

   _____
   _____
   _____
   _____
   _____
   _____

Fire and Emergency Medical Care

Workbook Activities

The following activities have been designed to help you. Your instructor may require you to complete some or all of these activities as a regular part of your fire fighter training program. You are encouraged to complete any activity that your instructor does not assign as a way to enhance your learning in the classroom.

Chapter Review

The following exercises provide an opportunity to refresh your knowledge of this chapter.

Matching

Match each of the terms in the left column to the appropriate definition in the right column.

_____ 1. Medical director   **A.** The first trained individual to arrive at the scene of an emergency to provide initial medical assistance

_____ 2. Basic life support (BLS)   **B.** An EMT who has the highest level of training in EMS, including cardiac monitoring, administering drugs, inserting advanced airways, and manual defibrillation

_____ 3. Medical control   **C.** The fire department provides medical first response, while another agency transports the patient to the hospital emergency room

_____ 4. Combination EMS system   **D.** Written documents, signed by the system's medical director, that outline specific directions, permissions, and sometimes prohibitions regarding patient care; also called protocols

_____ 5. First responder   **E.** Noninvasive emergency life-saving care that is used to treat airway obstruction, respiratory arrest, or cardiac arrest

_____ 6. EMT-Basic   **F.** Physician instructions that are given directly by radio or indirectly by protocol/guidelines as authorized by the medical director of the service

_____ 7. Standing orders   **G.** An EMT who can perform limited procedures that usually fall between those provided by an EMT-Basic and those provided by an EMT-Paramedic

_____ 8. EMT-Intermediate   **H.** The physician providing direction for patient care activities in the prehospital setting

_____ 9. EMT-Paramedic   **I.** An EMT who has training in basic emergency care skills, including oxygen therapy, bleeding control, CPR, automated external defibrillation, use of basic airway devices, and assisting patients with certain medications

_____ 10. Advanced life support   **J.** Advanced life-saving procedures, such as cardiac monitoring, administration of IV fluids and medications, and use of advanced airway adjuncts

CHAPTER 23

Multiple Choice
Read each item carefully, and then select the best response.

_____ 1. What are the levels of EMS certification, in order from lowest to highest level of training?
   A. First responder, EMT-Basic, EMT-Intermediate, EMT-Paramedic
   B. EMT-Basic, EMT-Paramedic, EMT-Intermediate, first responder
   C. First responder, EMT-Basic, EMT-Paramedic, EMT-Intermediate
   D. EMT-Basic, EMT-Intermediate, EMT-Paramedic, first responder

_____ 2. Basic life support services include
   A. administering oxygen, interpreting heart rhythms, and splinting.
   B. administering medications, oxygen, and intravenous fluids to treat shock.
   C. scene control, splinting, treating for shock, and defibrillating the heart.
   D. administering oxygen, controlling external bleeding, and lifting and moving patients.

_____ 3. Advanced life support services include
   A. endotracheal intubation and administration of intravenous fluids or medications.
   B. administration of intravenous fluids and interfacility transport of critical care patients.
   C. pediatric and critical care specialties.
   D. defibrillation, administration of medications, and blood diagnostics.

_____ 4. What are the two primary types of EMS delivery systems within fire departments?
   A. Paramedic and transport
   B. EMS and transport
   C. Combination EMS and public transport
   D. Combination and fire department EMS system

_____ 5. Standing orders are a type of medical control that
   A. is provided by an EMS physician who can be reached by radio during a call.
   B. directs the EMS providers to take specific actions when they encounter different situations.
   C. allows fire fighters to provide only rescue operations.
   D. is used only by the medical director.

_____ 6. What are the two levels of training for basic life support providers?
   A. EMT-Basic and EMT-Intermediate
   B. Fire Fighter I and EMT-Basic
   C. Fire Fighter I and first responder
   D. First responder and EMT-Basic

_____ 7. What are the two levels of training for advanced life support providers?
   A. EMT-Basic and EMT-Intermediate
   B. EMT-Intermediate and first responder
   C. EMT-Intermediate and EMT-Paramedic
   D. First responder and EMT-Paramedic

_____ 8. Basic life support providers can perform cardiac defibrillation by using
   A. an automated external defibrillator (AED).
   B. internal massage.
   C. medication.
   D. intravenous fluids.

_____ 9. The three main groups that interact with the EMS provider are
   A. fire fighters, hospital administrators, and patients.
   B. patients, public onlookers, and police officers.
   C. police officers, public onlookers, and medical directors.
   D. patients, hospital personnel, and medical directors.

_____ 10. Protecting the privacy of the people you serve is
   A. an ethical responsibility.
   B. seldom a concern.
   C. handled by the medical director.
   D. an insurance issue.

Vocabulary

Define the following terms using the space provided.

1. EMT-Paramedic:
   _____
   _____
   _____

2. Basic life support (BLS):
   _____
   _____
   _____

3. Advanced life support (ALS):
   _____
   _____
   _____

4. Combination EMS system:
   _____
   _____
   _____

5. Fire department EMS system:
   _____
   _____
   _____

6. Medical director:
   _____
   _____
   _____

Fill-in
Read each item carefully, and then complete the statement by filling in the missing word(s).

1. Strict _____ and effective _____ efforts have been quite successful in reducing the numbers of fires, enabling the fire service to take a greater role in providing emergency medical care.

2. From the caller's perspective, all calls are seen as a(n) _____. Therefore, an EMS provider must always remain supportive of the patient and the caller.

3. The _____ course is designed for people such as teachers and daycare providers who encounter medical emergencies as part of their jobs.

4. _____ is the only level of EMS provider that covers the causes and treatments of diseases.

5. In some fire departments, more than _____ of emergency calls are for emergency medical services.

6. EMS care is offered at both _____ and _____ life support levels.

7. The mission of the fire service is to _____ and protect _____.

8. _____ services do not include the administration of medications beyond assisting the patient with his or her prescribed medications.

9. _____ personnel operate as an extension of a physician and use _____, _____, and/or _____.

10. Some fire departments _____ their personnel in both fire suppression and EMS.

True/False
If you believe the statement to be more true than false, write the letter "T" in the space provided.
If you believe the statement to be more false than true, write the letter "F."

1. _____ In most departments, the number of fire calls is much greater than the number of EMS calls.
2. _____ The National Registry of Emergency Medical Technicians registers only EMT-Paramedics.
3. _____ Most fire departments provide some level of emergency medical services, although the degree of their involvement varies.
4. _____ Local protocols provide direction for the treatments administered and actions taken by EMS personnel.
5. _____ Urgent circumstances may require that EMS personnel contact the online medical control physician for online or direct treatment orders.
6. _____ All EMS providers in a system must be trained to work together and coordinate their activities.
7. _____ There are no advantages to having EMS systems located within the fire department.
8. _____ CME classes are unimportant to EMS providers.
9. _____ In a combination EMS system, the fire department provides medical first response and another agency operates the ambulances that transport the patients.

Short Answer
Complete this section with short written answers using the space provided.

1. Identify the responsibilities of the medical director.

2. List some of the advantages to having EMS systems located within the fire department.

3. Identify the duties and abilities of a first responder.

4. Identify the duties and abilities of the EMT-Basic.

5. Identify the duties and abilities of the EMT-Paramedic.

6. Describe fire department EMS systems.

7. Describe a combination EMS system.

_____
_____
_____

8. Describe appropriate interactions with patients.

_____
_____
_____

9. Describe the difference between offline (indirect) and online (direct) medical control.

_____
_____
_____

Word Fun

The following crossword puzzle is an activity provided to reinforce correct spelling and understanding of terminology associated with firefighting. Use the clues provided to complete the puzzle.

Clues

Across

1. Has the highest level of training in EMS, including cardiac monitoring, administering drugs, inserting advanced airways, and manual defibrillation.
3. Noninvasive, emergency life-saving care that is used to treat airway obstruction, respiratory arrest, or cardiac arrest.
5. Physician instructions that are given directly by radio or indirectly by protocols/guidelines as authorized by the medical director of the service.
6. Advanced life-saving procedures, such as cardiac monitoring, administration of IV fluids and medications, and use of advanced airway adjuncts.
7. Written documents, signed by the system's medical director, that outline specific directions regarding patient care.
8. Has training in basic emergency care skills, including oxygen therapy, bleeding control, CPR, automated external defibrillation, use of basic airway devices, and assisting patients with certain medications.
9. The first trained individual to arrive at the scene of an emergency to provide initial medical assistance.

Down

1. Can perform limited procedures that usually fall between those provided by an EMT-Basic and those provided by an EMT-Paramedic.
2. The fire department provides medical first response, while another agency transports the patient to the hospital emergency room.
4. The physician providing direction for patient care activities in the prehospital setting.

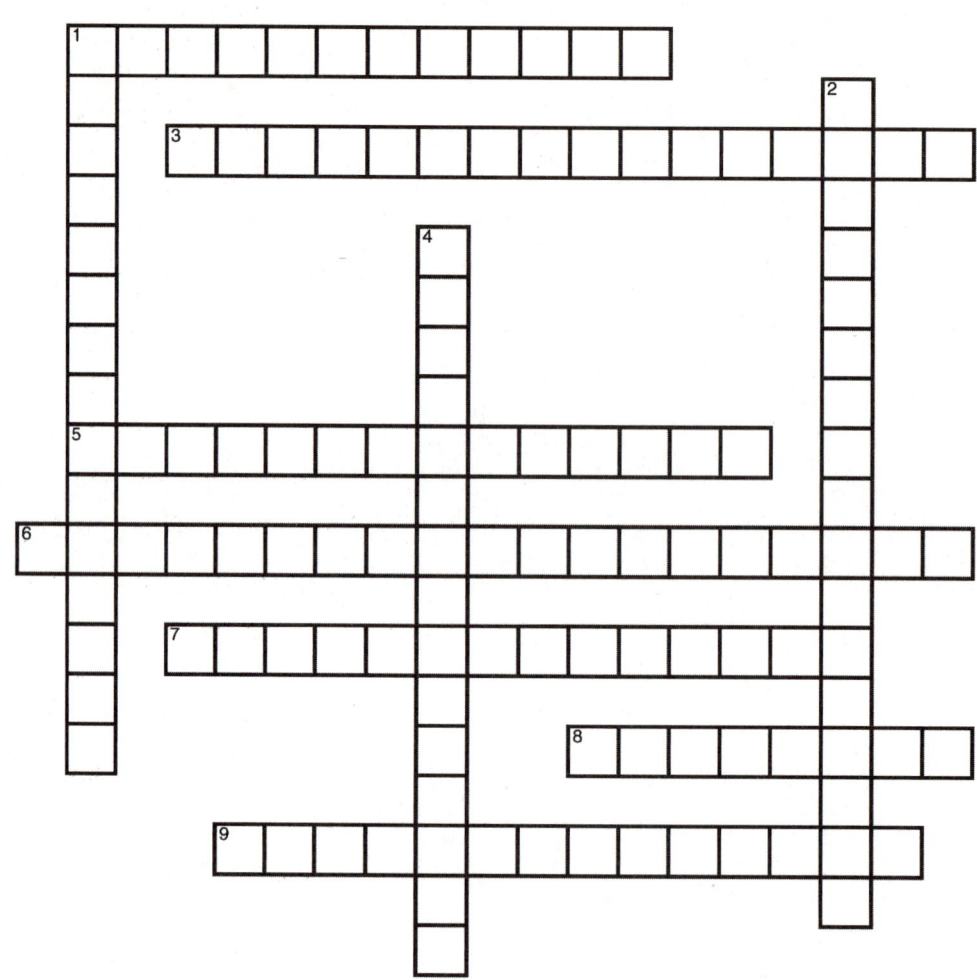

Fire Alarms

The following real case scenarios will give you an opportunity to explore the concerns associated with fire and emergency medical care. Read each scenario, and then answer each question in detail.

1. You have just completed your probationary period as a fire fighter, and you are interested in entering the emergency medical services. Your department offers a number of EMS training programs, including first responder, EMT-Basic, and EMT-Paramedic. You ask your Lieutenant for her opinion about where to start. She suggests that you enroll in the EMT-Basic course but also encourages you to research the training requirements and decide whether you can meet the demands of the class. What can you expect in terms of the time commitment and curriculum for an EMT-Basic course?

   _____
   _____
   _____
   _____
   _____
   _____

2. Your department is in the process of renewing its EMS contract with the local jurisdiction. A private EMS company is also bidding on providing these services. A family member asks you to explain the benefit of having the EMS system located within the fire department. How do you respond?

   _____
   _____
   _____
   _____
   _____
   _____

Emergency Medical Care

Workbook Activities

The following activities have been designed to help you. Your instructor may require you to complete some or all of these activities as a regular part of your fire fighter training program. You are encouraged to complete any activity that your instructor does not assign as a way to enhance your learning in the classroom.

Chapter Review

The following exercises provide an opportunity to refresh your knowledge of this chapter.

Matching

Match each of the terms in the left column to the appropriate definition in the right column.

_____ 1. Capillaries A. The fluid part of blood
_____ 2. Decapitation B. The key landmark for CPR and the Heimlich maneuver
_____ 3. Trachea C. Separation of the head from the body
_____ 4. Bruise D. A bandage placed directly on the wound
_____ 5. Plasma E. An injury in which a piece of skin is left hanging by a flap
_____ 6. Esophagus F. The windpipe
_____ 7. Avulsion G. The smallest blood vessels
_____ 8. Xiphoid process H. Voice box
_____ 9. Larynx I. The breastbone
_____ 10. Dressing J. An irregular cut or tear through the skin
_____ 11. Laceration K. A closed wound (contusion)
_____ 12. Sternum L. The tube that passes food from the throat to the stomach

Multiple Choice

Read each item carefully, and then select the best response.

_____ 1. The concept of using protective equipment to prevent exposure to infectious diseases is known as
   A. workplace safety equipment.
   B. personal protective equipment.
   C. body substance isolation.
   D. infection protection.

_____ 2. The immunizations recommended for medical care providers include the tetanus vaccine and
   A. hepatitis B vaccine.
   B. hepatitis C vaccine.
   C. HIV vaccine.
   D. SARS vaccine.

CHAPTER 24

_____ 3. The three critical components needed to sustain life in human beings are
   A. food, clothing, and shelter.
   B. food, water, and shelter.
   C. nutrients, clothing, and air.
   D. food, water, and oxygen.

_____ 4. Brain cells begin to die if they are deprived of oxygen and nutrients for
   A. two to three minutes.
   B. four to six minutes.
   C. seven to nine minutes.
   D. nine or more minutes.

_____ 5. What is the name of the thin flapper valve that allows air to enter the trachea, but that prevents food or water from doing so?
   A. Esophagus
   B. Epiglottis
   C. Larynx
   D. Alveoli

_____ 6. What is another name for the voice box?
   A. Esophagus
   B. Trachea
   C. Bronchi
   D. Larynx

_____ 7. The tiny air sacs in the lungs where the actual exchange of gases takes place are the
   A. alveoli.
   B. capillaries.
   C. bronchi.
   D. xiphoid.

_____ 8. What are the smallest branches of the circulatory system, where the exchange of oxygen and carbon dioxide takes place?
   A. Veins
   B. Arteries
   C. Capillaries
   D. Blood vessels

_____ 9. Opening the airway by lifting the victim's head backward and lifting the chin forward, bringing the entire lower jaw with it, is called the
   A. jaw-thrust technique.
   B. head tilt–chin lift.
   C. head tilt and shift.
   D. jaw and head lift.

_____ 10. A normal adult has a breathing rate of approximately
   A. 8 to 12 breaths per minute.
   B. 12 to 20 breaths per minute.
   C. 20 to 30 breaths per minute.
   D. 30 to 40 breaths per minute.

_____ 11. What is the most critical sign of inadequate breathing?
   A. Gasping
   B. Cyanosis
   C. Respiratory arrest
   D. Unconsciousness

_____ 12. The "look, listen, and feel" check for breathing should take no more than
   A. 5 seconds.
   B. 10 seconds.
   C. 20 seconds.
   D. 30 seconds.

_____ 13. If a victim is not breathing, you must breathe for him or her. This technique is known as
   A. rescue breathing.
   B. ventilation.
   C. the ABCs.
   D. victim recovery.

_____ 14. A "child" is defined as a person between
   A. 1 and 8 years of age.
   B. 5 and 12 years of age.
   C. 8 and 12 years of age.
   D. 8 and 18 years of age.

_____ 15. When assisting an infant's breathing, after the first two breaths, rescue breaths should follow every
   A. 2 seconds.
   B. 3 seconds.
   C. 5 seconds.
   D. 10 seconds.

_____ 16. What is the most common cause of airway obstruction?
   A. Food
   B. Small toys
   C. Dentures
   D. The tongue

_____ 17. What is the most effective way of expelling a foreign object that is causing airway obstruction?
   A. The Heimlich maneuver
   B. Patting or rubbing the back
   C. Coughing
   D. Throwing up

_____ 18. If the airway is completely obstructed, the victim will lose consciousness in
   A. less than one minute.
   B. one to two minutes.
   C. three to four minutes.
   D. five to six minutes.

_____ 19. To relieve an airway obstruction in an infant, use a combination of back slaps and
   A. the Heimlich maneuver.
   B. rescue breathing.
   C. tilts.
   D. chest thrusts.

_____ 20. What is the major artery in the neck?
   A. Carotid artery
   B. Radial artery
   C. Brachial artery
   D. Femoral artery

_____ 21. The pressure wave generated by the pumping action of the heart is called the
   A. heart rate.
   B. pulse.
   C. heart rhythm.
   D. arterial push.

_____ 22. Oxygen is carried from the lungs to the body and carbon dioxide back to the lungs by the
   A. plasma.
   B. platelets.
   C. white blood cells.
   D. red blood cells.

_____ 23. After a victim suffers cardiac arrest, brain damage begins within
   A. 2 to 3 minutes.
   B. 4 to 6 minutes.
   C. 7 to 9 minutes.
   D. 10 to 15 minutes.

_____ 24. During CPR on an adult, chest compressions should be at the rate of
   A. 30 compressions per minute.
   B. 60 compressions per minute.
   C. 100 compressions per minute.
   D. 120 compressions per minute.

_____ 25. If the heart cannot pump enough blood to supply the needs of the body, the victim will experience
   A. cardiogenic shock.
   B. pipe failure.
   C. anaphylactic shock.
   D. spinal shock.

_____ 26. Extreme allergic reactions to a foreign substance can cause
   A. fluid loss.
   B. cardiogenic shock.
   C. anaphylactic shock.
   D. spinal shock.

_____ 27. How much blood does the average adult's circulatory system contain?
   A. Approximately 6 pints
   B. Approximately 10 pints
   C. Approximately 12 pints
   D. Approximately 15 pints

_____ 28. What is the most serious type of bleeding?
   A. Venous bleeding
   B. Arterial bleeding
   C. Capillary bleeding
   D. External bleeding

_____ 29. A wound where the skin stays intact is called a(n)
   A. abrasion.
   B. open wound.
   C. laceration.
   D. closed wound.

_____ 30. Any call that involves multiple victims is termed a
   A. mass-casualty incident.
   B. multiple-casualty incident.
   C. triage.
   D. violent incident.

268 FUNDAMENTALS OF FIRE FIGHTER SKILLS

Labeling
Label the following diagrams with the correct terms.

1. The respiratory system.

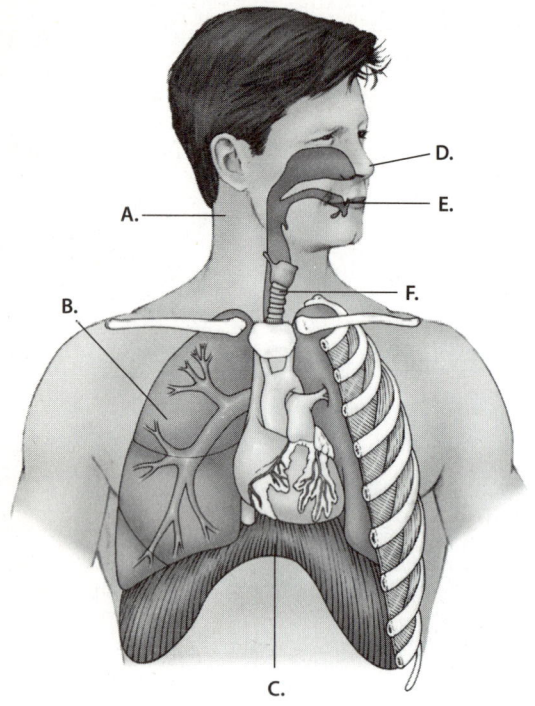

A. _____
B. _____
C. _____
D. _____
E. _____
F. _____

Figure 24-2

2. Anatomy of the respiratory system.

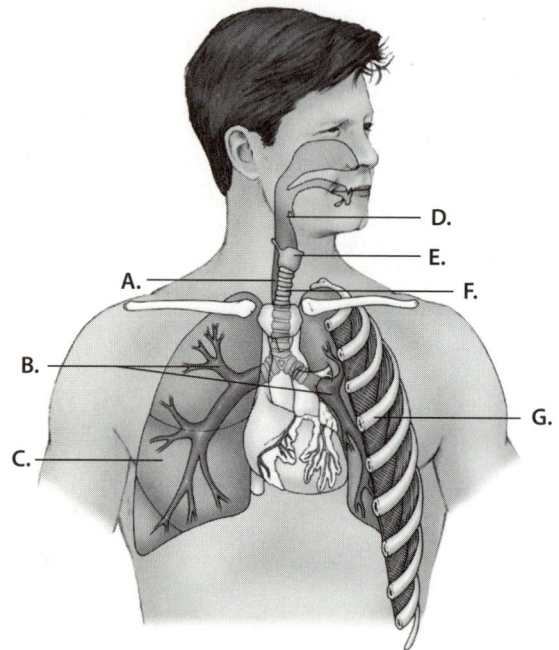

A. _____
B. _____
C. _____
D. _____
E. _____
F. _____
G. _____

Figure 24-4

3. The CPR chain of survival.

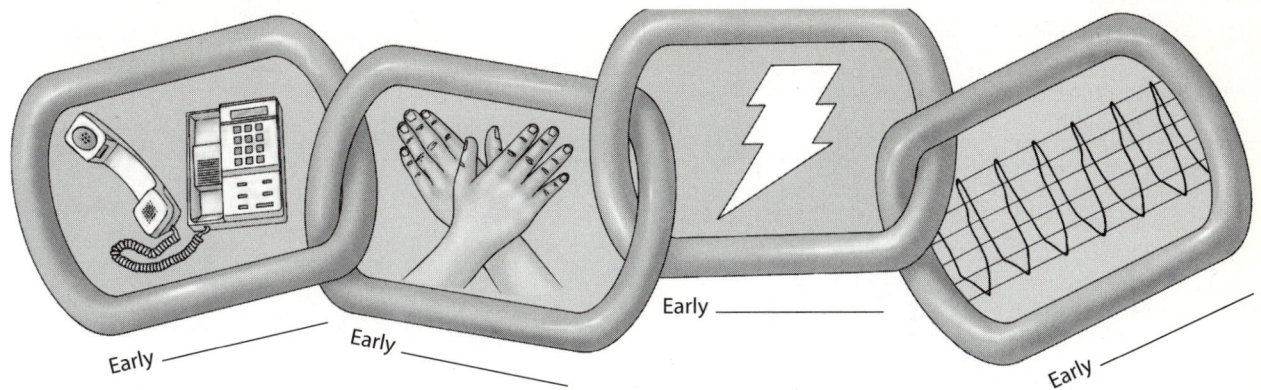

Early _____ Early _____ Early _____ Early _____

Figure 24-14

4. Types of external bleeding.

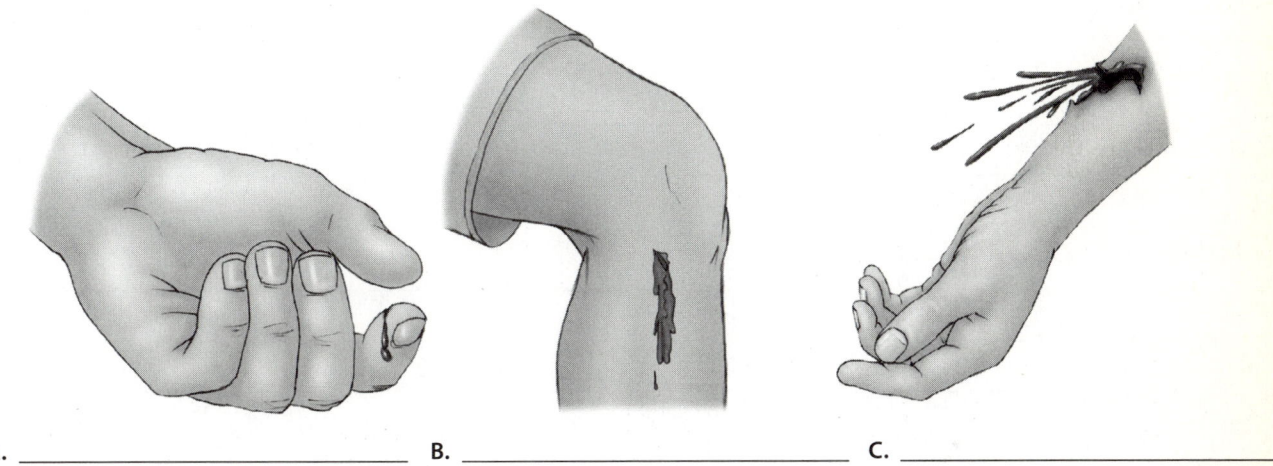

A. _____ B. _____ C. _____

Figure 24-20

5. Location of pressure points.

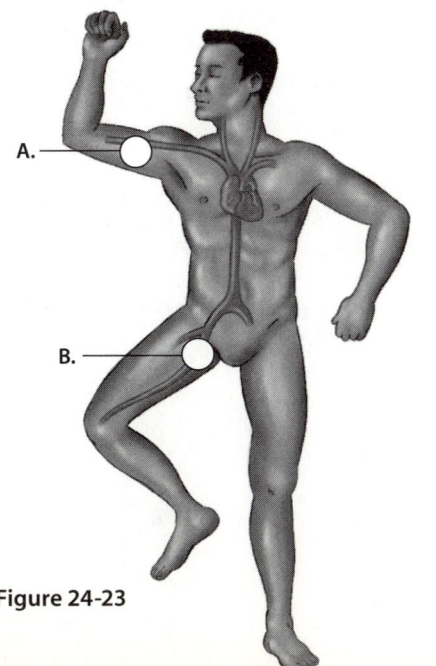

A. _____
B. _____

Figure 24-23

Vocabulary

Define the following terms using the space provided.

1. Cardiopulmonary resuscitation (CPR):

2. Recovery position:

3. Dependent lividity:

4. Shock:

5. Radial artery:

Fill-in
Read each item carefully, and then complete the statement by filling in the missing word(s).

1. _____ is the virus that can lead to acquired immune deficiency syndrome.
2. Based on the assumption that all victims are potential carriers of bloodborne pathogens, the CDC recommends all healthcare workers use _____.
3. The main purpose of the respiratory system is to provide _____ and to remove _____ from the body.
4. The _____ rely on the diaphragm for movement, as they consist of soft, spongy tissue with no muscles.
5. The first step in assessing a victim's airway is to check the victim's level of _____.
6. To open the airway of a victim with a suspected neck injury, use the _____ technique.
7. When a person's internal body temperature falls below 95 °F, the individual has severe _____.
8. The most common foreign object that causes an airway obstruction is _____.
9. When air is forced into the stomach instead of the lungs, _____ occurs.
10. If there is no pulse, you must correct the victim's circulation by performing external _____.
11. If performing CPR on an infant, give _____ rescue breath(s) after every 15 chest compressions.
12. More trauma victims die from _____ than from any other reason.

True/False
If you believe the statement to be more true than false, write the letter "T" in the space provided.
If you believe the statement to be more false than true, write the letter "F."

1. _____ Drug-resistant strains of tuberculosis can be transmitted through the air.
2. _____ Heavy leather gloves prevent the spread of infectious diseases.
3. _____ The "B" in the CPR ABCs stands for "bleeding."
4. _____ Once a brain cell has been destroyed, it cannot be healed or replaced.
5. _____ Unconscious victims will not be able to keep their airways open.
6. _____ Mouth-to-mask devices prevent the transmission of infectious diseases.
7. _____ Cardiac arrest occurs when the heart stops contracting and no blood is pumped through the blood vessels.
8. _____ Within 8 to 10 minutes after a cardiac arrest, the damage to the brain may be irreversible.
9. _____ Rigor mortis is an indication that a victim has been dead for more than a day.
10. _____ In adult CPR, after every 15 chest compressions, give two rescue breaths.
11. _____ If the patient's pupils constrict when exposed to light, he or she is receiving effective CPR.
12. _____ Living wills, advance directives, and DNR orders are legal documents that specify the patient's wishes regarding particular medical procedures.

Short Answer
Complete this section with short written answers using the space provided.

1. Identify the Centers for Disease Control and Prevention's recommended five steps for universal precautions.

   _____
   _____
   _____
   _____
   _____

2. List five of the major causes of respiratory arrest.

   _____
   _____
   _____
   _____
   _____

3. Identify and describe blood flow through the four chambers of the heart.

   _____
   _____
   _____
   _____

4. Identify the six acceptable criteria for discontinuing CPR on a victim.

   _____
   _____
   _____
   _____
   _____
   _____

5. Identify five of the possible signs and symptoms of shock.

6. List the seven steps used to combat or begin treatment of shock.

274 Fundamentals of Fire Fighter Skills

CLUES

Across
1. Either of the two lower chambers of the heart.
4. The smallest blood vessels.
7. The major vessel in the upper extremity that supplies blood to the arm.
12. A bandage.
13. A state of collapse of the cardiovascular system.
14. Loss of skin as a result of a body part being scraped across a rough surface.
15. Also called a contusion.

Down
2. The windpipe.
3. The fluid part of the blood that carries blood cells.
5. Microorganisms capable of causing disease.
6. The breastbone.
8. An injury in which a piece of skin is left hanging by a flap.
9. The organs that supply the body with oxygen and eliminate carbon dioxide from the blood.
10. The air sacs of the lungs, where the exchange of oxygen and carbon dioxide takes place.
11. The passages through which air enters and leaves the lungs.

Word Fun

The following crossword puzzle is an activity provided to reinforce correct spelling and understanding of terminology associated with firefighting. Use the clues provided to complete the puzzle.

Fire Alarms

The following real case scenarios will give you an opportunity to explore the concerns associated with emergency medical care. Read each scenario, and then answer each question in detail.

1. You and your company are rehabilitating after a structure fire. Your partner suddenly slaps his wrist and says he has been stung by a bee. Within minutes, he begins to show signs of labored breathing, complains of an itchy sensation, and has a red tint to his skin.

   a. What may be happening to him?

   _____
   _____
   _____

   b. How should you proceed?

   _____
   _____
   _____

2. Your company responds to a motor vehicle accident for extrication. Your officer directs you to use universal precautions. What are they?

   _____
   _____
   _____

Skill Drills

Skill Drill 24-1: Performing the Head Tilt–Chin Lift Technique
Test your knowledge of this skill drill by filling in the correct words in the photo captions.

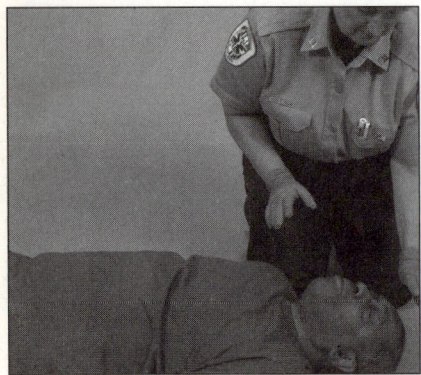

1. Position yourself at the side of the _____ victim.

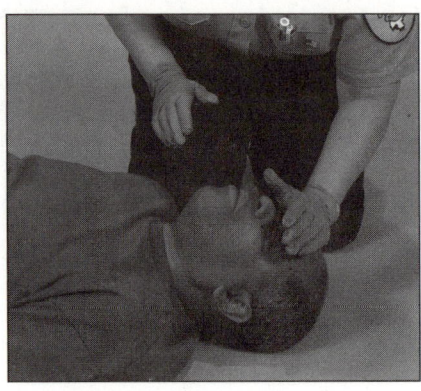

2. Place your hand closest to the victim's head on the _____.

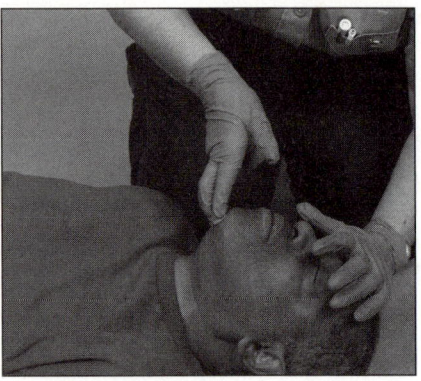

3. With your other hand, place _____ fingers on the underside of the victim's chin.

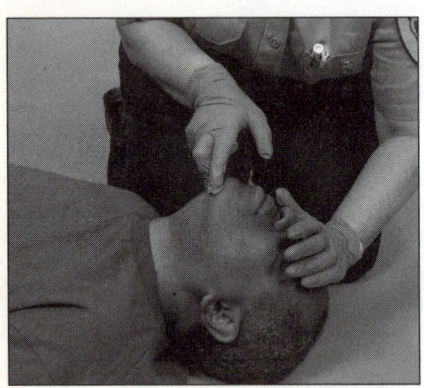

4. Simultaneously apply backward and _____ pressure to the victim's forehead and lift the jaw straight up. Do not depress the soft tissue below the chin.

Skill Drill 24-2: Performing the Jaw-Thrust Technique

Test your knowledge of this skill drill by placing the photos below in the correct order. Number the first step with a "1," the second step with a "2," and so on.

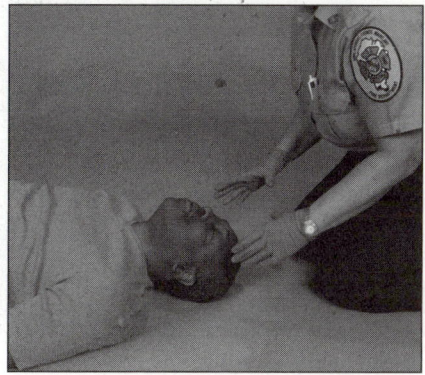
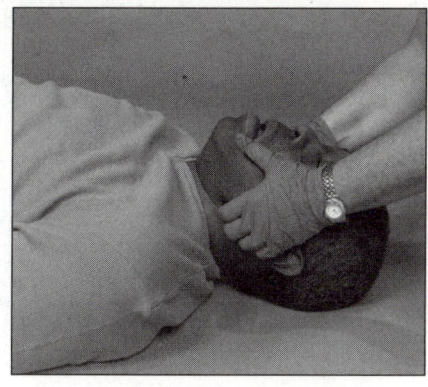
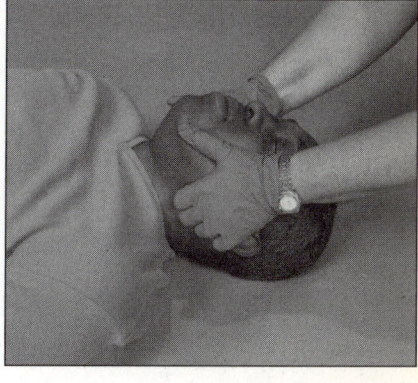

_____ Position yourself at the top of the victim's head.

_____ While holding the victim's head still, displace the jaw forward and open the victim's mouth with your thumb tips.

_____ Place the meaty portion of the base of your thumbs on the arches of the jaw, and hook the tips of your index fingers under the angle of the mandible, in the indent below the ear.

Skill Drill 24-4: Performing Mouth-to-Mask Rescue Breathing

Test your knowledge of this skill drill by filling in the correct words in the photo captions.

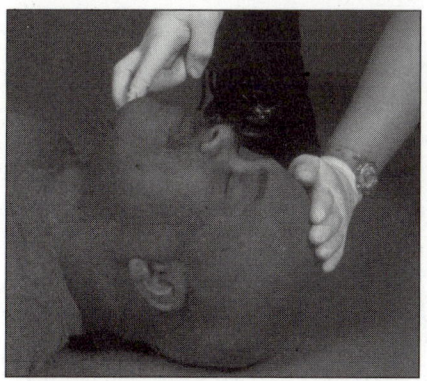
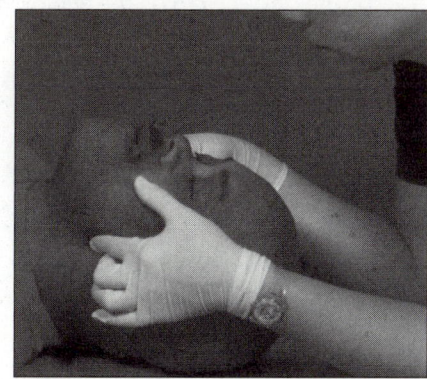
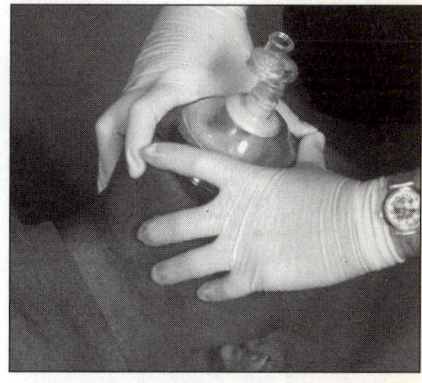

1. Open the victim's airway using the _____ technique.

2. If there is a suspected head injury, open the airway using the _____ technique.

3. Seal the _____ against the victim's face.

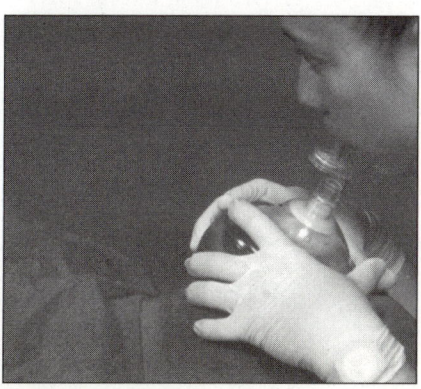

4. _____ through the mouthpiece.

Skill Drill 24-5: Performing Mouth-to-Barrier Device Rescue Breathing

Test your knowledge of this skill drill by placing the photos below in the correct order. Number the first step with a "1," the second step with a "2," and so on.

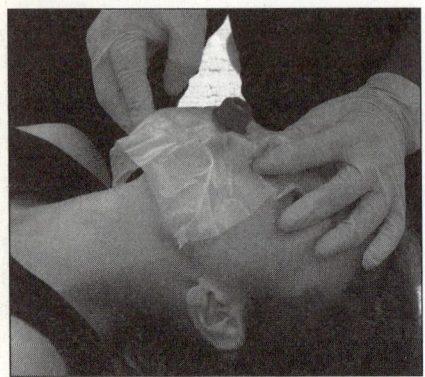 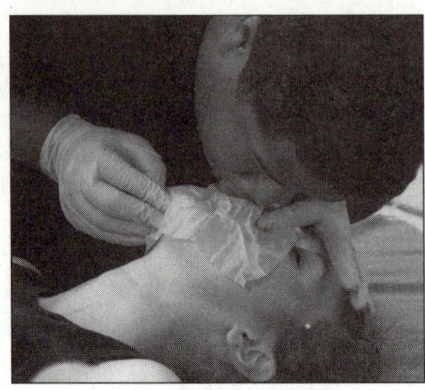 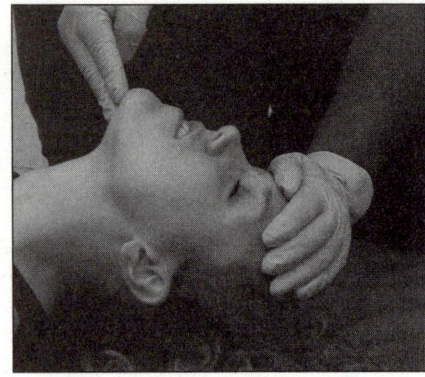

_____ Place the barrier device over the victim's mouth. Pinch the victim's nostrils together.

_____ Perform rescue breathing.

_____ Open the victim's airway using the head tilt–chin lift technique.

Skill Drill 24-7: Performing the Abdominal-Thrust Maneuver

Test your knowledge of this skill drill by filling in the correct words in the photo captions.

 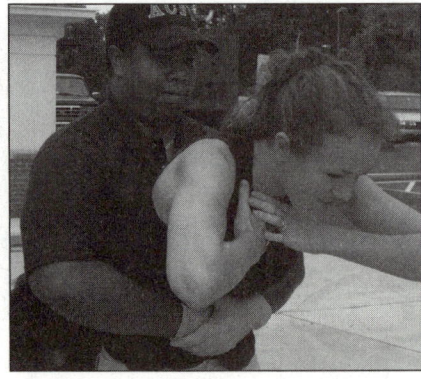

1. Stand behind the victim and deliver an abdominal thrust. Press into the victim's abdomen with a quick _____ and upward thrust.

2. Repeat the abdominal thrusts until either the foreign body is expelled or the victim becomes _____.

Skill Drill 24-8: Treating Airway Obstruction in an Unconscious Adult

Test your knowledge of this skill drill by placing the photos below in the correct order. Number the first step with a "1," the second step with a "2," and so on.

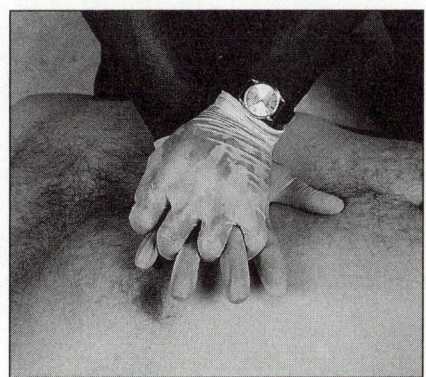

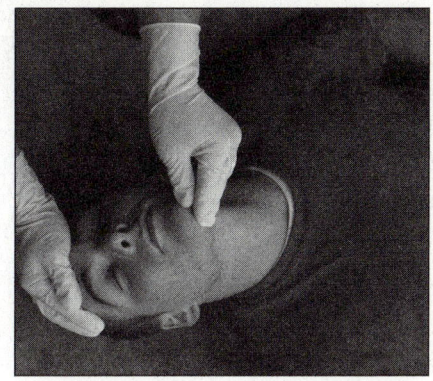

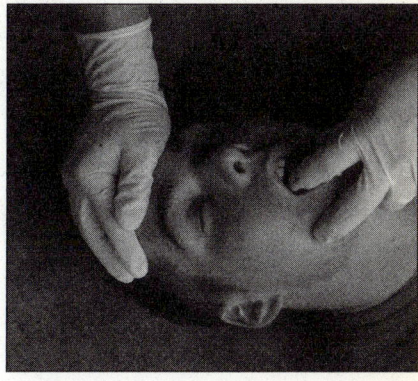

_____ Reposition the victim's head to improve the likelihood of opening the airway. Attempt rescue breathing again. If it is unsuccessful, perform 30 chest compressions.

_____ Open the victim's airway by using the head tilt–chin lift technique. If trauma to the spine is suspected, use the jaw-thrust technique. Attempt rescue breathing.

_____ Open the victim's mouth so you can check for obstructions. Use your finger to sweep the mouth clear only if you see an object. Attempt rescue breathing again. Repeat until the airway is cleared.

Skill Drill 24-11: Performing Chest Compressions

Test your knowledge of this skill drill by filling in the correct words in the photo captions.

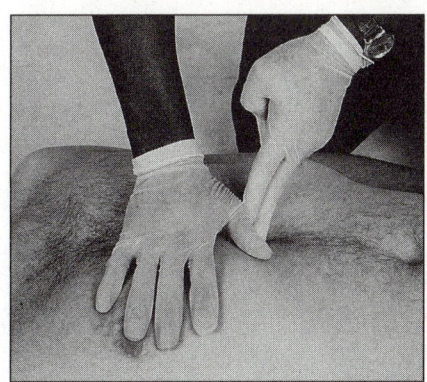

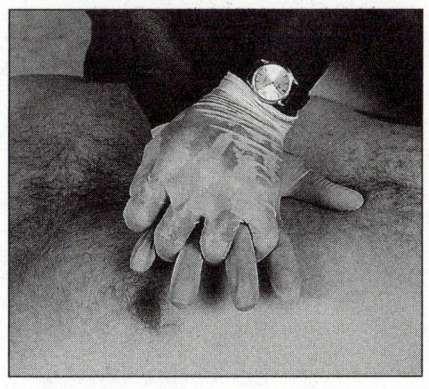

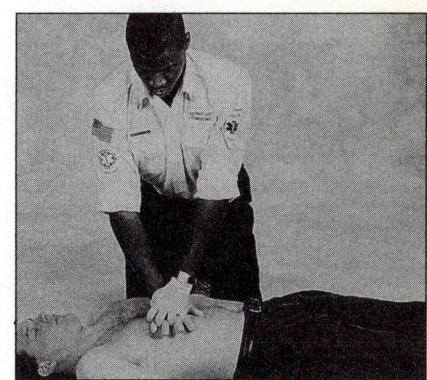

1. Place the heel of one hand on the victim's _____ between the nipples.

2. Place the _____ of your other hand over the first hand.

3. With your arms straight, lock your elbows, and position your shoulders directly over your _____. Depress the sternum 1 ½ to 2 inches using a direct downward movement.

Skill Drill 24-12: Performing One-Rescuer Adult CPR

Test your knowledge of this skill drill by placing the photos below in the correct order. Number the first step with a "1," the second step with a "2," and so on.

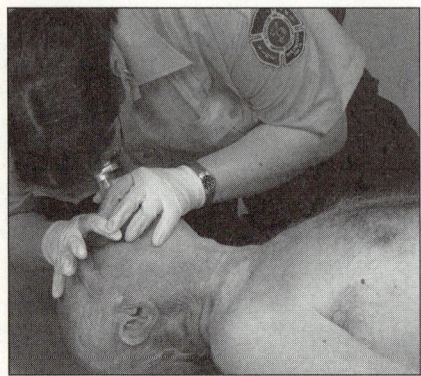

_____ If the victim is not breathing, give two breaths of 1 second each.

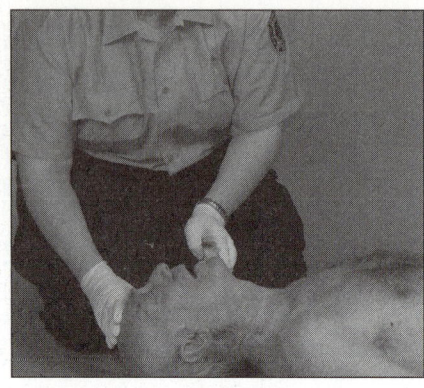

_____ Open the airway.

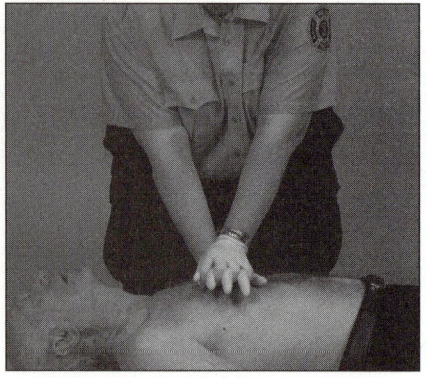

_____ If no pulse is found, apply the AED. If no AED is available, place your hands in the proper position for chest compressions. Give 30 compressions at a rate of about 100 per minute. Open the airway and give two ventilations of 1 second each. Perform five cycles of chest compressions and ventilation. Stop CPR and check for return of the victim's pulse. Depending on the victim's condition, continue CPR, continue rescue breathing only, or place the victim in the recovery position and monitor his or her breathing and pulse.

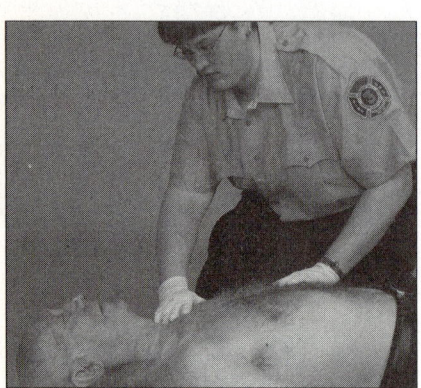

_____ Establish unresponsiveness and call for help.

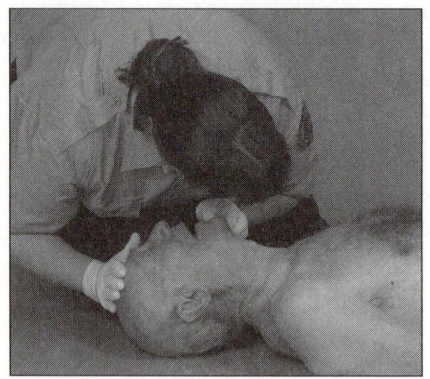

_____ Look, listen, and feel for breathing. If breathing is adequate, place the victim in the recovery position and monitor him or her.

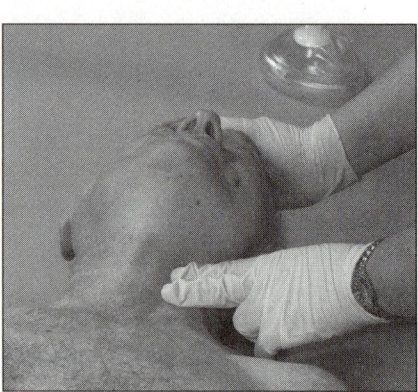

_____ Check for a carotid pulse.

Skill Drill 24-13: Performing Two-Rescuer Adult CPR

Test your knowledge of this skill drill by filling in the correct words in the photo captions.

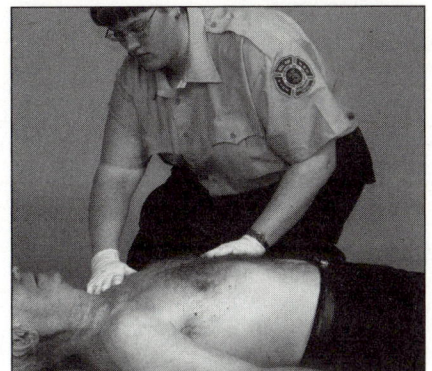

1. Establish the victim's unresponsiveness and take up your _____.

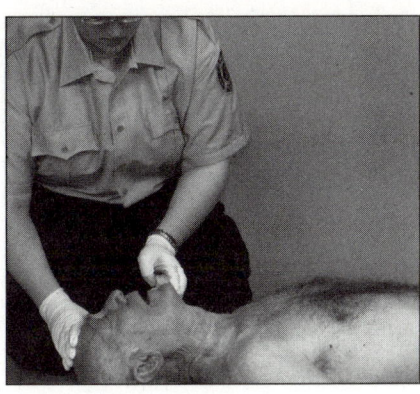

2. Open the _____.

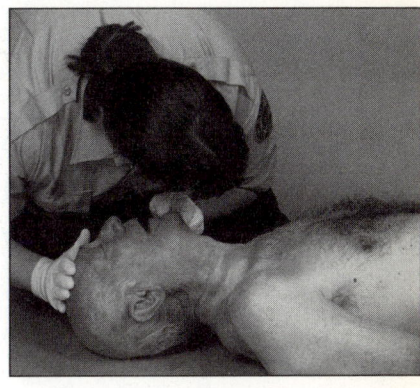

3. Look, listen, and _____ for breathing. If the victim's breathing is adequate, place the victim in the recovery position and monitor.

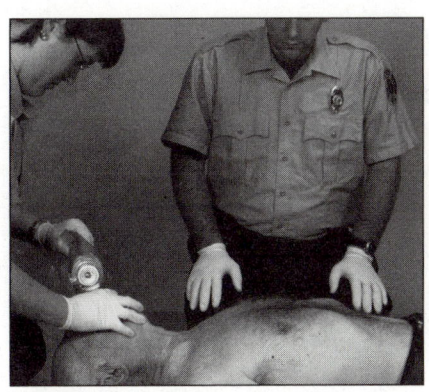

4. If the victim is not breathing, give _____ breaths of 1 second each.

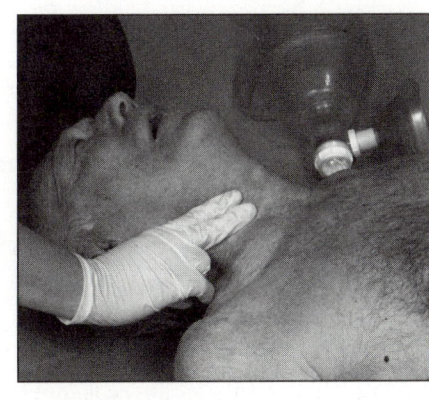

5. Check for a(n) _____ pulse. If no pulse is felt in 10 seconds, begin CPR.

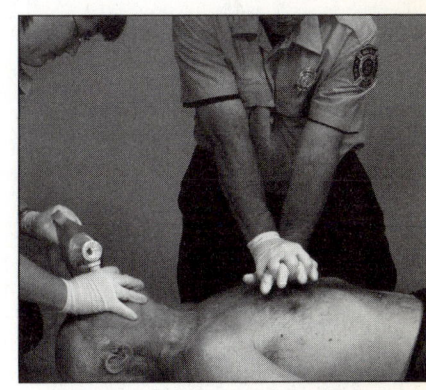

6. If there is no pulse but an AED is available, apply it now. If no AED is available, begin chest compressions at a rate of _____ per minute (30 compressions for each 2 ventilations). After every five cycles, switch rescuer positions to minimize fatigue. Keep the switch time to 5 to 10 seconds. Depending on the victim's condition, continue CPR, continue ventilations only, or place the victim in recovery position.

Vehicle Rescue and Extrication

Workbook Activities

The following activities have been designed to help you. Your instructor may require you to complete some or all of these activities as a regular part of your fire fighter training program. You are encouraged to complete any activity that your instructor does not assign as a way to enhance your learning in the classroom.

Chapter Review

The following exercises provide an opportunity to refresh your knowledge of this chapter.

Matching

Match each of the terms in the left column to the appropriate definition in the right column.

_____ 1. Wedges
_____ 2. Bulkhead
_____ 3. Posts
_____ 4. Conventional vehicles
_____ 5. Cribbing
_____ 6. Step chocks
_____ 7. Air bags
_____ 8. Air chisel
_____ 9. Tempered glass
_____ 10. Laminated glass

A. Short lengths of usually hardwood timber used to stabilize vehicles
B. One of the vertical support members of a vehicle that holds up the roof and forms the upright columns of the occupant cage
C. Specialized cribbing assemblies
D. The wall that separates the engine compartment from the passenger compartment
E. Used to snug loose cribbing under the load or when using lift air bags to fill the void between the crib and the object as it is raised
F. A vehicle that uses an internal combustion engine
G. Glass used to make car windshields
H. Glass used for side and rear car windows
I. A tool that is efficient at cutting sheet metal and some plastics
J. Pneumatic-filled bladders made out of rubber or synthetic material

Multiple Choice

Read each item carefully, and then select the best response.

_____ 1. Vehicles that are powered by compressed natural gas are known as
   A. electric-powered vehicles.
   B. hybrid vehicles.
   C. alternative-powered vehicles.
   D. conventional vehicles.

_____ 2. Which posts are located closest to the front of the vehicle?
   A. "A" posts
   B. "B" posts
   C. "C" posts
   D. "D" posts

CHAPTER 25

_____ 3. Cribbing protects the vehicle from
   A. electrical hazards.
   B. excessive exposure.
   C. rolling.
   D. other transportation.

_____ 4. The suspension system of most vehicles can be stabilized with
   A. cribbing.
   B. straps.
   C. wedges
   D. step chocks.

_____ 5. When using rescue-lift air bags, use _____ to fill the void between the crib and the vehicle.
   A. step chocks
   B. wedges
   C. posts
   D. bulkheads

_____ 6. What is the most common type of rescue-lift air bag?
   A. Low-pressure lift air bag
   B. Medium-pressure lift air bag
   C. High-pressure lift air bag
   D. Dual-pressure lift air bag

_____ 7. Cribbing, rescue-lift air bags, and step blocks are all types of
   A. stabilization devices.
   B. patient extractors.
   C. bracing tools.
   D. prying tools.

_____ 8. Axes, bolt cutters, hacksaws, and manual hydraulic cutters are all tools that can be used for
   A. prying.
   B. stabilizing.
   C. windshield removal.
   D. cutting.

_____ 9. A vehicle windshield is made of
   A. tempered glass.
   B. laminated windshield glass.
   C. block glass.
   D. sheet glass.

_____ 10. The rear window of a vehicle is made of
   A. tempered glass.
   B. laminated windshield glass.
   C. block glass.
   D. sheet glass.

_____ 11. What are the most efficient and widely used tools for opening jammed doors?
   A. Cutting tools
   B. Manual hydraulic tools
   C. Powered hydraulic tools
   D. Prying tools

12. As soon as you have secured access to the victim, what is your next step as a rescuer?
    A. Remove the excess glass.
    B. Remove the extraneous materials.
    C. Begin to provide emergency medical care.
    D. Begin to communicate with the victims.

13. What is the simplest way to displace a seat backward?
    A. Cut the material out of the bottom of the seat.
    B. Remove the backseat.
    C. Tilt the seat backward.
    D. Move the seat backward in its tracks.

14. Between the layers of glass that make the vehicle windshield is
    A. an empty space.
    B. a thin layer of flexible plastic.
    C. an epoxy glue.
    D. an invisible wire mesh.

15. When removing the roof of a vehicle, it is essential to remove the
    A. windshield.
    B. "C" posts.
    C. doors.
    D. bulkhead.

16. What is the first step of the dash displacement procedure?
    A. Open both front doors.
    B. Remove the bulkhead.
    C. Remove the steering wheel.
    D. Cut the "A" post.

17. What is the first step in displacing the roof of a vehicle?
    A. Remove the "A" posts.
    B. Cut the "C" posts.
    C. Ensure the safety of the rescuers.
    D. Remove the glass.

18. The victim needs to be stabilized and packaged in preparation for removal in the
    A. stabilization phase.
    B. extrication phase.
    C. size-up phase.
    D. rescue phase.

19. When removing a windshield with an axe, one rescuer begins
    A. at the top, in the middle.
    B. at the driver side.
    C. at the passenger side.
    D. at the bottom.

20. Steering wheels can be cut using
    A. a hacksaw.
    B. a bolt cutter.
    C. a hydraulic cutter.
    D. all of the above.

Labeling
Label the following diagram with the correct terms.

1. The anatomy of a vehicle.

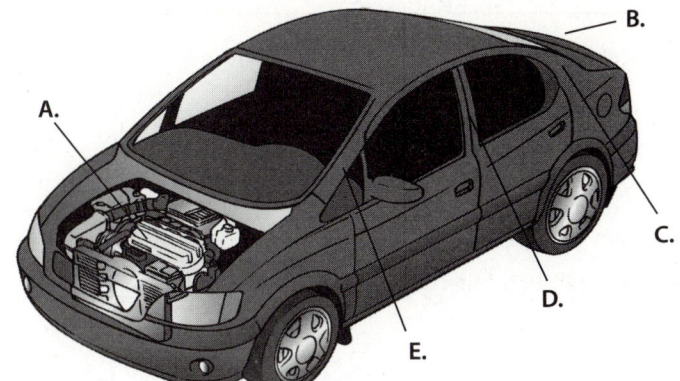

A. _____
B. _____
C. _____
D. _____
E. _____

Figure 25-2

Vocabulary
Define the following terms using the space provided.

1. Unibody:

2. Hybrid vehicle:

3. Platform frame:

4. Firewall:

5. Post:

Fill-in

Read each item carefully, and then complete the statement by filling in the missing word(s).

1. A fire in a natural gas-powered vehicle poses the threat of a(n) _____.
2. The right side of the vehicle is where the _____ seat is located.
3. The _____ posts are located between the front and rear doors of a vehicle.
4. The first step in the extrication process is _____.
5. After arriving at the scene of a motor vehicle collision, it is important to assess the _____ present and to determine the _____ of the incident.
6. Traffic hazards are best handled by the appropriate _____ agency.
7. The three types of commonly used pneumatic rescue-lift air bags are low, medium, and high _____.
8. The simplest way to access a victim of a crash is to open a(n) _____.
9. If the door cannot be opened or glass removal will not provide access to the victim, the most common technique for gaining access is _____ displacement.
10. One method of displacing the roof is to cut the _____ posts and fold the roof back toward the rear of the vehicle.

True/False

If you believe the statement to be more true than false, write the letter "T" in the space provided.
If you believe the statement to be more false than true, write the letter "F."

1. _____ Hybrid vehicles contain the hazards associated with conventional gas-powered vehicles.
2. _____ The incident commander will usually perform a size-up of the scene by conducting a 360-degree walk around the scene.
3. _____ Downed power lines can create a mechanical hazard.
4. _____ The steps of scene stabilization consist of reducing, removing, or mitigating the hazards at the incident scene.
5. _____ Unstable vehicles pose a more serious threat to rescuers than do stabilized vehicles.
6. _____ Wedges should be the same width as the cribbing used in stabilization efforts.
7. _____ Rescue-lift air bags are among the best pieces of equipment used to shore a vehicle by themselves.
8. _____ When it is necessary to force a door to gain access to a victim, choose the door closest to the victim.
9. _____ The purpose of disentangling the victim is to remove those parts of the vehicle that are trapping the victim.

Short Answer

Complete this section with short written answers using the space provided.

1. Identify five safety tips for using rescue-lift air bags.

   _____
   _____
   _____
   _____
   _____

2. Identify and provide an example of each of the four general functions of gaining access and disentangling a victim.

   _____
   _____
   _____
   _____

3. Identify five vehicle air bag safety tips.

   _____
   _____
   _____
   _____
   _____

Word Fun

The following crossword puzzle is an activity provided to reinforce correct spelling and understanding of terminology associated with firefighting. Use the clues provided to complete the puzzle.

Clues

Across

3. One of the vertical support members of a vehicle that holds up the roof and forms the upright columns of the occupant cage.
4. Specialized cribbing assemblies used to stabilize vehicles.
6. The wall that separates the engine compartment from the passenger compartment.
8. The most commonly used frame construction in vehicles.
9. A type of window that breaks into small cubical shapes.

Down

1. A vehicle that uses an internal combustion engine and is fueled with either gasoline or diesel fuel.
2. An extrication tool consisting of air bags, filler hoses, air regulators, control valves, and a supply of compressed air.
5. An extrication tool used to lift vehicles involved in accidents.
7. Used to snug loose cribbing under the load or when using lift air bags to fill the void between the crib and the object as it is raised.

Fire Alarms

The following real case scenarios will give you an opportunity to explore the concerns associated with vehicle rescue and extrication. Read each scenario, and then answer each question in detail.

1. Your company has been dispatched to a motor vehicle accident involving two vehicles with trapped occupants. It is 3:00 P.M. on a bright, clear day; both vehicles are on their wheels. The IC orders your company to stabilize both vehicles.

    a. Which equipment will you use for vehicle stabilization?

    _____
    _____
    _____
    _____
    _____
    _____

    b. How will you stabilize these vehicles?

    _____
    _____
    _____
    _____
    _____
    _____
    _____

2. You have been dispatched to a vehicle accident on the interstate at 1:15 A.M. on a rainy night. Your company is first on the scene, and SOPs state your first actions are to protect the scene from traffic and then do a size-up.

    a. How will you protect the scene from approaching traffic?

    _____
    _____
    _____
    _____
    _____
    _____

    b. How will you proceed with your scene size-up?

    _____
    _____
    _____
    _____
    _____
    _____
    _____

Skill Drills

Skill Drill 25-1: Performing a Scene Size-up at a Motor Vehicle Crash
Test your knowledge of this skill drill by filling in the correct words in the photo captions.

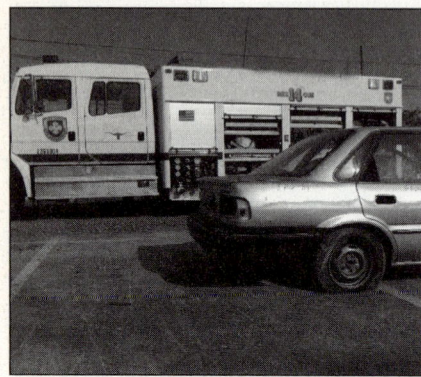

1. Position the emergency vehicle so as to _____ the scene. Size up the scene from inside this vehicle. Transmit the initial report over the radio. Establish command. Perform a scene size-up outside the vehicle. Check for overhead hazards.

2. Approach the crash scene vehicles and examine the space under the vehicles. Perform a _____-degree walk-around of the entire scene. Assess the hazards. Determine the number of people involved. Determine the severity of the victims' injuries. Determine the level of entanglement. Assess the resources available, and call for additional units if needed. Give an updated report. Establish a secure work area. Establish a staging resource area. Direct the placement of arriving apparatus. Assign personnel and tasks.

Skill Drill 25-4: Breaking Tempered Glass

Test your knowledge of this skill drill by placing the photos below in the correct order. Number the first step with a "1," the second step with a "2," and so on.

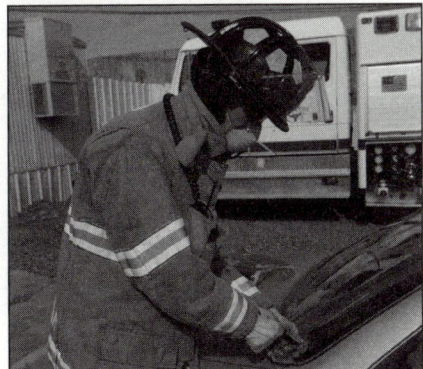

_____ Warn personnel with the verbal command "Breaking glass," unless a stop/freeze call is made, while you continue with glass removal starting with a window farthest from the victim. Place the tool in the lower corner of the window and apply pressure until the spring is activated, or strike the lower corner with the axe.

_____ Remove any loose glass around the window opening with a tool. Follow this procedure until all glass has been removed.

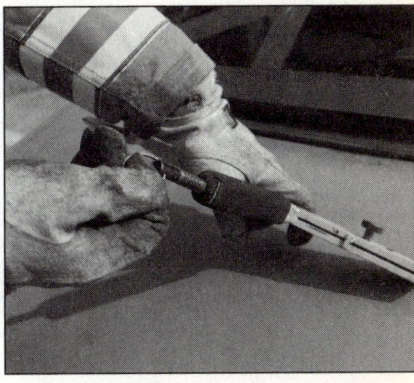

_____ All fire fighters should be in proper PPE, including dust mask, safety glasses, or goggles. Attempt to lower windows as far as possible before breaking glass, and then select a spring-loaded center punch, a multipurpose tool, or axe.

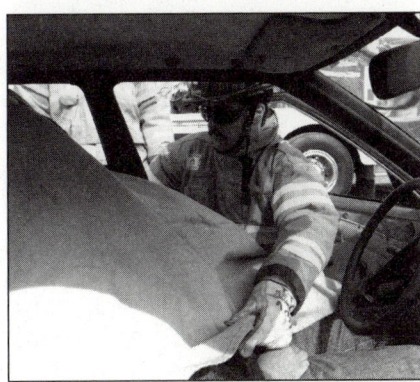

_____ Ensure that the victim and other fire fighters are properly protected.

Skill Drill 25-8: Displacing the Dashboard of a Vehicle
Test your knowledge of this skill drill by filling in the correct words in the photo captions.

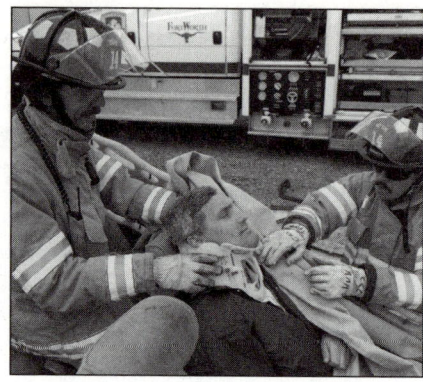

1. Don PPE. Retrieve the required tools. Set up any power units. Couple hoses and cords if needed. Check the equipment for readiness. Enter and secure the work area. Assess the vehicle for hazards, including _____.

2. Communicate with the victim and ensure that both the victim and rescuers are _____ from hazards. Operate tools according to the manufacturer's recommendations.

3. Make relief cuts at the bottom of the "A" post. Use extra cribbing under the base end of the tool to keep the spreader from _____. Guard against the tool kicking out.

4. Add _____ as the dashboard is being lifted. Use good body mechanics when lifting. Maintain control of all tools at all times. Notify command that the dashboard has been displaced. Return the tools to the staging area upon completion of the tasks.

Skill Drill 25-9: Removing the Roof of a Vehicle

Test your knowledge of this skill drill by placing the photos below in the correct order. Number the first step with a "1," the second step with a "2," and so on.

_____ Control the vehicle roof and protect against sharp edges at all times. Remove the roof to a safe location. Notify command when the roof is removed. Return the tools to the staging area upon completion of the tasks.

_____ Operate the required tools to remove the roof. Operate tools according to the manufacturer's instructions. Use good body mechanics.

_____ Don PPE. Retrieve the required tools. Set up any power units. Couple hoses and cords if needed. Check the equipment for readiness. Enter and secure the work area. Assess the vehicle for hazards, including passive restraint devices. Communicate with the victim; ensure that the victim and rescuers are protected from hazards. Remove any remaining glass.

Assisting Special Rescue Teams

Workbook Activities

The following activities have been designed to help you. Your instructor may require you to complete some or all of these activities as a regular part of your fire fighter training program. You are encouraged to complete any activity that your instructor does not assign as a way to enhance your learning in the classroom.

Chapter Review

The following exercises provide an opportunity to refresh your knowledge of this chapter. All questions in this chapter are Fire Fighter II level.

Matching

Match each of the terms in the left column to the appropriate definition in the right column.

_____ 1. Placards
_____ 2. Shoring
_____ 3. Warm zone
_____ 4. Spoil pile
_____ 5. Hot zone
_____ 6. IDLH
_____ 7. Low-angle operations
_____ 8. PFD
_____ 9. SAR system
_____ 10. High-angle operations

A. When fire fighters are dependent on the ground for their primary support, and the rope system is a secondary means of support

B. An emergency breathing system that utilizes an air line running from the rescuers to a fixed air supply located outside of the confined space

C. Signage required to be placed on all four sides of vehicles that identifies the hazardous materials contents being transported

D. Allows the body to float in water

E. The area between the hot zone and the cold zone

F. The area immediately surrounding an incident site that is directly dangerous to life and health

G. A method of supporting a trench wall to prevent its collapse

H. Flammable, toxic, or oxygen-deficient atmospheres that pose immediate or possible adverse health effects

I. The unstable pile of dirt removed from an evacuation

J. The slope of the ground is greater than 45 degrees, and fire fighters are dependent on life safety rope for suppor

CHAPTER 26

Multiple Choice

Read each item carefully, and then select the best response.

_____ 1. Which level of training allows an individual to work in the warm zone and directly assist those conducting the rescue operation?
   A. Awareness level
   B. Operations level
   C. Technician level
   D. Incident commander level

_____ 2. All emergency service personnel at a rescue situation must
   A. constantly assess and reassess the scene.
   B. communicate with the victim(s).
   C. report directly to the incident commander.
   D. be prepared to assist with the technical rescue team.

_____ 3. A technical rescue team will usually respond with a rescue squad,
   A. medic unit, and safety officer.
   B. paramedic, and incident commander.
   C. logistics team, and operational team.
   D. medic unit, engine company, and chief.

_____ 4. When responding to an industrial facility, the IC should make contact with the
   A. business owner.
   B. property owner.
   C. responsible party.
   D. city office or administration.

_____ 5. The outer perimeter of an incident scene that is established to keep the public and media out of the staging area is called the
   A. hot zone.
   B. warm zone.
   C. cold zone.
   D. public zone.

_____ 6. What is the most common method of establishing the control zones for an emergency incident site?
   A. Barricades
   B. Pylons
   C. Chalk or paint lines
   D. Fire line tape

_____ 7. During a rescue incident, emergency medical care should be initiated as soon as
   A. the technical rescue team is clear of the scene.
   B. access is made to the victim.
   C. the medical team arrives on scene.
   D. the incident commander indicates that medical treatment is required.

_____ 8. In a vehicle accident, disentanglement is the process of
   A. cutting a vehicle away from the victim.
   B. removing the victim from the vehicle.
   C. cutting and removing the doors of the vehicle.
   D. establishing medical control of the victim.

_____ 9. If a victim's general condition is deteriorating and time will not permit meticulous splinting and dressing procedures, victims may be removed by
   A. the medical team.
   B. rapid extrication.
   C. complete immobilization.
   D. special extraction teams.

_____ 10. The only time the victim should be moved prior to completion of initial care, assessment, stabilization, and treatment is when there is immediate danger to
   A. the victim's or emergency responder's life.
   B. the surrounding areas.
   C. a rescue team member's life.
   D. the entire rescue operations team.

_____ 11. The overriding objective for each rescue, transfer, and removal is to complete the process as
   A. quickly as possible.
   B. safely and efficiently as possible.
   C. a team.
   D. an integrated team.

_____ 12. Once the victim has been removed from the hazard area, who transports the victim to an appropriate medical facility?
   A. The technical rescue team
   B. The operations team
   C. EMS
   D. The safety officer

_____ 13. In a hazardous materials incident, the site clean-up should be completed by
   A. contracted cleaning crews.
   B. government agencies.
   C. the property owner.
   D. trained disposal crews.

_____ 14. In an industrial setting, securing the scene is the responsibility of the
   A. facility supervisor.
   B. incident commander.
   C. emergency response team.
   D. property owner.

_____ 15. To ensure the continuity of quality care and proper transfer of responsibility, there must be
   A. a complete team debriefing.
   B. adequate reporting and accurate records.
   C. an incident analysis report.
   D. a critical incident stress management (CISM) intervention.

_____ 16. The fire fighter should start compiling the facts about an incident from the
   A. captain.
   B. scene size-up.
   C. initial dispatch of the rescue call.
   D. technical rescue team specialists.

_____ 17. Shutting off the utilities in the area where the rescuers will be working is a responsibility of the
   A. incident commander.
   B. safety officer.
   C. logistics officer.
   D. shift captain.

_____ 18. To assist fire fighters in remaining visible to one another in a crowd or in wilderness locations, they should wear
   A. reflective tape.
   B. strobe lights.
   C. reflective vests.
   D. guidelines.

_____ 19. What is the single most important process that any rescuer needs to follow to ensure his or her safety?
   A. Buddy system
   B. Tagout system
   C. Accountability system
   D. Two-in/two-out rule

_____ 20. During a technical rescue incident, whose orders do fire fighters follow?
   A. Company officer
   B. Incident commander
   C. Battalion chief
   D. Rescue captain

_____ 21. The lack of oxygen and the presence of poisonous gases are the greatest hazards associated with a
   A. vehicle or machinery rescue.
   B. high-angle rescue.
   C. hazardous materials rescue.
   D. confined-space rescue.

_____ 22. What are the most versatile and widely used technical rescue skills?
   A. Disentanglement skills
   B. Medical skills
   C. Rope skills
   D. Hazardous materials knowledge

_____ 23. A collapse that occurs after an initial collapse is called a
   A. follow-up collapse.
   B. supportive collapse.
   C. shoring collapse.
   D. secondary collapse.

_____ 24. In search and rescue, removing a victim from a hostile environment is classified as a
   A. search.
   B. rescue.
   C. recovery.
   D. removal.

_____ 25. During elevator and escalator rescue, some activities should be attempted only by
   A. technical rescue team members.
   B. emergency responders.
   C. professionally trained service technicians.
   D. veteran department members.

Vocabulary

Define the following terms using the space provided.

1. Hazardous materials:
   _____
   _____
   _____

2. Technical rescue incident:
   _____
   _____
   _____

3. Lockout/tagout system:
   _____
   _____
   _____

Fill-in
Read each item carefully, and then complete the statement by filling in the missing word(s).

1. The _____ level of training provides an emphasis on recognizing the hazards, securing the scene, and calling for appropriate assistance.
2. To assist in more efficient communication with other rescuers, it is important to know the _____ used in the field.
3. A rescue area is an area that surrounds the incident site and whose size is _____ to the hazards that exist.
4. Scene control activities are sometimes assigned to _____ personnel.
5. The process of preparing the victim for transport is called _____.
6. It is extremely important that hazardous materials incident victims are _____ prior to transport.
7. Once the rescue is complete, the scene must be _____ by the rescue crew to ensure that no one else becomes injured.
8. Natural gas and liquefied petroleum gas are nontoxic, but are classified as _____ because they displace breathing air.
9. To ensure the safety of the rescuers there must be a(n) _____ in place.
10. If you have the role of assisting a technical rescue team, _____ with the team is probably the most important thing you can do.
11. Information gathered _____ to the technical rescue team's arrival will save valuable time during the actual rescue.
12. _____ collapse is the sudden and unplanned collapse of part or all of a structure.

True/False

If you believe the statement to be more true than false, write the letter "T" in the space provided.
If you believe the statement to be more false than true, write the letter "F."

1. _____ Rescue situations have many hidden hazards.
2. _____ Many fire departments are run like a military organization.
3. _____ Rescue efforts often require a small focused group of individuals to complete the operation.
4. _____ The most important part of any rescue is the arrival and size-up of the incident scene.
5. _____ Tagout procedures are used for personnel accountability.
6. _____ During a rescue, a team member should remain with the victim to direct the rescuers performing disentanglement.
7. _____ The best way to prepare for the next rescue call is to review the last one.
8. _____ To assist a victim in remaining calm, communicate calmly, slowly, and clearly.
9. _____ Any machine that is involved in a machinery rescue should be considered electrically charged.
10. _____ Without a solid command structure, most large-scale rescue efforts are doomed to failure.
11. _____ During water rescue incidents, all responders within 10 feet of the water must wear an approved personal flotation device.

Chapter 26: Assisting Special Rescue Teams

Short Answer
Complete this section with short written answers using the space provided.

1. List the five guidelines that a fire fighter should follow when assisting rescue team members.

2. Identify the paramilitary guidelines for which a fire fighter must have a strong appreciation to understand the command and control concept of fire departments.

3. Identify the components of the acronym "FAILURE" used to describe why rescuers fail.

4. Identify the 10 steps of the special rescue sequence.

5. List five considerations during size-up.

Word Fun

The following crossword puzzle is an activity provided to reinforce correct spelling and understanding of terminology associated with firefighting. Use the clues provided to complete the puzzle.

Clues

Across

7. A collapse that occurs following a trench, excavation, or structural collapse.
8. The pile of dirt that has been removed from an excavation.
9. The training level that allows responders to directly assist those conducting the rescue operation and to use certain rescue skills and procedures.
12. A condition in which a victim is trapped and is unable to extricate himself or herself.
13. A safe area.
14. A rope rescue operation where the rescuers depend on life safety rope rather than a fixed support surface such as the ground.

Down

1. The level of training with an emphasis on recognizing the hazards, securing the scene, and calling for appropriate assistance.
2. Preparing the victim for movement as a unit.
3. The area that contains the decontamination stations.
4. The training level that provides a high level of competency for rescuers who will be directly involved in the rescue operation itself.
5. The absolute closure of a pipe, line, or duct.
6. A space with limited or restricted access that is not meant for continuous occupancy.
10. A method of supporting a trench wall or building components to prevent collapse.
11. The area immediately surrounding an incident site that is directly dangerous to life and health.

Fire Alarms

The following real case scenarios will give you an opportunity to explore the concerns associated with assisting special rescue teams. Read each scenario, and then answer each question in detail.

1. Your company has been dispatched to a call of wires down, with an individual unconscious and in contact with the downed wires. There are several citizens in the immediate area. Your company officer has ordered you to secure the scene and establish a barrier. How will you accomplish these orders?

   _____
   _____
   _____
   _____
   _____
   _____

2. Your ladder has been dispatched to an incident involving an individual trapped in industrial machinery. Your company officer has ordered you to assemble the tools that may be needed to release the victim from the machinery while he performs a scene size-up. Which tools will you assemble for use?

   _____
   _____
   _____
   _____
   _____
   _____

3. Your company is dispatched to a trench collapse where a worker is trapped. How will you make a safe approach to the collapse zone?

   _____
   _____
   _____
   _____
   _____
   _____

Hazardous Materials: Overview

Workbook Activities

The following activities have been designed to help you. Your instructor may require you to complete some or all of these activities as a regular part of your fire fighter training program. You are encouraged to complete any activity that your instructor does not assign as a way to enhance your learning in the classroom.

Chapter Review

The following exercises provide an opportunity to refresh your knowledge of this chapter.

Matching

Match each of the terms in the left column to the appropriate definition in the right column.

_____ 1. Awareness level
_____ 2. Technician level
_____ 3. Hazardous material
_____ 4. EPA
_____ 5. Hazardous waste
_____ 6. NFPA
_____ 7. Target hazard
_____ 8. SARA
_____ 9. SERC
_____ 10. MSDS

A. Training that provides the ability to enter heavily contaminated areas using the highest levels of protection
B. The impure substance left after manufacturing
C. A federal agency that ensures safe manufacturing, use, transportation, and disposal of hazardous substances
D. A facility that presents a high potential for loss of life
E. The body that develops and maintains nationally recognized minimum consensus standards on many areas of fire safety and hazardous materials
F. Training that provides the ability to recognize a potential hazardous emergency and isolate the area
G. Any material that poses an unreasonable risk of damage or injury to persons, property, or the environment if not properly controlled
H. Liaison between local and state levels of emergency response authorities
I. A detailed profile of a single chemical or mixture of chemicals provided by the manufacturer or supplier of a chemical
J. Law that affects how fire departments respond in a hazardous materials emergency

CHAPTER 27

Multiple Choice
Read each item carefully, and then select the best response.

_____ 1. A material that poses an unreasonable risk to the health and safety of the public and/or the environment if it is not controlled properly during handling, processing, and disposal is called a
   A. hazardous waste.
   B. hazardous material.
   C. hazardous target.
   D. hazardous substance.

_____ 2. Which of the following is a nongovernment agency that issues fire response standards?
   A. CANUTEC
   B. NFPA
   C. OSHA
   D. EPA

_____ 3. In the United States, the federal document containing the hazardous materials response competencies is known as
   A. NFPA.
   B. EPCRA.
   C. SARA.
   D. HAZWOPER.

_____ 4. In the United States, which federal government agency enforces and publicizes laws and regulations that govern the transportation of goods by highways, rail, and air?
   A. Environmental Protection Agency (EPA)
   B. Occupational Safety and Health Agency (OSHA)
   C. State Emergency Response Commission (SERC)
   D. Department of Transportation (DOT)

_____ 5. What act requires a business that handles chemicals to report storage type, quantity, and storage methods to the fire department and the local emergency planning committee?
   A. Superfund Amendments and Reauthorization Act
   B. Local Emergency Planning Committee Act
   C. Emergency Planning and Community Right to Know Act
   D. Occupational Safety and Health Act

_____ 6. Which of the following is a group that gathers information about hazardous materials and disseminates that information to the public?
   A. LEPC
   B. NFPA
   C. SARA
   D. EPA

_____ 7. Which of the following is a state group that acts as a liaison between local- and state-level response authorities?
   A. SARA
   B. EPA
   C. SERC
   D. MSDS

_____ 8. Which response level is trained to recognize a hazardous materials emergency and call for assistance?
   A. Awareness
   B. Operations
   C. Technician
   D. Specialist

_____ 9. Which response level is trained to take defensive actions?
   A. Awareness
   B. Operations
   C. Technician
   D. Specialist

_____ 10. Which response level is trained to take offensive actions?
   A. Awareness
   B. Operations
   C. Technician
   D. Specialist

Vocabulary
Define the following terms using the space provided.

1. Local emergency planning committee (LEPC):

2. Material safety data sheet (MSDS):

3. Specialist level:

4. HAZWOPER:

5. Operations level:

Fill-in
Read each item carefully, and then complete the statement by filling in the missing word(s).

1. The _____ regulates and governs issues relating to hazardous materials in the environment.
2. A(n) _____ is a detailed profile of a single chemical or mixture of chemicals, provided by the manufacturer and/or supplier of a chemical.
3. Hazardous materials incidents are _____ complicated than most structural firefighting incidents.
4. _____ activities enable agencies to develop logical and appropriate response procedures for anticipated incidents.
5. Awareness level skills are _____ and not defensive.
6. Hazardous materials _____ assume command of a hazardous materials incident beyond the operations level.
7. _____ is the *Standard for Competence of Responders to Hazardous Materials/Weapons of Mass Destruction Incidents*.
8. The federal document containing the hazardous materials response competencies is known as _____.
9. _____ is the material that remains after a manufacturing plant has used some chemicals, and they are no longer pure.
10. States that have adopted OSHA safety and health regulations are called _____ states.

True/False
If you believe the statement to be more true than false, write the letter "T" in the space provided.
If you believe the statement to be more false than true, write the letter "F."

1. _____ The ability to recognize a potential hazardous materials incident is critical to ensuring one's safety.
2. _____ The Emergency Planning and Community Right to Know Act was one of the first laws to affect how fire departments respond in a hazardous materials emergency.
3. _____ Each state has a State Emergency Response Commission (SERC) that acts as a liaison between local and state levels of authority.
4. _____ The actions taken at hazardous materials incidents are largely dictated by the chemicals involved.
5. _____ Fires require a less straightforward response than do hazardous materials incidents.
6. _____ When approaching a hazardous materials event, you should make a conscious effort to change your response perspective.
7. _____ Response agencies should not preplan target hazards owing to the health issues involved in such planning.
8. _____ The goal of a fire fighter is to favorably change the outcome of a hazardous materials incident.
9. _____ The SARA regulates and governs issues relating to hazardous materials and the environment.
10. _____ The EPA's version of HAZWOPER is in Title 40, Protection of the Environment, Part 311, Worker Safety.

Short Answer
Complete this section with short written answers using the space provided.

1. Identify the five levels of hazardous materials training and competencies, according to NFPA 472.

   _____
   _____
   _____
   _____
   _____

2. Discuss the Superfund Amendments and Reauthorization Act.

   _____
   _____
   _____

Word Fun

The following crossword puzzle is an activity provided to reinforce correct spelling and understanding of terminology associated with firefighting. Use the clues provided to complete the puzzle.

Clues

Across

2. The federal agency that regulates worker safety and, in some cases, responder safety.
3. Any material that poses an unreasonable risk of damage or injury to persons, property, or the environment if it is not properly controlled.
5. A facility that presents a high potential for loss of life or serious impact to the community resulting from fire, explosion, or chemical release.
7. The association that maintains nationally recognized minimum consensus standards on many areas of fire safety.
8. A detailed profile of a chemical, provided by the manufacturer and/or supplier of that chemical.
9. The federal agency that ensures safe manufacturing, use, transportation, and disposal of hazardous materials.

Down

1. A substance that remains after a process or manufacturing plant has used some of the material and it is no longer pure.
2. The level of training that should allow responders to be able to recognize and isolate hazardous materials, and deny entry to other responders.
4. The level of training that provides first responders with the ability to recognize a potential hazardous materials emergency.
5. Responders trained to this level are able to enter heavily contaminated areas and take offensive actions.
6. The federal department that publicizes and enforces rules and regulations that relate to the transportation of many hazardous materials.

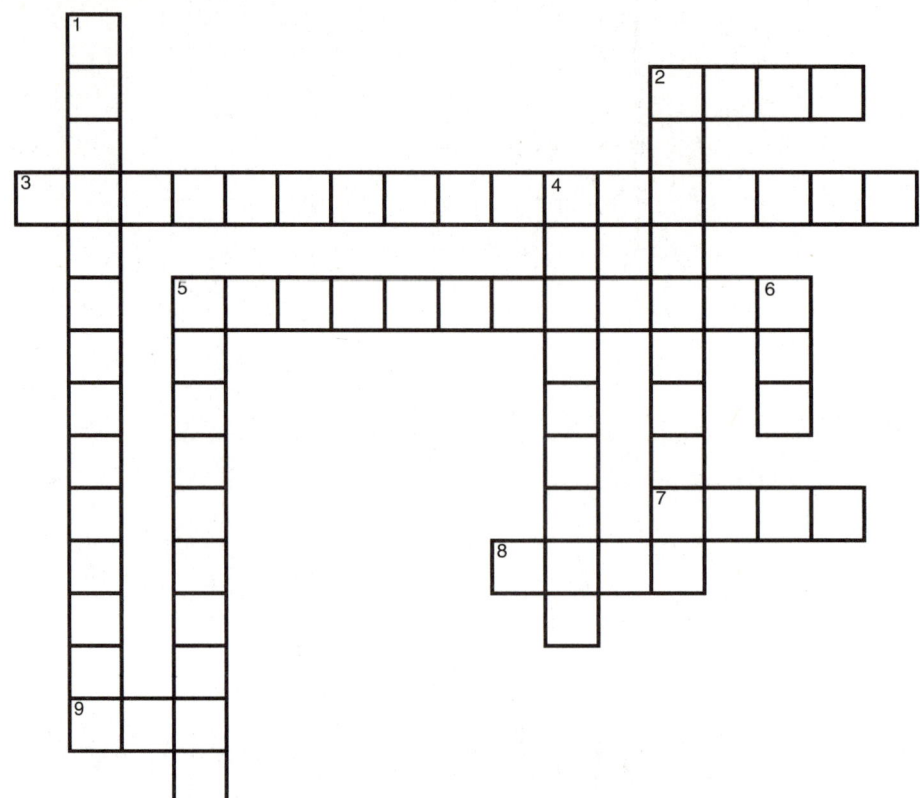

Fire Alarms

The following real case scenarios will give you an opportunity to explore the concerns associated with hazardous materials. Read each scenario, and then answer each question in detail.

1. You are participating in a hazardous materials awareness class, and the instructor asks you to define the term "hazardous materials." What is your response?

   _____
   _____
   _____
   _____
   _____
   _____

2. You have just completed your hazardous materials training. Your Lieutenant gives you an assignment to complete a mock preincident plan for a hazardous materials response to an occupancy near your station. How should you proceed?

   _____
   _____
   _____
   _____
   _____
   _____

Hazardous Materials: Properties and Effects

Workbook Activities

The following activities have been designed to help you. Your instructor may require you to complete some or all of these activities as a regular part of your fire fighter training program. You are encouraged to complete any activity that your instructor does not assign as a way to enhance your learning in the classroom.

Chapter Review

The following exercises provide an opportunity to refresh your knowledge of this chapter.

Matching

Match each of the terms in the left column to the appropriate definition in the right column.

_____ 1. Vapor
_____ 2. Lower flammable limit
_____ 3. Corrosivity
_____ 4. Toxicology
_____ 5. Flammable range
_____ 6. Vapor density
_____ 7. Pulmonary edema
_____ 8. Vapor pressure
_____ 9. Carcinogen
_____ 10. Expansion ratio

A. The ability of a material to cause damage upon skin contact
B. The boundaries of a fuel/air mixture necessary for a combustible material to burn properly
C. Fluid build-up in the lungs
D. The weight of a gas as compared to an equal volume of dry air
E. The gas phase of a substance
F. The minimum amount of gaseous fuel that must be present in the air mixture for the mixture to be flammable or explosive
G. A cancer-causing agent
H. The study of the adverse effects of chemical or physical agents on living organisms
I. The pressure exerted by a liquid's vapor until the liquid and vapor reach an equilibrium
J. A description of a volume increase that occurs when a liquid changes to a gas

Multiple Choice

Read each item carefully, and then select the best response.

_____ 1. The characteristics of a chemical that are measurable are
   A. physical properties.
   B. chemical properties.
   C. states of matter.
   D. radiation agents.

_____ 2. The first step in understanding the hazard of any chemical involves identifying
   A. physical properties.
   B. chemical properties.
   C. states of matter.
   D. radiation agents.

CHAPTER 28

___ 3. The expansion ratio is a description of the volume increase that occurs when a material changes from
   A. a liquid to a solid.
   B. a solid to a gas.
   C. a solid to a liquid.
   D. a liquid to a gas.

___ 4. The ability of a chemical to undergo a change in its chemical make-up, usually with a release of some form of energy, is a
   A. property change.
   B. physical change.
   C. chemical change.
   D. change of state.

___ 5. Steel rusting and wood burning are examples of
   A. physical changes.
   B. chemical changes.
   C. vaporization.
   D. ionization.

___ 6. The temperature at which a liquid changes into a gas is the
   A. flash point.
   B. vaporization point.
   C. boiling point.
   D. gas point.

___ 7. The weight of an airborne concentration as compared to an equal volume of dry air is the
   A. vapor density.
   B. vapor ratio.
   C. flammable range.
   D. explosive ratio.

___ 8. Air has a set vapor density value of
   A. 0.59.
   B. 1.0.
   C. 2.4.
   D. 3.8.

___ 9. The vapor pressure at the standard atmospheric pressure of 20°C can be expressed in pounds per square inch, atmospheres, and millimeters of mercury as follows:
   A. 14.7 psi = 1 atm = 760 torr = 1 mm Hg.
   B. 1 psi = 0.59 atm = 760 torr = 10 mm Hg.
   C. 10 psi = 1 atm = 550 torr = 0.59 mm Hg.
   D. 14.7 psi = 100 atm = 30 torr = 100 mm Hg.

___ 10. The ability of a substance to dissolve in water is known as its
   A. expansion ratio.
   B. dissolvability.
   C. water solubility.
   D. dispersement value.

11. pH is an expression of the concentration of
    A. hydrogen ions in a given substance.
    B. acid ions in a given substance.
    C. oxygen ions in a given substance.
    D. base ions in a given substance.

12. Common acids have pH values that are
    A. equal to zero.
    B. greater than 7.
    C. equal to 7.
    D. less than 7.

13. Bases have pH values that are
    A. equal to zero.
    B. greater than 7.
    C. equal to 7.
    D. less than 7.

14. The hazardous chemical compounds released when a material decomposes under heat are known as
    A. carcinogens.
    B. alpha particles.
    C. toxic products of combustion.
    D. beta particles.

15. The nucleus of a radioactive isotope includes an unstable configuration of
    A. protons and neutrons.
    B. electrons and protons.
    C. electrons and neutrons.
    D. protons, electrons, and neutrons.

16. Which is the least penetrating of the three types of radiation?
    A. Alpha particles
    B. Beta particles
    C. Gamma radiation
    D. Neutrons

17. When a person or object transfers contamination to another person or object by direct contact, what is this process called?
    A. Contamination by association
    B. Secondary exposure
    C. Direct contamination
    D. Secondary contamination

18. Exposure to which of the following substances prevents the body from using oxygen?
    A. Chlorine
    B. Cyanide
    C. Lewisite
    D. Sarin

19. Phosgene is a
    A. nerve agent.
    B. blistering agent.
    C. choking agent.
    D. blood agent.

20. Which type of exposure occurs when harmful substances are brought into the body through the respiratory system?
    A. Ingestion exposure
    B. Inhalation exposure
    C. Absorption exposure
    D. Injection exposure

_____ 21. Adverse health effects caused by long-term exposure to a substance are termed a(n)
   A. acute health hazards.
   B. chronic health hazards.
   C. long-term disablers.
   D. overexposure.

_____ 22. Which type of chemical causes a substantial proportion of exposed people to develop an allergic reaction in normal tissue after repeated exposure to that chemical?
   A. Sensitizer
   B. Irritant
   C. Convulsant
   D. Contaminant

Labeling
Label the following diagrams with the correct terms.

1. Vapor density.

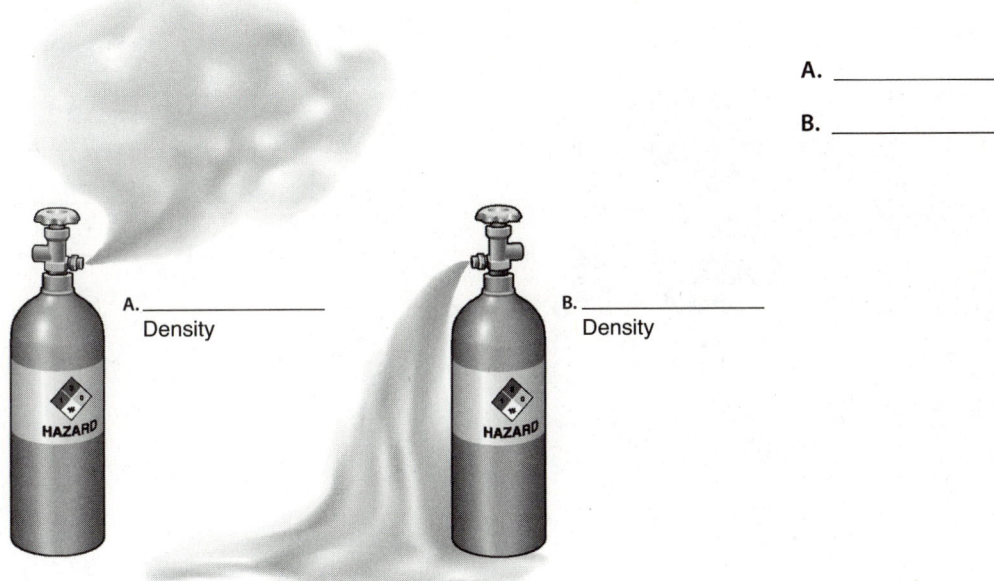

A. _____
   Density

B. _____
   Density

A. _____

B. _____

Figure 28-3

2. Radiation.

A. _____
B. _____
C. _____
D. _____
E. _____
F. _____

Figure 28-6

3. Four ways a chemical substance can enter the body.

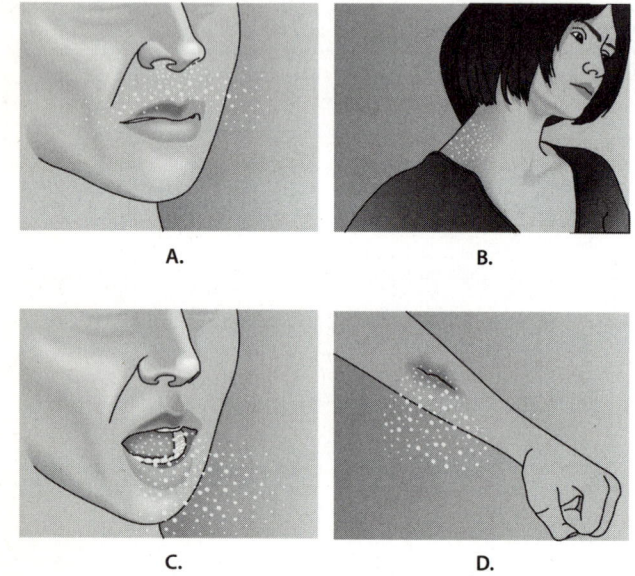

A. _____
B. _____
C. _____
D. _____

Figure 28-8

Vocabulary

Define the following terms using the space provided.

1. Flammable vapor:

   _____
   _____
   _____

2. HEPA filter:

   _____
   _____
   _____

3. Radiation:

   _____
   _____
   _____

4. Contamination:

   _____
   _____
   _____

5. Weapons of mass destruction (WMD):

   _____
   _____
   _____

Fundamentals of Fire Fighter Skills

Fill-in
Read each item carefully, and then complete the statement by filling in the missing word(s).

1. If the state of matter and physical properties of a chemical are known, a fire fighter can _____ what the substance will do if it escapes its container.
2. Chemicals can undergo a(n) _____ change when subjected to outside influences such as heat, cold, and pressure.
3. Standard atmospheric pressure at sea level is _____ pounds per square inch.
4. The flash point of gasoline is _____.
5. _____ is the minimum temperature at which a liquid or a solid emits vapor sufficient to form an ignitable mixture with air.
6. The _____ temperature is the minimum temperature at which a substance will ignite without an external ignition source.
7. Most flammable liquids will _____ on water.
8. The periodic table illustrates all the known _____ that make up every known compound.
9. _____ particles can break chemical bonds, creating ions; therefore they are considered ionizing radiation.
10. Chemicals that are capable of causing seizures are classified as _____.

True/False
If you believe the statement to be more true than false, write the letter "T" in the space provided.
If you believe the statement to be more false than true, write the letter "F."

1. _____ A physical change is essentially a change in state; a chemical change results in an alteration of the chemical nature of the material.
2. _____ Water has an expansion rate of 100:1 and a boiling point of 100 °F.
3. _____ Diesel fuel has a higher flash point than does gasoline.
4. _____ The wider the flammable range, the more dangerous the material.
5. _____ Vapor pressure directly correlates to the speed at which a material will evaporate once it is released from its container.
6. _____ Radioactive isotopes can be detected by the noise and odors they give off.
7. _____ The nucleus of an atom is made up of protons, neutrons, and electrons.
8. _____ A hazard is a material capable of posing an unreasonable risk to health, safety, or the environment.
9. _____ Nerve agents attack the central nervous system.
10. _____ A chemical brought into the body through an open cut is an injection exposure.

Short Answer

Complete this section with short written answers using the space provided.

1. Identify the five types of possible hazardous material incidents represented by the mnemonic "TRACEMP."

2. Identify the nerve agent signs and symptoms represented by the mnemonic "SLUDGEM."

3. Identify and define the four ways through which chemical substances can enter the human body.

4. The health hazards posed by radiation are a function of what two factors?

5. What does "HA HA MINCE" stand for?

Word Fun

The following crossword puzzle is an activity provided to reinforce correct spelling and understanding of terminology associated with firefighting. Use the clues provided to complete the puzzle.

Clues

Across

7. The least penetrating of the three common types of radiation emitted by radioactive material.
8. A cancer-causing agent.
12. Materials with pH values less than 7.
13. Exposure to a hazardous material by breathing it into the lungs.
14. Exposure to a hazardous material by swallowing it.
15. The process of transferring a hazardous material from its source to people.

Down

1. The gas phase of a substance.
2. A nerve agent that is primarily a vapor hazard.
3. Materials with a pH value greater than 7.
4. High-energy, short-wavelength electromagnetic radiation.
5. Hazardous materials entering cuts or other breaches in the skin.
6. The contamination process by which a contaminant is carried out of the hot zone and contaminates additional areas or people.
9. Toxic substances that attack the central nervous system in humans.
10. A particle ejected from the nucleus of an unstable atom.
11. Penetrating particles found in the nucleus of the atom that are removed through nuclear fusion or fission.

Fire Alarms

The following real case scenarios will give you an opportunity to explore the concerns associated with hazardous materials. Read each scenario, and then answer each question in detail.

1. You are dispatched to the local hospital for a hazardous materials incident. When your engine arrives at the scene, a lab technician states that she believes that one of the medical containers is leaking radioactive material. How should you proceed?

   _____
   _____
   _____
   _____
   _____
   _____

2. It is 2:00 in the afternoon on a Saturday when your engine is dispatched to a shopping mall for a hazardous materials incident. Approximately 50 people have been exposed to a substance that is causing pain and burning to their skin and eyes. Last week, the U.S. terror alert system was activated to level red and a warning was posted for possible attacks on shopping malls and other highly populated buildings. How should you proceed?

   _____
   _____
   _____
   _____
   _____
   _____

Hazardous Materials: Recognizing and Identifying the Hazards

Workbook Activities

The following activities have been designed to help you. Your instructor may require you to complete some or all of these activities as a regular part of your fire fighter training program. You are encouraged to complete any activity that your instructor does not assign as a way to enhance your learning in the classroom.

Chapter Review

The following exercises provide an opportunity to refresh your knowledge of this chapter.

Matching

Match each of the terms in the left column to the appropriate definition in the right column.

_____ 1. Bill of lading
_____ 2. Waybill
_____ 3. Signal words
_____ 4. Dewar containers
_____ 5. Cylinder
_____ 6. Consist
_____ 7. Bungs
_____ 8. Drums
_____ 9. Vent pipes
_____ 10. Totes

A. The list of every car on a train
B. Barrel-like containers
C. Containers designed to preserve the temperature of the cold liquid held inside
D. Portable tanks characterized by a unique style of construction
E. One or more small openings in closed-head drums
F. Shipping papers for trains
G. Information on a pesticide label that indicates the relative toxicity of the material
H. Shipping papers for roads and highways
I. Inverted J-shaped tubes that allow for pressure relief from the pipeline
J. A portable compressed-gas container

Multiple Choice

Read each item carefully, and then select the best response.

_____ 1. Liquid bulk storage containers have an internal capacity of more than
   A. 500 gallons.
   B. 250 gallons.
   C. 182 gallons.
   D. 119 gallons.

_____ 2. Solid bulk storage containers have an internal capacity of more than
   A. 1118 pounds.
   B. 1084 pounds.
   C. 919 pounds.
   D. 882 pounds.

CHAPTER 29

_____ 3. When large-volume horizontal tanks are stored above ground, they are referred to as
   A. OSTs.
   B. ASTs.
   C. USTs.
   D. GSTs.

_____ 4. Which high-pressure vessels have internal pressures of several hundred pounds per square inch and carry liquefied propane?
   A. IM-101s
   B. IM-102s
   C. IMO-Type 10 containers
   D. IMO-Type 5 containers

_____ 5. Drum bungs can be removed using a
   A. bung wrench.
   B. drum ratchet.
   C. cinching wrench.
   D. drum ring.

_____ 6. Solids and powders are often stored in
   A. drums.
   B. boxes.
   C. bags.
   D. carboys.

_____ 7. A glass, plastic, or steel container that holds 5 to 15 gallons of product is a
   A. carboy.
   B. bottle.
   C. drum.
   D. cylinder.

_____ 8. Propane cylinders contain a liquefied gas and have low pressures of approximately
   A. 50–110 psi.
   B. 150–200 psi.
   C. 200–300 psi.
   D. 300–400 psi.

_____ 9. Gaseous substances that have been chilled until they liquefy are classified as
   A. cryogenic gases.
   B. crystals.
   C. Dewar liquids.
   D. cryogenic liquids.

_____ 10. One of the most common chemical tankers is a gasoline tanker, also known as a(n)
   A. MC-331 pressure cargo tanker.
   B. MC-307 chemical hauler.
   C. MC-306 flammable liquid tanker.
   D. tube trailer.

_____ 11. Which types of containers are generally V-shaped with rounded sides and are used to carry grain or fertilizers?
   A. Consist tankers
   B. Dry bulk cargo tankers
   C. Carboys
   D. ASTs

_____ 12. Fire fighters should be able to recognize the three basic railcar configurations of
   A. nonpressurized, pressurized, and special use.
   B. dry bulk, liquid, and hazardous materials.
   C. agricultural, mechanical, and products.
   D. contained, unpackaged, and hazardous materials.

_____ 13. What are the 10 ¾-inch diamond-shaped indicators that must be placed on all four sides of hazardous materials transportation called?
   A. MSDS markers
   B. Labels
   C. Placards
   D. NAERG tags

_____ 14. Within the NFPA hazard identification system, which number is used to identify materials that will not burn?
   A. 4
   B. 2
   C. 1
   D. 0

_____ 15. Within the NFPA hazard identification system, which number is used to identify materials that can cause death after a short exposure?
   A. 4
   B. 2
   C. 1
   D. 0

_____ 16. Shipping papers on a marine vessel are referred to as
   A. waybills.
   B. dangerous cargo manifest.
   C. consists.
   D. freight bills.

_____ 17. Which DOT packaging group designation is used to represent the highest level of danger?
   A. Packaging group I
   B. Packaging group II
   C. Packaging group III
   D. Packaging group V

_____ 18. If a radiation incident is expected at a fixed facility, which person should be contacted for information?
   A. The safety officer
   B. The incident commander
   C. The radiation safety officer
   D. The shift supervisor

_____ 19. Which type of packaging has an inner containment vessel of glass, plastic, or metal and rubber or vermiculite packaging materials?
   A. Type A
   B. Type B
   C. Type C
   D. Type D

_____ 20. What is the type of clandestine lab most commonly encountered by fire fighters?
   A. Paint lab
   B. Drug lab
   C. Chemical lab
   D. Biological lab

Labeling

Label the following diagrams with the correct terms.

1. Chemical transport vehicles.

Figure 29-5 A. _____

Figure 29-5 B. _____

Figure 29-5 C. _____

Figure 29-5 D. _____

Figure 29-5 E. _____

Figure 29-5 F. _____

Figure 29-5 G. _____

Figure 29-5 H. _____

Vocabulary

Define the following terms using the space provided.

1. Shipping papers:
   _____
   _____
   _____

2. Secondary containment:
   _____
   _____
   _____

3. Hazardous materials:
   _____
   _____
   _____

4. Pipeline right-of-way:
   _____
   _____
   _____

5. Placards and labels:
   _____
   _____
   _____

Fill-in

Read each item carefully, and then complete the statement by filling in the missing word(s).

1. Scene _____ is especially important in all hazardous materials incidents.
2. IM-_____ containers primarily carry flammable liquids and corrosives.
3. MC-_____ corrosives tankers are used for transporting concentrated nitric acids and other corrosive substances.
4. Compressed gases such as hydrogen, oxygen, and methane are carried by _____ trailers.
5. Large-diameter _____ transport natural gas, diesel fuel, and other products from delivery terminals to distribution facilities.
6. Within the NFPA hazard identification system, special hazards that react with water are identified by _____.
7. The _____ describes the chemical hazards posed by a particular substance and provides guidance about personal protective equipment employees need to use to protect themselves from workplace hazards.
8. A common source of information about a particular chemical is the _____ specific to that substance.
9. The _____ marking system has been developed primarily to identify detonation, fire, and special hazards.
10. Shipping papers for railroad transportation are called _____; the list of every car on the train is called a(n) _____.

True/False

If you believe the statement to be more true than false, write the letter "T" in the space provided.
If you believe the statement to be more false than true, write the letter "F."

1. _____ Hazardous materials incidents can occur almost anywhere.
2. _____ Intermodal tanks are both shipping and storage vehicles.
3. _____ Nonbulk storage vessels can also be used as intermodal tanks.
4. _____ Hazardous materials can be transported in cardboard drums or paper bags.
5. _____ The Department of Transportation's marking system is characterized by a system of signs, colors, and numbers.
6. _____ OX is used to represent compressed oxygen in the NFPA hazard identification system.
7. _____ ACID is used to represent acid in the NFPA hazard identification system.
8. _____ More than 4 billion tons of hazardous materials are shipped annually in the United States.
9. _____ An MSDS will usually include a responsible-party contact.
10. _____ CHEMRESPECT is a free service that connects fire fighters with chemical manufacturers, chemists, and other product specialists who can help during a chemical incident.

Short Answer

Complete this section with short written answers using the space provided.

1. List five pieces of specific information that are included on a pesticide bag label.

2. Identify the nine ERG chemical families.

3. Identify and describe the four colored sections of the ERG.

4. Describe the parts and purpose of the NFPA 704 hazard identification system.

5. List five pieces of information that are normally included on a material safety data sheet.

CLUES

Across

1. A glass, plastic, or steel container, ranging in volume from 5 gallons to 15 gallons.
2. Inverted J-shaped tubes that allow for pressure relief or natural venting of the pipeline for maintenance and repairs.
4. Smaller versions of placards, which are placed on four sides of individual boxes and smaller packages.
9. Any device or structure that prevents environmental contamination when the primary container or its appurtenances fail.
11. Containers designed to preserve the temperature of the cold liquid held inside them.
12. The list of every car on a train.
13. The shipping papers on an airplane.
16. A document that usually includes the names and addresses of both the shipper and the receiver, as well as a list of shipped materials along with their quantity and weight.
17. Portable tanks, usually holding a few hundred gallons of product, characterized by a unique style of construction.
18. Shipping papers for roads and highways.

Down

1. A portable compressed-gas container.
3. A length of pipe, including pumps, valves, flanges, control devices, strainers, and/or similar equipment, for conveying fluids and gases.
5. Shipping papers for roads and highways.
6. Information on a pesticide label that indicates the relative toxicity of the material.
7. Barrel-like containers built to DOT specification 5P.
8. Bulk containers that can be shipped by all modes of transportation—air, sea, or land.
10. High-volume transportation devices made up of several individual compressed-gas cylinders banded together and affixed to a trailer.
14. Shipping papers for trains.
15. One or more small openings in closed-head drums.

Word Fun

The following crossword puzzle is an activity provided to reinforce correct spelling and understanding of terminology associated with firefighting. Use the clues provided to complete the puzzle.

Fire Alarms

The following real case scenarios will give you an opportunity to explore the concerns associated with hazardous materials. Read each scenario, and then answer each question in detail.

1. Your engine company is first on the scene of a vehicle accident on the four-lane highway leading into town. You see an MC-307 laying on its side with liquid leaking from the valving. Your company officer orders you to gather information for a call to CHEMTREC.

    a. What information will you collect for the phone call?
    _____
    _____
    _____
    _____
    _____
    _____

    b. What is the CHEMTREC emergency phone number?
    _____
    _____
    _____
    _____
    _____
    _____

2. Your engine company has responded to an explosion at a local abortion clinic. You are instructed to be aware that a secondary device might potentially be present.

    a. What is a secondary device?
    _____
    _____
    _____
    _____
    _____
    _____

    b. What are indicators of a potential secondary device?
    _____
    _____
    _____
    _____
    _____
    _____

Hazardous Materials:
Implementing a Response

Workbook Activities

The following activities have been designed to help you. Your instructor may require you to complete some or all of these activities as a regular part of your fire fighter training program. You are encouraged to complete any activity that your instructor does not assign as a way to enhance your learning in the classroom.

Chapter Review

The following exercises provide an opportunity to refresh your knowledge of this chapter.

Matching

Match each of the terms in the left column to the appropriate definition in the right column.

_____ 1. Material safety data sheet (MSDS)

_____ 2. Defensive objectives

_____ 3. Level I

_____ 4. Level II

_____ 5. Level III

A. Actions that do not involve the actual stopping of the leak or release of a hazardous material

B. A form provided by manufacturers and compounders (blenders) of chemicals containing information about a potentially hazardous material

C. Highest level of threat

D. Lowest level of threat

E. Level at which a hazardous materials response is needed

Multiple Choice

Read each item carefully, and then select the best response.

_____ 1. When planning an initial hazardous materials incident response, what is the first priority?
   A. Consider the effect on the environment.
   B. Consider the safety of the victims.
   C. Consider the equipment and personnel needed to mediate the incident.
   D. Consider the safety of the responding personnel.

_____ 2. Planning a response begins with the
   A. size-up.
   B. initial call for help.
   C. incident commander's orders.
   D. review of standard operating procedures.

_____ 3. Responders to a hazardous materials incident need to know the
   A. type of material involved.
   B. general operating guidelines.
   C. short- and long-term effects of the hazardous material.
   D. duration of the incident.

CHAPTER 30

_____ 4. The determination of which personal protective equipment is needed is based on the
   A. hazardous material involved.
   B. level of training of the responder.
   C. direction of the incident commander.
   D. standard operating procedures of the department.

_____ 5. What is the main hub of the incident management system?
   A. The hot zone
   B. The command post
   C. The staging area
   D. The logistics tent

_____ 6. When/where do secondary attacks take place?
   A. As responders treat victims
   B. At the firehouse
   C. At the police station
   D. Never

_____ 7. Responders to hazardous materials incidents need to consider
   A. the size of the container.
   B. the nature and amount of the material released.
   C. the area exposed to the material.
   D. A, B, and C.

_____ 8. Litmus paper is used to determine
   A. time at which the contamination occurred.
   B. pH.
   C. weather.
   D. location of the contamination.

_____ 9. Victims removed from contaminated zones must be
   A. searched.
   B. confined.
   C. decontaminated.
   D. arrested.

_____ 10. Defensive actions include
   A. plugging.
   B. patching.
   C. overpacking.
   D. diking.

Vocabulary
Define the following terms using the space provided.

1. Defensive objectives:
   _____
   _____
   _____

2. Material safety data sheet (MSDS):
   _____
   _____
   _____

3. Hazardous materials safety officer:
   _____
   _____
   _____

4. Hot zone entry team:
   _____
   _____
   _____

5. Decontamination team:
   _____
   _____
   _____

Chapter 30: Hazardous Materials: Implementing a Response

Fill-in
Read each item carefully, and then complete the statement by filling in the missing word(s).

1. When a hazardous materials incident is detected, there should be an initial call for additional _____.
2. During a hazardous materials incident, no _____ action should be taken until the identity of the hazardous material involved is confirmed.
3. The _____ of the affected area near the location of the spill or leak are important factors in planning the response to an incident.
4. At the operational level of training, all response objectives should be primarily _____ in purpose.
5. The methods of decontamination are dictated by the _____.
6. _____ paper can be used to determine the concentration of an acid or a base by reporting the hazardous material's pH.
7. The basic incident management system consists of five sections: _____, _____, _____, _____, and _____.
8. When choosing a site for the ICP, the _____ margin of safety must be used.
9. A primary terrorist attack may purposely injure members of the public to draw _____ into the scene.
10. Monitoring devices such as wind direction and weather forecasting equipment are critical resources for the _____ in formulating response plans.

True/False
If you believe the statement to be more true than false, write the letter "T" in the space provided.
If you believe the statement to be more false than true, write the letter "F."

1. _____ A predetermined list of contact names, agencies, and numbers should be established and maintained by each and every fire fighter.
2. _____ When dealing with a hazardous material, a variety of sources of information should be compared for consistency.
3. _____ If information regarding the hazardous material is unknown or is unconfirmed, the responders should prepare and approach the incident by assuming that it involves the normal hazardous materials present in the area.
4. _____ Monitoring and portable detection devices assist the incident commander in determining the hot, warm, and cold zones and the evacuation distances required.
5. _____ The safety of responders is paramount to maintaining an effective response to any hazardous materials incident.

Short Answer

Complete this section with short written answers using the space provided.

1. Identify 10 pieces of information that could be reported to agencies to assist in their preparation for a response to a hazardous materials incident.

   _____
   _____
   _____
   _____
   _____
   _____
   _____
   _____
   _____
   _____

2. Identify the three defensive objectives.

   _____
   _____
   _____

3. Identify and briefly describe the three hazardous materials incident levels.

   _____
   _____
   _____

4. List the special technical groups that may develop under the Operations section during an incident involving hazardous materials.

   _____
   _____
   _____

Fire Alarms

The following real case scenarios will give you an opportunity to explore the concerns associated with hazardous materials. Read each scenario, and then answer each question in detail.

1. It is 4:40 in the afternoon when your engine company is dispatched to an industrial plant for a hazardous materials spill. Upon arrival, you see a vapor cloud coming from the rear of the building. The plant has been evacuated. After talking with witnesses who were inside the plant, you learn that the vapor cloud is anhydrous ammonia. A careless forklift operator punctured an unknown-sized storage tank. You are now in the process of gathering the facts and reporting pertinent information to the appropriate agencies. How should you proceed?

___
___
___
___
___
___

2. You are on the scene of a large hazardous material release. Multiple agencies are en route to the site, and it appears that the incident will entail a lengthy response. Recognizing the severity of the incident, the incident commander tells your engine company to secure a location at which to establish a formal command post. How should you proceed?

___
___
___
___
___
___

Hazardous Materials:
Personal Protective Equipment, Scene Safety, and Scene Control

Workbook Activities

The following activities have been designed to help you. Your instructor may require you to complete some or all of these activities as a regular part of your fire fighter training program. You are encouraged to complete any activity that your instructor does not assign as a way to enhance your learning in the classroom.

Chapter Review

The following exercises provide an opportunity to refresh your knowledge of this chapter.

Matching

Match each of the terms in the left column to the appropriate definition in the right column.

_____ 1. Cold zone
_____ 2. Degradation
_____ 3. TLV/C
_____ 4. Hot zone
_____ 5. TLV/TWA
_____ 6. Warm zone
_____ 7. APRs
_____ 8. SARs
_____ 9. Level A
_____ 10. Level C

A. The decontamination corridor is located here
B. The maximum concentration of hazardous material to which a worker should not be exposed, even for an instant
C. The safe area that houses the command post at an incident
D. The airborne concentration of a material to which a worker can be exposed for 8 hours a day, 40 hours a week, and not suffer any ill effects
E. The physical destruction of clothing following chemical exposure
F. Devices worn to filter particulates and contaminants from the air
G. The area immediately surrounding a hazardous materials incident
H. Worn when the airborne substance is known, criteria for APR are met, and skin and eye exposure is unlikely
I. Useful during extended operations such as decontamination, clean-up, and remedial work
J. Used when the hazardous material identified requires the highest level of protection for skin, eyes, and respiration

CHAPTER 31

Multiple Choice

Read each item carefully, and then select the best response.

_____ 1. The level of PPE required for responding to a hazardous materials incident should be approved by the
   A. incident commander.
   B. safety officer.
   C. hazardous material technician.
   D. crew captain.

_____ 2. The process by which a hazardous chemical moves through closures, seams, or porous materials is called
   A. penetration.
   B. degradation.
   C. permeation.
   D. vaporization.

_____ 3. The physical destruction of clothing due to chemical exposure is called
   A. penetration.
   B. degradation.
   C. permeation.
   D. vaporization.

_____ 4. Chemical resistance, flexibility, abrasion, temperature resistance, shelf life, and sizing criteria are requirements that need to be considered when selecting
   A. entry tools.
   B. respirators.
   C. testing equipment.
   D. chemical-protective materials.

_____ 5. Air-purifying respirators should be worn in atmospheres where the type and quantity of contaminants are
   A. unknown.
   B. known.
   C. suspected.
   D. indistinguishable.

_____ 6. Chemical-protective clothing is rated for its effectiveness against chemical permeation, including how quickly it protects the fire fighter and
   A. how well it fits the fire fighter.
   B. how many times the suit can be used.
   C. to what degree it protects the fire fighter.
   D. how visible the fire fighter is when wearing the suit.

_____ 7. The principal dangers of hazardous materials are toxicity, flammability, and
   A. reactivity.
   B. instability.
   C. tolerance.
   D. transportability.

_____ 8. If a person's body temperature falls below 95°F (35°C), he or she may experience
   A. hypothermia.
   B. death.
   C. hyperthermia.
   D. cold exhaustion.

_____ 9. One of the primary objectives of a medical surveillance program is to determine
   A. the intensity of the response at an incident.
   B. the concentration of the chemicals at an incident.
   C. the time of duty at an incident.
   D. any changes in the functioning of body systems.

_____ 10. After team members undergo decontamination, they should
   A. prepare for reassignment.
   B. remove all layers of their protective uniforms.
   C. have all vital signs checked.
   D. report to the incident commander.

_____ 11. The first step in gaining control of a hazardous materials incident is to isolate the problem and
   A. equip the cold zone.
   B. keep people away.
   C. establish a backup team.
   D. identify the hazardous materials involved.

_____ 12. Designated areas at a hazardous materials incident based on safety and the degree of hazard are called
   A. control zones.
   B. hot zones.
   C. warm zones.
   D. cold zones.

_____ 13. What is the area immediately around and adjacent to the incident called?
   A. Control zone
   B. Hot zone
   C. Warm zone
   D. Cold zone

_____ 14. What is the area where personnel and equipment are staged before they enter and after they leave the hot zone called?
   A. Control zone
   B. Hot zone
   C. Warm zone
   D. Cold zone

_____ 15. What is the safe area in which personnel do not need to wear any special protective clothing for safe operation called?
   A. Control zone
   B. Hot zone
   C. Warm zone
   D. Cold zone

Chapter 31: Hazardous Materials: Personal Protective Equipment, Scene Safety, and Scene Control

Vocabulary

Define the following terms using the space provided.

1. Immediately dangerous to life and health (IDLH):

   _____
   _____
   _____

2. Heat stroke:

   _____
   _____
   _____

3. Heat exhaustion:

   _____
   _____
   _____

4. Backup personnel:

   _____
   _____
   _____

5. High temperature–protective clothing:

   _____
   _____
   _____

Fill-in

Read each item carefully, and then complete the statement by filling in the missing word(s).

1. Work uniforms offer the _____ amount of protection in a hazardous materials emergency.
2. _____ are most likely to penetrate material.
3. _____-protective clothing is designed to prevent chemicals from coming in contact with the body and may have varying degrees of resistance.
4. An encapsulated suit is a(n) _____-piece garment that completely encloses the wearer.
5. The high absorbency rate of the _____ make them more susceptible than normal skin to contaminants and one of the fastest means of exposure.
6. _____ burns are often much deeper and more destructive than acid burns.
7. Wet clothing extracts heat from the body as many as _____ times faster than dry clothing.
8. The layer of clothing next to the skin, especially the _____, should always be kept dry.
9. Fire fighters should be encouraged to drink _____ of water before donning any protective clothing.
10. _____ is enhanced by abrasions, cuts, heat, and moisture.

True/False

If you believe the statement to be more true than false, write the letter "T" in the space provided.
If you believe the statement to be more false than true, write the letter "F."

1. _____ All members of the responding team must know the shielding capabilities and limitations of their personal protective clothing.
2. _____ A hazardous materials incident may require different levels of PPE.
3. _____ Tyvek provides satisfactory protection from all chemicals.
4. _____ When possible, approach a hazardous materials incident cautiously from downwind of the site.
5. _____ The backup personnel remain on standby in the cold zone awaiting orders to prepare for follow-up duties.
6. _____ The decontamination team must be in place before anyone enters the hot zone.
7. _____ There are several ways to isolate the hazard area and create the control zones.
8. _____ All personnel must be fully briefed before they approach the hazard area or enter the cold zone.
9. _____ The warm zone contains control points for access corridors as well as the decontamination corridor.
10. _____ An incident that involves a gaseous contaminant will require a larger hot zone than one involving a liquid leak.

Short Answer
Complete this section with short written answers using the space provided.

1. Identify and define the three basic atmospheres at a hazardous materials emergency according to the exposure guidelines.

   _____
   _____
   _____

2. Identify and provide a brief description of the four levels of protective clothing.

   _____
   _____
   _____
   _____

3. Identify and provide a brief description of the three zones at a hazardous materials incident.

   _____
   _____
   _____

344 FUNDAMENTALS OF FIRE FIGHTER SKILLS

CLUES

Across
2. A safe area at a hazardous materials incident.
4. The area immediately surrounding a hazardous materials spill/incident site that is directly dangerous to life and health.
6. A severe and sometimes fatal condition resulting from the failure of the temperature-regulating capacity of the body.
7. The level of PPE used when the type and atmospheric concentration of substances have been identified and the PPE needs to provide a high level of respiratory protection but less skin protection.
9. A totally encapsulating suit that provides the highest level of protection.
10. The physical destruction or decomposition of clothing material due to chemical exposure.
11. Individuals who remove or rescue those working in the hot zone in the event of an emergency.

Down
1. The movement of a hazardous liquid chemical through zippers, stitched seams, flaps, pinholes, or other imperfections in a material.
2. Areas at a hazardous materials incident that are designated based on their safety and the degree of hazard present there.
3. A mild form of shock caused when the circulatory system begins to fail as a result of the body's inadequate effort to give off excessive heat.
5. The process by which a hazardous liquid chemical moves through clothing material.
8. Primarily a work uniform that includes coveralls and affords minimal protection.
9. Consists of standard work clothing with the addition of chemical-protective clothing, chemically resistant gloves, and a form of respirator protection.

Word Fun

The following crossword puzzle is an activity provided to reinforce correct spelling and understanding of terminology associated with firefighting. Use the clues provided to complete the puzzle.

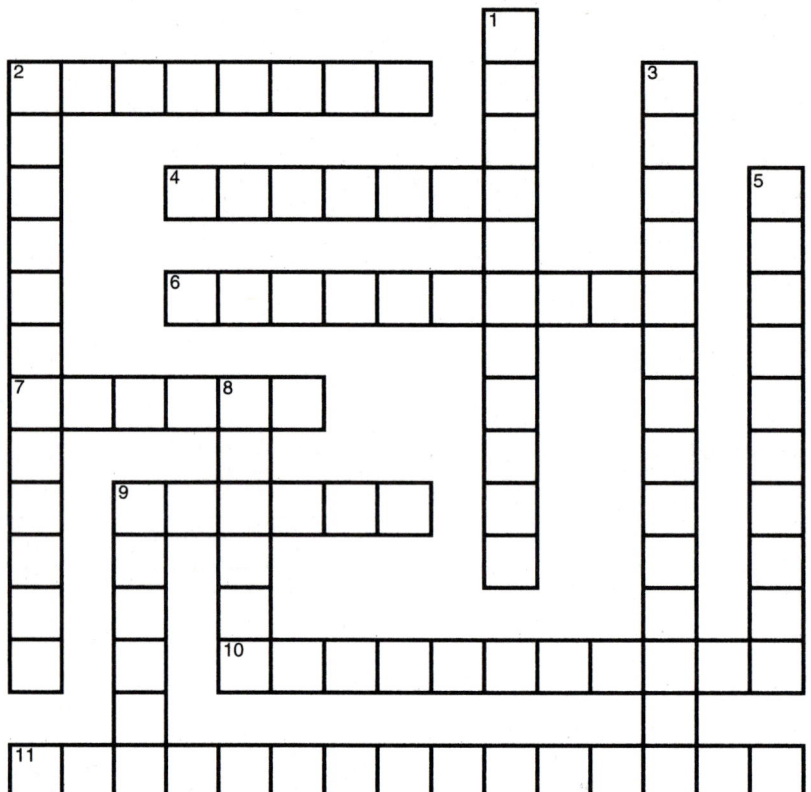

Chapter 31: Hazardous Materials: Personal Protective Equipment, Scene Safety, and Scene Control   345

Fire Alarms
The following real case scenarios will give you an opportunity to explore the concerns associated with hazardous materials. Read each scenario, and then answer each question in detail.

1. During hazardous materials response training, you are assigned to an entry team wearing Level B nonencapsulated personal protective clothing. What is the recommended PPE for Level B protection?

   _____
   _____
   _____
   _____
   _____
   _____

2. After 45 minutes of training while wearing your Level B PPE, you begin feeling dizzy and sweat profusely. You are also feeling weak and notice some blurring of your vision.

   a. What is the probable cause of your sudden illness?

   _____
   _____
   _____
   _____
   _____
   _____

   b. Which actions should be taken?

   _____
   _____
   _____
   _____
   _____
   _____

Skill Drills

Skill Drill 31-1: Donning a Level B Encapsulated Chemical-Protective Clothing Ensemble

Test your knowledge of this skill drill by placing the photos below in the correct order. Number the first step with a "1," the second step with a "2," and so on.

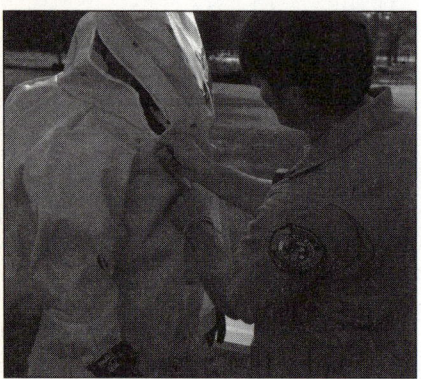

_____ Instruct the assistant to close the chemical suit by closing the zipper and sealing the splash flap.

_____ Review hand signals and indicate that you are okay.

_____ Stand up and don SCBA and the SCBA face piece, but do not connect the regulator to the face piece.

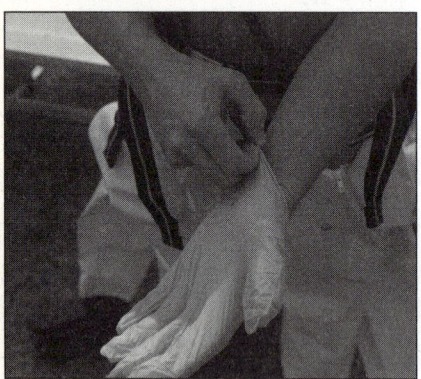

_____ Don the inner gloves.

_____ With assistance, complete donning the suit by placing both arms in the suit, pulling the expanded back piece over the SCBA, and placing the chemical suit over the head. Instruct the assistant to connect the regulator to the SCBA face piece and ensure air flow.

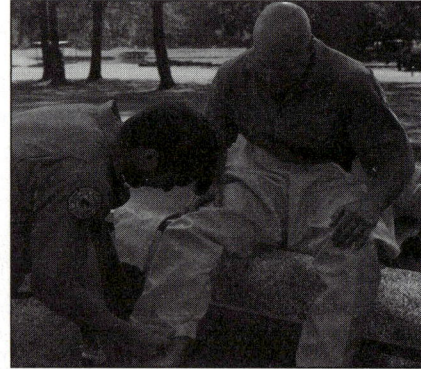

_____ Conduct a pre-entry briefing, medical monitoring, and equipment inspection. While seated, pull on the suit to waist level; pull on the chemical boots over the top of the chemical suit. Pull the suit boot covers over the tops of the boots.

_____ Place the helmet on the head.

Skill Drill 31-2: Doffing a Level B Encapsulated Chemical-Protective Clothing Ensemble
Test your knowledge of this skill drill by filling in the correct words in the photo captions.

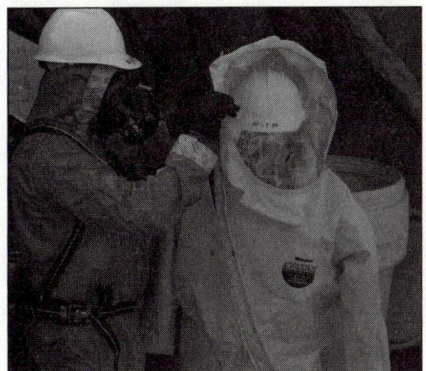

1. After completing decontamination, proceed to the _____. Remove the hands and arms from the suit gloves and sleeves, and cross the arms in front inside the suit.

2. Instruct the assistant to open the _____ and open the suit zipper.

3. Instruct the assistant to begin at the head and roll the suit _____ and _____ from you until the suit is below waist level.

4. Sit and instruct the assistant to complete rolling down the suit and remove the outer boots and suit. Rotate on the bench to the direction that will allow you to place your _____ on a dry, clean area.

5. Stand and doff the SCBA using the _____ method. Keep the face piece in place while the SCBA frame is placed on the ground.

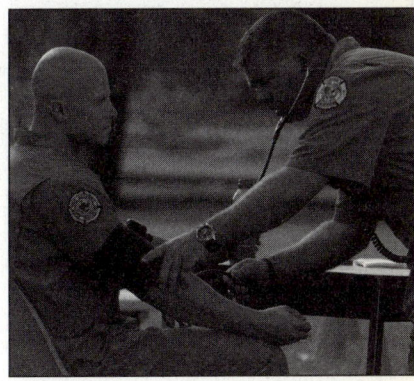

6. Take a deep breath, doff the SCBA mask, and walk away from the clean area. Go to rehabilitation area for medical monitoring, rehydration, and _____.

Skill Drill 31-5: Donning a Level C Chemical-Protective Clothing Ensemble

Test your knowledge of this skill drill by placing the photos below in the correct order. Number the first step with a "1," the second step with a "2," and so on.

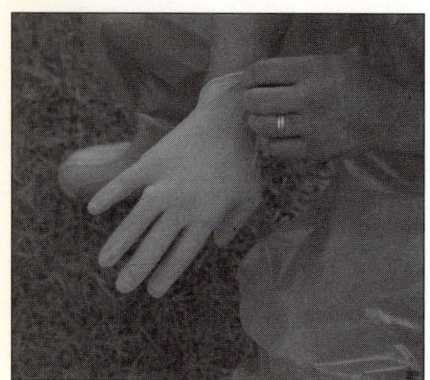

_____ Don the inner glove.

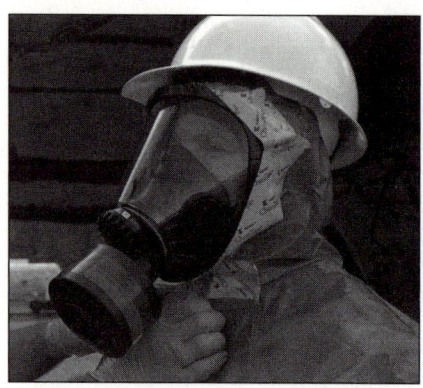

_____ Stand up and don the APR/PAPR and APR/PAPR face piece. With assistance, pull the hood over the head and the APR/PAPR face piece. Place the helmet on the head.

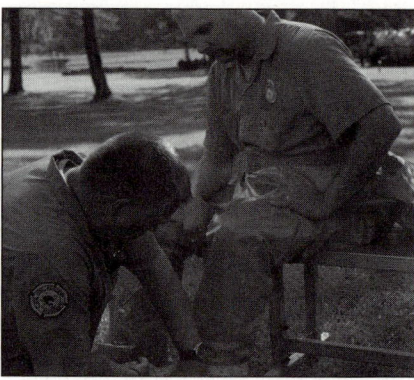

_____ Conduct a pre-entry briefing, medical monitoring, and equipment inspection. While seated, pull on the suit to waist level; pull on the chemical boots over the top of the chemical suit. Pull the suit boot covers over the tops of the boots.

_____ With assistance, complete donning the suit by placing both arms in the suit and pulling the suit over the shoulders. The assistant closes the chemical suit by closing the zipper and sealing the splash flap.

Skill Drill 31-6: Doffing a Level C Chemical-Protective Clothing Ensemble
Test your knowledge of this skill drill by filling in the correct words in the photo caption.

1. After completing _____, proceed to the clean area. The assistant opens the chemical splash flap and suit zipper. Remove the hands and arms from the gloves (except the inner gloves) and sleeves, and cross the arms in front _____ the suit. You are seated and the assistant completes rolling down the suit and takes the _____ and suit away. The assistant helps remove the inner gloves. _____ on the bench to the direction that will allow you to place your feet on a dry, clean area. Remove APR/PAPR. Go to the _____ area for medical monitoring, rehydration, and personal decontamination shower.

Hazardous Materials: Response Priorities and Actions

Workbook Activities

The following activities have been designed to help you. Your instructor may require you to complete some or all of these activities as a regular part of your fire fighter training program. You are encouraged to complete any activity that your instructor does not assign as a way to enhance your learning in the classroom.

Chapter Review

The following exercises provide an opportunity to refresh your knowledge of this chapter.

Matching

Match each of the terms in the left column to the appropriate definition in the right column.

_____ 1. Vapor dispersion   A. The process of keeping a hazardous material on the site or within the immediate area of the release

_____ 2. Confinement   B. Actions relating to stopping a hazardous materials container from leaking, such as patching, plugging, or righting the container

_____ 3. Containment   C. The process of lowering the concentration of vapors by spreading them out

_____ 4. Diversion   D. The process of controlling vapors by covering the product with foam or by reducing the temperature of the material

_____ 5. Dilution   E. The process of adding some substance to a product to weaken its concentration

_____ 6. Vapor suppression   F. Redirecting spilled material to an area where it will have less impact

_____ 7. Retention   G. The process of applying a material that will soak up and hold the hazardous material

_____ 8. Diking   H. Used when a liquid is flowing in a natural channel or depression and its progress can be stopped by blocking the channel

_____ 9. Damming   I. The placement of impervious materials to form a barrier that will keep a hazardous material in liquid form from entering an area

_____ 10. Absorption   J. The process of creating an area to hold hazardous materials

CHAPTER 32

Multiple Choice
Read each item carefully, and then select the best response.

_____ 1. Which of the following is one of the first response priorities at a hazardous materials incident?
   A. Contacting the property owners
   B. Evacuating the exposed area
   C. Containing the hazardous materials
   D. Alerting the appropriate responding agencies

_____ 2. To determine how far to extend evacuation distances, hazardous materials technicians should
   A. contact product specialists.
   B. refer to the ERG.
   C. get direction from the incident commander.
   D. use detection and monitoring equipment.

_____ 3. Which of the following factors is a major concern when considering evacuation?
   A. Potential for exposure to the material
   B. Distance to a safe area
   C. Time of day
   D. Amount of property involved in the incident

_____ 4. Which method of safeguarding people in a hazardous area involves keeping them in a safe atmosphere?
   A. Staying indoors
   B. Duck and cover
   C. Shelter-in-place
   D. Containment

_____ 5. During the initial size-up at a hazardous materials incident, the first decision concerns
   A. the amount of property affected.
   B. personnel safety.
   C. the number of people involved.
   D. the type of material involved.

_____ 6. The process of attempting to keep the hazardous material on the site or within the immediate area of the release is known as
   A. confinement.
   B. containment.
   C. exposure.
   D. suppression.

_____ 7. Most flammable and combustible liquid fires can be extinguished by the use of
   A. water.
   B. carbon monoxide.
   C. foam.
   D. dilution.

_____ 8. As fire fighters approach a hazardous materials incident, they should look for
   A. a means of egress.
   B. damage to property or surfaces.
   C. natural control points.
   D. plumes of smoke.

_____ 9. Why is the technique of absorption difficult for operational personnel to implement?
   A. It creates an extensive clean-up process.
   B. It requires the appropriate material matches.
   C. It involves a large number of personnel.
   D. It generally involves being in close proximity to the spill.

_____ 10. The process of creating an area to hold hazardous materials is called
   A. retention.
   B. diking.
   C. damming.
   D. diversion.

_____ 11. The addition of another liquid to weaken the concentration of a hazardous material is called
   A. dispersion.
   B. dilution.
   C. extension.
   D. liquidation.

_____ 12. The process of lowering the concentration of vapors by spreading them out is called
   A. vapor suppression.
   B. vapor release.
   C. vapor evacuation.
   D. vapor dispersion.

_____ 13. When the imminent danger has passed and clean-up and the return to normalcy have begun, the incident has reached the
   A. debriefing phase.
   B. clean-up phase.
   C. recovery phase.
   D. termination phase.

_____ 14. Which phase of the incident includes the compilation of all records necessary for documentation of the incident?
   A. Administration phase
   B. Recovery phase
   C. Wrap-up phase
   D. Size-up phase

_____ 15. Who makes the decision to terminate a hazardous materials incident?
   A. Safety officer
   B. Incident commander
   C. Operations officer
   D. Planning officer

Vocabulary

Define the following terms using the space provided.

1. Exposures (hazardous materials):
   _____
   _____
   _____

2. Shelter-in-place:
   _____
   _____
   _____

3. Recovery phase:
   _____
   _____
   _____

Fill-in

Read each item carefully, and then complete the statement by filling in the missing word(s).

1. The protection of _____ is the first priority in any emergency response situation.
2. Before an evacuation order is given, a(n) _____ and suitable shelter are established.
3. The _____ of the hazardous material is a major factor in the decision whether to evacuate.
4. If fire fighters expose themselves to _____ risk, injury, exposure, or contamination, they only complicate the problem.
5. Only a responder at the _____ level would patch or plug a container.
6. _____ is the process of creating an area to hold hazardous materials.
7. A(n) _____ is placed across a small stream or ditch to completely stop the flow of materials through the channel.
8. Incidents involving pressurized-gas _____ may involve fires and/or releases of their contents.
9. Most flammable and combustible _____ can be extinguished by the use of foam.
10. Victim search is _____ if a hazardous material exposure is not survivable.

True/False

If you believe the statement to be more true than false, write the letter "T" in the space provided.
If you believe the statement to be more false than true, write the letter "F."

1. _____ In a hazardous materials incident, all emergency response personnel must first recognize and identify which hazardous materials may be present.

2. _____ The duration of the hazardous materials incident is a factor in determining whether shelter-in-place is a viable option.

3. _____ All exposures need to be protected in the same way.

4. _____ Firefighting foams should be sprayed directly on the burning material and surface.

5. _____ In some cases. the incident commander may decide to withdraw to a safe distance and let the hazardous materials incident run its course.

6. _____ The recovery phase and clean-up will likely require amounts of resources and equipment that are far beyond the capabilities of local responders.

7. _____ MC 307/DOT 407 cargo tanks are certified to carry chemicals that are transported at high pressure.

8. _____ Many chemical processes, or piped systems that carry chemicals, have a way to remotely shut down a system or isolate a valve.

9. _____ Dilution can be used only when the identity and properties of the hazardous material are known with certainty.

10. _____ A retention technique is used to redirect the flow of a liquid away from an area.

Short Answer

Complete this section with short written answers using the space provided.

1. List three of the types of firefighting foams.

   _____

   _____

   _____

Word Fun

The following crossword puzzle is an activity provided to reinforce correct spelling and understanding of terminology associated with firefighting. Use the clues provided to complete the puzzle.

Clues

Across

2. Actions relating to stopping a hazardous materials container from leaking.
4. Redirecting spilled or leaking material to an area where it will have less impact.
9. The stage of a hazardous materials incident when clean-up and the return to normalcy have begun.
10. The process of adding some substance to a product to weaken its concentration.

Down

1. The placement of materials to form a barrier that will keep a hazardous material in liquid form from entering an area, or to hold the material in an area.
3. The process of keeping a hazardous material on the site or within the immediate area of the release.
5. The process of purposefully collecting hazardous materials in a defined area.
6. The process of applying a material that will soak up and hold a hazardous material in a sponge-like manner for collection and disposal.
7. People or areas of the environment that are subject to damage or injury from hazardous materials.
8. A process used when liquid is flowing in a natural channel or depression and its progress can be stopped by blocking the channel.

Chapter 32: Hazardous Materials: Response Priorities and Actions

Fire Alarms
The following real case scenarios will give you an opportunity to explore the concerns associated with hazardous materials. Read each scenario, and then answer each question in detail.

1. You have been dispatched to a diesel fuel spill on the state route south of town. On arrival at the site, you find 10 gallons of diesel fuel spilled across the roadway. Your company officer directs you to absorb the fuel with on-board absorbent. How will you absorb the spilled hazardous material?

   _____
   _____
   _____
   _____
   _____
   _____

2. Before you can begin absorbing the spilled diesel fuel, the combustible material is ignited by a road flare placed too close to the spill.

   a. What can you use to extinguish the burning combustible fuel?

   _____
   _____
   _____
   _____
   _____
   _____

   b. How should the extinguishing agent be applied?

   _____
   _____
   _____
   _____
   _____
   _____

Hazardous Materials: Decontamination Techniques

Workbook Activities

The following activities have been designed to help you. Your instructor may require you to complete some or all of these activities as a regular part of your fire fighter training program. You are encouraged to complete any activity that your instructor does not assign as a way to enhance your learning in the classroom.

Chapter Review

The following exercises provide an opportunity to refresh your knowledge of this chapter.

Matching

Match each of the terms in the left column to the appropriate definition in the right column.

_____ 1. Decontamination
_____ 2. Emulsification
_____ 3. Removal
_____ 4. Absorption
_____ 5. Solidification
_____ 6. Disinfection
_____ 7. Adsorption
_____ 8. Dilution
_____ 9. Vapor dispersion
_____ 10. Vacuuming

A. The process used to destroy recognized pathogenic microorganisms

B. The process of mixing a spongy material into a spilled liquid and picking up the mixture together

C. The process of chemically treating a hazardous liquid to turn it into a solid material, making the material easier to handle

D. The process of changing the chemical properties of a hazardous material, thereby reducing its harmful effects

E. The process of adding a material to a contaminant, which then adheres to the surface of the material for collection

F. The process of removing any form of contaminant from a person, an object, or the environment

G. A mode of decontamination that applies specifically to contaminated soil that is taken away from the scene

H. The removal of dusts, particles, and some liquids by sucking them up into a container

I. Uses plain water or a soap-and-water mixture to lower the concentration of a hazardous material while flushing it off a contaminated person or object

J. The process of separating and diminishing harmful vapors

CHAPTER 33

Multiple Choice
Read each item carefully, and then select the best response.

_____ 1. What is the situation called when a contaminated person comes into direct contact with another person or object?
   A. Cross-contamination
   B. Dispersion
   C. Transference
   D. Integration

_____ 2. Which agency is responsible for laws governing the disposal of absorbent materials?
   A. Fire department
   B. Government
   C. Department of Transportation
   D. Emergency response team

_____ 3. Which method of decontamination is used during incidents involving unknown agents and large groups of people?
   A. Emergency decontamination
   B. Group decontamination
   C. Gross decontamination
   D. Mass decontamination

_____ 4. Which decontamination procedure mixes a spongy material with a liquid hazardous material?
   A. Absorption
   B. Adsorption
   C. Dilution
   D. Vapor dispersal

_____ 5. Which of the following is a two-step removal process for items that cannot be properly decontaminated?
   A. Disinfection
   B. Solidification
   C. Disposal
   D. Rapid mass decontamination

_____ 6. During decontamination, what is usually the last item of clothing removed?
   A. Shoes
   B. SCBA mask
   C. Inner gloves
   D. Face shield

_____ 7. Removed equipment should be placed
   A. on the contaminated side of the corridor.
   B. in the hot zone.
   C. in the cold zone.
   D. in the hazardous materials truck.

_____ 8. After personnel are thoroughly decontaminated, they should proceed to
   A. the rehabilitation area.
   B. EMS personnel.
   C. the incident commander.
   D. the Operations section.

_____ 9. All personal clothing should be
   A. diluted.
   B. solidified.
   C. bagged and tagged.
   D. burned.

_____ 10. Hazardous materials that have been emulsified should be
   A. diluted.
   B. solidified.
   C. bagged and tagged.
   D. disposed of properly.

Chapter 33: Hazardous Materials: Decontamination Techniques

Vocabulary
Define the following terms using the space provided.

1. Decontamination team:

2. Contamination:

3. Adsorption:

4. Solidification:

5. Emulsification:

Fill-in

Read each item carefully, and then complete the statement by filling in the missing word(s).

1. The _____ is a controlled area, usually within the warm zone, where decontamination procedures take place.

2. During gross decontamination, runoff water should be controlled because it is likely to contain _____.

3. The process of separating and diminishing harmful vapors is known as _____.

4. The mode of decontamination that applies specifically to contaminated soil that can be taken away from the scene is called _____.

5. Fire fighters tend to use _____ as the first decontamination method.

6. A water spray is commonly used to _____ vapors.

7. The opposite of absorption is _____.

8. _____ is performed after gross decontamination and is a more thorough cleaning process.

9. Do not allow the water runoff from emergency decontamination to flow into _____, _____, or _____.

10. Whenever possible, _____ the hazardous material before beginning decontamination.

True/False

If you believe the statement to be more true than false, write the letter "T" in the space provided.
If you believe the statement to be more false than true, write the letter "F."

1. _____ Emergency medical responders are responsible for establishing a decontamination corridor for the initial emergency response crews and victims.

2. _____ During gross decontamination, hospital staff use low-pressure, high-volume water flow to rinse off and dilute contaminants.

3. _____ Vacuuming is the removal of dusts, particles, and some liquids by sucking them into a container.

4. _____ Personnel leaving the hot zone should place used tools in a tool drop area near the decontamination corridor.

5. _____ Contact lenses can trap contaminants and need to be removed during decontamination.

Short Answer

Complete this section with short written answers using the space provided.

1. Identify and provide a brief description of the four major categories of decontamination.

_____
_____
_____
_____

Word Fun

The following crossword puzzle is an activity provided to reinforce correct spelling and understanding of terminology associated with firefighting. Use the clues provided to complete the puzzle.

CLUES

Across

8. The process of chemically treating a hazardous liquid to turn it into a solid material, making the material easier to handle.
9. The process of mixing a spongy material into a spilled liquid hazardous material and picking up the contaminated mixture together.
10. The initial phase of the decontamination process, which consists of removing the outer clothing and generally flushing most contaminants from a victim.
11. Contamination that occurs when a person or object comes into direct contact with a contaminated person or object.

Down

1. A two-step removal process for contaminated items that cannot be properly decontaminated.
2. The process used to destroy recognized disease-carrying (pathogenic) microorganisms.
3. The process of changing the chemical properties of a hazardous material, thereby reducing its harmful effects.
4. A decontamination method that uses water or a soap-and-water mixture to lower the concentration of a hazardous material while flushing it from protective clothing and equipment.
5. The process of transferring a hazardous material from its source to people.
6. The physical or chemical process of removing any form of contaminant from a person.
7. The process of adding a material such as sand to a contaminant, which then adheres to the surface of the material for collection.

Fire Alarms

The following real case scenarios will give you an opportunity to explore the concerns associated with hazardous materials. Read each scenario, and then answer each question in detail.

1. Your engine company is dispatched to a pesticide spill at a local hardware store. On arrival, you see several store patrons covered with liquid and powder. Your officer orders your company to set up emergency decontamination for the contaminated patrons. How will you proceed?

2. Your Lieutenant has given you the chance to prepare a short presentation on alternative decontamination procedures. Which topics will you discuss?

Skill Drills

Skill Drill 33-3: Performing Responder Decontamination
Test your knowledge of this skill drill by placing the photos below in the correct order. Number the first step with a "1," the second step with a "2," and so on.

_____ Perform gross decontamination.

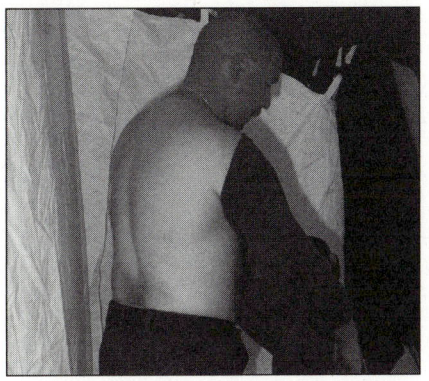

_____ Remove personal clothing. Bag and tag all personal clothing.

_____ Remove outer hazardous materials protective clothing and isolate PPE.

_____ Shower and wash the body. Dry off the body and put on clean clothing.

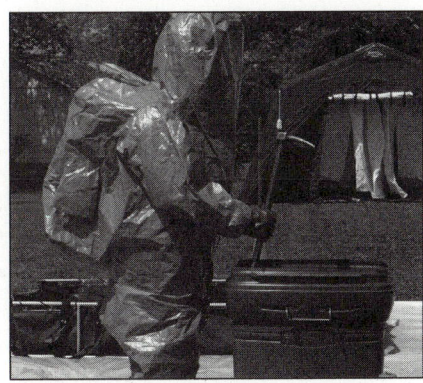

_____ Drop any tools and equipment.

_____ Remove SCBA.

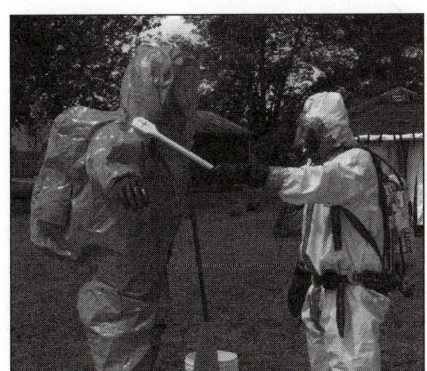

_____ Wash and rinse the fire fighter one to three times.

Terrorism Awareness

Workbook Activities

The following activities have been designed to help you. Your instructor may require you to complete some or all of these activities as a regular part of your fire fighter training program. You are encouraged to complete any activity that your instructor does not assign as a way to enhance your learning in the classroom.

Chapter Review

The following exercises provide an opportunity to refresh your knowledge of this chapter.

Matching

Match each of the terms in the left column to the appropriate definition in the right column.

_____ 1. Phosgene

_____ 2. Secondary device

_____ 3. Soman

_____ 4. Decontamination

_____ 5. ANFO

_____ 6. Sulfur mustard

_____ 7. Cyanide

_____ 8. Radiological agents

_____ 9. Nerve agent

_____ 10. Choking agent

_____ 11. Chlorine

_____ 12. Sarin

_____ 13. Triage

_____ 14. Incubation period

_____ 15. Weapon of mass destruction (WMD)

A. A nerve gas that is both a contact and a vapor hazard that has the odor of camphor

B. The physical or chemical process of removing any form of contaminant from a person, an object, or the environment

C. A yellowish gas that has many industrial uses but also damages the lungs when inhaled

D. The process of sorting victims based on the severity of injury and medical needs to establish treatment and transportation priorities

E. A chemical agent that causes severe pulmonary damage

F. A chemical designed to inhibit breathing, which is typically designed to incapacitate rather than kill

G. Toxic substances that attack the central nervous system in humans

H. Time period between the initial infection by an organism and the development of symptoms

I. A weapon intended to cause mass casualties, damage, and chaos

J. Materials that emit radioactivity

K. An explosive material containing ammonium nitrate fertilizer and fuel oil

L. A nerve agent that is primarily a vapor hazard

M. An explosive device designed to injure emergency responders who have responded to an initial event

N. A clear, yellow, or amber oily liquid with a faint sweet odor of mustard or garlic that may be dispersed in an aerosol form

O. A highly toxic chemical agent that attacks the circulatory system

CHAPTER 34

Multiple Choice

Read each item carefully, and then select the best response.

_____ 1. A terrorist threat requires fire fighters to work closely with
  A. local, state, and federal law enforcement agencies.
  B. emergency management agencies.
  C. the military.
  D. all of the above.

_____ 2. Bombing a store that sells fur coats would be an example of
  A. ecoterrorism.
  B. cyberterrorism.
  C. agroterrorism.
  D. religious terrorism.

_____ 3. Disrupting or deleting government or banking computer systems is an example of
  A. ecoterrorism.
  B. cyberterrorism.
  C. agroterrorism.
  D. religious terrorism.

_____ 4. Attacking a food industry or supply is an example of
  A. ecoterrorism.
  B. cyberterrorism.
  C. agroterrorism.
  D. religious terrorism.

_____ 5. An IED is an explosive device that is contained in a package. IED is an acronym for
  A. improvised explosive device.
  B. internal explosive device.
  C. imploding explosive device.
  D. illuminating explosive device.

_____ 6. At an incident where there is potential terrorist or secondary device activity, the fire department should be part of a joint command structure commonly referred to as
  A. a team command.
  B. an emergency response team.
  C. a unified command.
  D. a united command.

_____ 7. During a bomb disposal, where does the rapid intervention team stand by to provide immediate assistance?
  A. Bomb disposal containment area
  B. Forward staging area
  C. Incident command center
  D. Response area

_____ 8. Before anyone is allowed to enter a building involved in an explosion, what must happen?
  A. The utilities must be disconnected.
  B. All emergency response teams must arrive.
  C. Team members must review preincident plans.
  D. The stability of the building must be evaluated.

_____ 9. Many of the chemicals classified as weapons of mass destruction (WMD) are
   A. expensive.
   B. easy to obtain.
   C. restricted under the Anti-terrorist Act.
   D. kept away from ordinary citizen contact.

_____ 10. Which name is given to the time period between the actual infection and the appearance of symptoms?
   A. Growth period
   B. Dispersing period
   C. Incubation period
   D. Implementation period

_____ 11. What are the three types of radiation?
   A. Internal, external, and dispersement
   B. Alpha particles, beta particles, and gamma rays
   C. Alpha, beta, and gamma particles
   D. Alpha particles, beta particles, and sigma rays

_____ 12. For what purpose is a personal dosimeter used?
   A. To record personal exposure to contaminants
   B. To document personal exposure to contaminants
   C. To measure the amount of radioactive exposure
   D. To measure the active agents in the area

_____ 13. Fire fighters responding to a potential or known terrorist incident should use the same approach as they would a(n)
   A. structural fire.
   B. EMS incident.
   C. rescue incident.
   D. hazardous materials incident.

_____ 14. Fire fighters and emergency responders must remember that a terrorist incident is also a(n)
   A. crime scene.
   B. opportunity to improve working relations between departments.
   C. implementation of advanced rescue techniques.
   D. all of the above.

_____ 15. What is the process of sorting victims based on the severity of their injuries and medical needs to establish treatment and transportation priorities called?
   A. EMS
   B. Decon
   C. Triage
   D. Beta

Vocabulary

Define the following terms using the space provided.

1. V-agent:
   _____
   _____
   _____

2. Plague:
   _____
   _____
   _____

3. Smallpox:
   _____
   _____
   _____

4. Tabun:
   _____
   _____
   _____

5. Universal precautions:
   _____
   _____
   _____

6. Forward staging area:
   _____
   _____
   _____

7. Radiation dispersal device:
   _____
   _____
   _____

Fill-in
Read each item carefully, and then complete the statement by filling in the missing word(s).

1. _____ can be described as the unlawful use of violence or threats of violence to intimidate or coerce a person or group to further political or social objectives.
2. The most common improvised explosive device is the _____.
3. _____ are toxic substances used to attack the central nervous system and were first developed in Germany before World War II.
4. _____ is a mnemonic used to remember the symptoms of possible nerve agent exposure.
5. The most common dispersal method for chemical agents is _____.
6. The time period between the actual infection and the appearance of symptoms is known as the _____.
7. _____ release energy in the form of electromagnetic waves or energy particles that cannot be detected by the senses.
8. Decontamination should occur as soon as possible to prevent further _____ of a contaminant and to reduce the possibility of spreading the contamination.
9. Decontamination of a large number of victims or emergency responders is referred to as _____.
10. If contamination is suspected, a plan must ensure that it does not spread beyond a(n) _____.

True/False
If you believe the statement to be more true than false, write the letter "T" in the space provided.
If you believe the statement to be more false than true, write the letter "F."

1. _____ Anthrax and the plague are examples of nerve agents.
2. _____ Emergency responders are decontaminated after they leave the contaminated area.
3. _____ Gamma rays are the least harmful of the three types of radiation.
4. _____ Beta particles are also active nerve agents.
5. _____ A personal dosimeter is used to measure the amount of radioactive exposure.
6. _____ Soman is a highly infectious disease that kills approximately 30 percent of those infected with it.
7. _____ Universal precautions must be enacted during acts of cyberterrorism.
8. _____ Fire fighters must become familiar with potential terrorist targets and actions because they are often involved in the initial response and handling of a terrorist incident.
9. _____ The first emergency response units to arrive should establish an outer perimeter to control access to and from the scene.
10. _____ Exposure to high levels of radiation can cause vomiting and digestive system damage within a short time.

Short Answer

Complete this section with short written answers using the space provided.

1. What motivates terrorists?

2. Describe ecoterrorism, cyberterrorism, and agroterrorism.

3. Describe the issues that fire fighters must consider following a large explosion.

4. Describe why responding to a terrorist incident puts fire fighters and emergency personnel at increased risk.

5. Identify the three types of radiation and ways to limit exposure to each.

Word Fun

The following crossword puzzle is an activity provided to reinforce correct spelling and understanding of terminology associated with firefighting. Use the clues provided to complete the puzzle.

Clues

Across

1. Disease-causing bacteria, viruses, and other agents that attack the human body.
5. Terrorism directed against causes that radical environmentalists think would damage the earth or its creatures.
8. A blister-forming agent that is an oily, colorless-to-dark brown liquid with an odor of geraniums.
10. An infectious disease spread by the bacteria *Bacillus anthracis*; it is typically found around farms, infecting livestock.
11. The process of sorting victims based on the severity of injury and medical needs to establish treatment and transportation priorities.
14. Devices that measure the amount of radioactive exposure to an individual.

Down

2. A chemical designed to inhibit breathing, and typically intended to incapacitate rather than kill.
3. A toxic substance that attacks the central nervous system in humans.
4. A nerve agent that is primarily a vapor hazard.
6. The time period between the initial infection by an organism and the development of symptoms.
7. The intentional act of electronically attacking government or private computer systems.
9. An infectious disease caused by the bacterium *Yersinia pestis*; it is commonly found on rodents.
12. A chemical agent that causes severe pulmonary damage.
13. A type of radiation that can travel significant distances, penetrating most materials and passing through the body.

Fire Alarms

The following real case scenarios will give you an opportunity to explore the concerns associated with terrorism awareness. Read each scenario, and then answer each question in detail.

1. Your fire department contracts with a small farming community for fire protection and EMS. The government has just announced in a press release that it has uncovered some information that agroterrorists are planning a chemical attack on the U.S. food supply. During your shift meeting, your Lieutenant reviewed the department's response guidelines to terrorist incidents. What is the fire department's role in protecting communities from terrorism?

   _____
   _____
   _____
   _____
   _____
   _____

2. It is 11:30 on a Saturday morning when your engine is dispatched to a car explosion in front of a government building near the center of your community. Upon arrival, you find a heavily damaged passenger vehicle that is on fire. The building's widows have been blown out in the front and the structural status is unknown. You observe approximately 10 people who are injured. How should you proceed?

   _____
   _____
   _____
   _____
   _____
   _____

Fire Prevention and Public Education

Workbook Activities

The following activities have been designed to help you. Your instructor may require you to complete some or all of these activities as a regular part of your fire fighter training program. You are encouraged to complete any activity that your instructor does not assign as a way to enhance your learning in the classroom.

Chapter Review

The following exercises provide an opportunity to refresh your knowledge of this chapter.

Matching
Match each of the terms in the left column to the appropriate definition in the right column.

_____ 1. Fire prevention   A. Teaches techniques to reduce fire deaths and injuries
_____ 2. Fire codes   B. Activities intended to prevent the outbreak of fires
_____ 3. Public education   C. Regulations adopted to ensure a minimum level of fire safety
_____ 4. Stop, Drop, and Roll program   D. Teaches residents how to safely get out of their homes in a fire or other emergency
_____ 5. EDITH   E. Instructs people what to do if clothing catches fire

Multiple Choice
Read each item carefully, and then select the best response.

_____ 1. Activities that are intended to prevent the outbreak of fires or to limit the damage if a fire does occur are referred to as
   A. fire prevention.
   B. fire codes.
   C. fire regulations.
   D. public awareness.

_____ 2. The main objectives of fire prevention activities are to limit life loss, to prevent injuries, and
   A. to provide education.
   B. to minimize property damage.
   C. to provide an emergency response.
   D. to avoid regulation infraction.

_____ 3. Regulations that have been legally adopted by a government body with the authority to pass laws and enforce safety regulations are called
   A. fire bylaws.
   B. jurisdictional laws.
   C. jurisdictional regulations.
   D. fire codes.

CHAPTER 35

_____ 4. Which of the following is a set of documents produced by the National Fire Protection Association that is intended to address a wide range of issues relating to fire and safety?
   A. National Fire Codes
   B. National Training Standards
   C. Jurisdictional Regulations
   D. Recommended Occupant's Practices

_____ 5. A voluntary inspection of a private dwelling is called a
   A. fire department visitation.
   B. public fire safety inspection.
   C. legal requirement.
   D. home fire safety survey.

_____ 6. Helping people to understand how to prevent fires from occurring and teaching them how to react if a fire does occur are the goals of
   A. local governments.
   B. fire departments.
   C. public fire safety education.
   D. teachers.

_____ 7. The process of trying to establish the cause of a fire through careful investigation and analysis of available evidence is
   A. called fire investigation.
   B. called fire cause determination.
   C. the responsibility of law enforcement agencies.
   D. regulated through fire codes.

_____ 8. Accumulated trash and a visible house number should be checked during the
   A. walk-by assessment.
   B. interior survey.
   C. exterior survey.
   D. home escape plan.

_____ 9. Kitchen fires are responsible for
   A. 8 percent of all residential fires.
   B. 22 percent of all residential fires.
   C. 34 percent of all residential fires.
   D. 44 percent of all residential fires.

_____ 10. The primary causes of fires in living room areas are electrical equipment and
   A. smoking.
   B. fireplaces.
   C. children playing with matches.
   D. heating devices.

Vocabulary

Define the following terms using the space provided.

1. Fire code:

   _____
   _____
   _____

2. Fire prevention:

   _____
   _____
   _____

Fill-in

Read each item carefully, and then complete the statement by filling in the missing word(s).

1. Fire codes are closely related to _____ codes.
2. Citizens have a(n) _____ obligation to comply with fire codes.
3. In many cases, the fire code does not apply to the _____ of a private dwelling.
4. Home surveys should be conducted in a(n) _____ fashion for both the inside and the outside of the home.
5. The most common causes of fires in _____ are defective wiring, improper use of candles, and children playing with matches.
6. Your highest priority as a fire fighter should always be to _____ fires.
7. Every fire fighter should know how to conduct a(n) _____ safety survey in a private dwelling.
8. Generally, fire codes apply to all _____, new or old.
9. Teach students to use the back of a(n) _____ to sense the temperature of a door.
10. Stress the importance of keeping _____ in working order.

True/False

If you believe the statement to be more true than false, write the letter "T" in the space provided.
If you believe the statement to be more false than true, write the letter "F."

1. _____ The fire code is enforced through a legal process similar to the way traffic regulations are enforced.

2. _____ After a jurisdiction adopts a fire code, it must be able to enforce that code.

3. _____ Residential fire safety surveys can be conducted without the occupant's permission.

4. _____ Good housekeeping is one of the most important issues when addressing fire safety in garages and basements.

5. _____ Remind students to avoid installing smoke alarms in kitchens and garages, or near fireplaces, windows, and exit doors.

6. _____ Test smoke alarms once a year using the test button.

7. _____ According to the NFPA, the top five causes of fire are cooking equipment, heating equipment, intentional, electrical distribution, and open flames.

8. _____ As part of your EDITH presentation, stress the importance of keeping bedroom doors open during sleeping hours.

9. _____ Having working smoke alarms on each level of a house reduces the risk of death from fire by 50 percent.

10. _____ Every kitchen should be equipped with an approved ABC-rated fire extinguisher.

Short Answer
Complete this section with short written answers using the space provided.

1. Provide three examples of public fire safety education programs.

2. List five important smoke alarm tips.

3. Identify five recommendations for kitchen safety.

Fire Alarms

The following real case scenarios will give you an opportunity to explore the concerns associated with fire prevention and public education. Read each scenario, and then answer each question in detail.

1. Your Captain tells you that a group of 10 teenagers will visit the fire station for a tour during the next shift. He tells you to prepare an appropriate presentation and give them a tour of the station. How do you proceed?

   _____
   _____
   _____
   _____
   _____
   _____

2. You are in the middle of a home safety survey of a private dwelling. So far, the inspection is going well. You enter the basement and notice that the occupant is storing five full gasoline cans next to the gas water heater. How should you proceed?

   _____
   _____
   _____
   _____
   _____
   _____
   _____

Fire Detection, Protection, and Suppression Systems

Workbook Activities

The following activities have been designed to help you. Your instructor may require you to complete some or all of these activities as a regular part of your fire fighter training program. You are encouraged to complete any activity that your instructor does not assign as a way to enhance your learning in the classroom.

Chapter Review

The following exercises provide an opportunity to refresh your knowledge of this chapter.

Matching

Match each of the terms in the left column to the appropriate definition in the right column.

_____ 1. ESFR sprinkler head — **A.** A fire alarm signal caused by malfunction or improper operation of a fire alarm system or component

_____ 2. Accelerator — **B.** A sprinkler control valve with a valve stem that moves in and out as the valve is opened or closed

_____ 3. Gas detector — **C.** A device that measures the concentration of dangerous gases

_____ 4. Dry-pipe valve — **D.** A device within a piping system that allows water to flow in only one direction

_____ 5. Clapper valve — **E.** A sprinkler system in which the pipes are normally filled with water

_____ 6. Line detector — **F.** The activation of a fire alarm system when there is no fire or emergency condition

_____ 7. Nuisance alarm — **G.** The valve assembly on a dry sprinkler system that prevents water from entering the system until the air pressure is released

_____ 8. Deluge head — **H.** A sprinkler head usually marked SSU

_____ 9. Temporal-3 pattern — **I.** A sprinkler head designed to react quickly and suppress a fire in its early stages

_____ 10. Wet sprinkler system — **J.** A device that increases the removal of the air from a dry-pipe or preaction sprinkler system

_____ 11. Cross-zoned system — **K.** Wire or tubing that can be strung along the ceiling of large open areas to detect an increase in heat

_____ 12. OS&Y valve — **L.** A sprinkler head that has no release mechanism

_____ 13. Upright sprinkler head — **M.** A fire alarm system that requires activation of two separate detection devices before initiating an alarm

_____ 14. False alarm — **N.** A standard fire alarm audible signal for alerting occupants of a building

CHAPTER 36

Multiple Choice
Read each item carefully, and then select the best response.

_____ 1. The most current codes require new homes to have a smoke alarm
   A. on every floor.
   B. in every room.
   C. in every bedroom and on every floor level.
   D. in every bedroom, hallway, and floor level.

_____ 2. Which type of detectors are triggered by the invisible products of combustion?
   A. Ionization smoke detectors
   B. Photoelectric smoke detectors
   C. Heat detectors
   D. Spot detectors

_____ 3. Many buildings have an additional fire alarm control display panel in the front of the building called a
   A. remote alarm station.
   B. remote control panel.
   C. remote annunciator.
   D. remote visual board.

_____ 4. Fire alarm systems are activated by the
   A. remote annunciator.
   B. ESFR device.
   C. alarm initiation device.
   D. line detector.

_____ 5. Which type of fire alarm requires two steps before the alarm will activate?
   A. Single-action pull-station
   B. Double-action pull-station
   C. Protected pull-station
   D. Tamper alarm

_____ 6. A smoke detector is designed to sense the presence of
   A. smoke.
   B. heat.
   C. fire.
   D. toxic gases.

_____ 7. Which type of detectors use wire or tubing strung along the ceiling of large open areas to detect an increase in heat?
   A. Spot detectors
   B. Heat detectors
   C. Beam detectors
   D. Line detectors

_____ 8. Which type of detectors detect the electromagnetic light waves produced by a flame?
   A. Beam detectors
   B. Line detectors
   C. Air sampling detectors
   D. Flame detectors

_____ 9. Which type of detectors are calibrated to detect the presence of a specific gas that is created by combustion?
   A. Gas detectors
   B. Combustion detectors
   C. Beam detectors
   D. Rate-of-calibration detectors

_____ 10. The activation of a single smoke detector plus the activation of a second smoke detector is characteristic of a
   A. double-pull alarm system.
   B. verification system.
   C. cross-zoned system.
   D. nuisance system.

_____ 11. The temporal-3 pattern is a(n)
   A. verification system.
   B. standardized audio pattern.
   C. alarm activation system.
   D. photoelectric detector system.

_____ 12. Which type of sprinkler head has a glass bulb filled with glycerin to hold the cap in place?
   A. Fusible-link sprinkler head
   B. Chemical-pellet sprinkler head
   C. Deluge head
   D. Frangible-bulb sprinkler head

_____ 13. Which type of sprinkler head is triggered by the melting of a metal alloy at a specific temperature?
   A. Fusible-link sprinkler head
   B. Frangible-bulb sprinkler head
   C. ESFR sprinkler head
   D. Pendant sprinkler head

_____ 14. Which type of valve is mounted on the outside wall of a building?
   A. PIV
   B. OS&Y valve
   C. WPIV
   D. Support control valve

_____ 15. To allow the fire department's engine to pump water into the sprinkler system, each sprinkler system should also have a
   A. primary feeder.
   B. secondary feeder.
   C. pumper outlet.
   D. fire department connection.

_____ 16. Most modern sprinkler systems are connected to the building's fire alarm system to alert the occupants by a pressure switch or a
   A. tamper switch.
   B. flow switch.
   C. clapper switch.
   D. trigger switch.

_____ 17. In most cases, the entire sprinkler system can be shut down by
   A. closing the main control valve.
   B. using the remote annunciator panel.
   C. deactivating the alarm.
   D. using a sprinkler wedge.

_____ 18. A network of pipes and outlets for fire hoses built into a structure to provide water for firefighting purposes is called a
   A. residential pipe system.
   B. grid system.
   C. standpipe system.
   D. closed flow system.

_____ 19. The network of pipes and outlets for fire hoses built into a structure and designed for use by the building occupants is designated as belonging to
   A. Class I.
   B. Class II.
   C. Class III.
   D. Class IV.

_____ 20. Which of the following is the name for a special extinguishing system that operates by discharging a gaseous agent into the atmosphere at a concentration that will extinguish a fire?
   A. Wet chemical extinguishing system
   B. Clean agent extinguishing system
   C. Dry chemical extinguishing system
   D. Carbon dioxide extinguishing system

Vocabulary

Define the following terms using the space provided.

1. Verification system:
   _____
   _____
   _____

2. Zoned system:
   _____
   _____
   _____

3. Deluge sprinkler system:
   _____
   _____
   _____

4. Fire department connection:
   _____
   _____
   _____

5. Outside stem and yoke valve:
   _____
   _____
   _____

6. Post indicator valve:
   _____
   _____
   _____

Fill-in
Read each item carefully, and then complete the statement by filling in the missing word(s).

1. _____ detectors are triggered by the visible products of combustion.
2. The fire alarm control panel serves as the _____ of the fire alarm system.
3. The fire alarm control panel should monitor the entire alarm system to detect any _____.
4. A(n) _____ detector is a type of photoelectric smoke detector used to protect large open areas.
5. _____ detectors will be activated if the temperature of the surrounding air rises more than a set amount in a given time period.
6. A(n) _____ alarm occurs when an alarm system is activated by a condition that is not really an emergency.
7. A(n) _____ system sends a signal directly to the fire department or to a monitoring location via telephone or radio signal.
8. _____ sprinkler heads are designed for horizontal mounting and projecting out from a wall.
9. The network of pipes that delivers water through the sprinkler system is the sprinkler _____.
10. The Class _____ standpipe is designed for use by fire department personnel only.

True/False

If you believe the statement to be more true than false, write the letter "T" in the space provided.
If you believe the statement to be more false than true, write the letter "F."

1. _____ Smoke alarms can be either hard-wired to a 110-volt electrical system or battery operated.
2. _____ A photoelectric detector has a small amount of radioactive material inside a chamber.
3. _____ An activated alarm sounds throughout a building.
4. _____ Heat detectors provide reliable life-safety protection.
5. _____ Bimetallic strips are made to respond to a rapid increase in temperature.
6. _____ Nuisance alarms are caused by individuals who deliberately activate fire alarms when there is no fire.
7. _____ Fire alarm systems can control doors and elevators.
8. _____ A central station is operated by the fire department.
9. _____ An activated sprinkler head in an automatic sprinkler system triggers the water-motor gong.
10. _____ All sprinkler systems should be equipped with a method for sounding an alarm whenever there is water flowing in the pipes.

Short Answer

Complete this section with short written answers using the space provided.

1. List the three basic components of a fire alarm system.

2. List the five fire department notification systems.

3. Identify the three fire suppression systems.

4. Identify the four categories of sprinkler systems and provide a brief description of each category.

5. Identify the three categories of standpipes, with a description of their intended use.

Word Fun

The following crossword puzzle is an activity provided to reinforce correct spelling and understanding of terminology associated with firefighting. Use the clues provided to complete the puzzle.

Clues

Across

1. A fire alarm signal caused by a device reacting properly to a condition that is not a true fire emergency.
9. An alarm system that provides no information at the alarm control panel indicating where the activated alarm is located.
10. An electrical switch that is activated by water moving through a pipe on a sprinkler system.
11. A device that detects smoke and sends a signal to a fire alarm control panel.

Down

2. A fire alarm system that sounds an alarm only in the building where it was activated. No signal is sent out of the building.
3. A switch on a sprinkler valve that transmits a signal to the fire alarm control panel if the normal position of the valve is changed.
4. A liquefied gas extinguishing agent that puts out the fire by chemically interrupting the combustion reaction between fuel and oxygen.
5. A device that accelerates the removal of the air from a dry-pipe or preaction sprinkler system.
6. A sensing device that detects the radiant energy emitted by a flame.
7. A fire alarm system design that divides a building or facility into zones and has audible notification devices that can be used to identify the area where an alarm originated.
8. A sprinkler head that has no release mechanism; the orifice is always open.

Fire Alarms

The following real case scenarios will give you an opportunity to explore the concerns associated with fire protection, suppression, and detection systems. Read each scenario, and then answer each question in detail.

1. It is Sunday morning and you have completed checking the apparatus. Your Lieutenant calls the crew together for a practice drill. She tells you that the drill will familiarize you with the fire suppression system at the new city courthouse. The fire suppression system at the courthouse has a supplied wet sprinkler system with fire department connections. Why is it important for fire fighters to have a basic understanding of fire suppression systems?

   _____
   _____
   _____
   _____
   _____
   _____

2. It is 10:00 on a Wednesday morning when your engine is dispatched to deal with an alarm activation at an office building. You and the crew check the annunciator panel. The annunciator indicates that the carbon dioxide extinguishing system has been activated in the computer server room. One of the managers of the company meets you at the door and reports that light smoke was seen in the server room. What are the special hazards of a carbon dioxide extinguishing system?

   _____
   _____
   _____
   _____
   _____
   _____

Fire Cause Determination

Workbook Activities

The following activities have been designed to help you. Your instructor may require you to complete some or all of these activities as a regular part of your fire fighter training program. You are encouraged to complete any activity that your instructor does not assign as a way to enhance your learning in the classroom.

Chapter Review

The following exercises provide an opportunity to refresh your knowledge of this chapter.

Matching

Match each of the terms in the left column to the appropriate definition in the right column.

_____ 1. Pyromaniac

_____ 2. Arsonist

_____ 3. Circumstantial evidence

_____ 4. Physical evidence

_____ 5. Trace evidence

_____ 6. Incendiary fires

_____ 7. Ignitable liquid

_____ 8. Direct evidence

_____ 9. Contaminated

_____ 10. Undetermined

A. The means by which alleged facts are proven by deduction or inference from other facts that were observed first-hand

B. A pathological fire-setter

C. Intentionally set fires

D. Evidence of a minute quantity that is conveyed from one place to another

E. A person who deliberately sets a fire to destroy property with criminal intent

F. Items that can be examined in a laboratory and presented in court to prove or demonstrate a point

G. Fire cause classification that includes fires for which the cause has not or cannot be proven

H. A term used to describe evidence that may have been altered from its original state

I. Evidence that is reported first-hand

J. Classification for liquid fuels including both flammable and combustible liquids

CHAPTER 37

Multiple Choice
Read each item carefully, and then select the best response.

_____ 1. In most jurisdictions, the _____ determines the cause of a fire.
   A. law enforcement agency
   B. chief of the fire department
   C. property owner
   D. fire investigator

_____ 2. An arson investigation must determine not only who was responsible for starting the fire, but also
   A. why the person started the fire.
   B. when the person started the fire.
   C. the property damage caused by the fire.
   D. the cause and origin of the fire.

_____ 3. The cause of a fire can be classified as either incendiary or
   A. malicious intent.
   B. criminal intent.
   C. undetermined.
   D. accidental.

_____ 4. Most fires, fire deaths, and injuries occur in
   A. residential occupancies.
   B. industrial settings.
   C. small businesses.
   D. gas or fueling stations.

_____ 5. One of the first steps in a fire investigation is identifying the
   A. fuel supply.
   B. accelerants.
   C. point of origin.
   D. evidence.

_____ 6. A charred V-pattern on a wall indicates that fire spread
   A. along the floor before reaching the wall.
   B. up and out from an unknown material at the base of the V.
   C. along the ceiling before reaching the wall.
   D. slowly.

_____ 7. An inverted V-pattern on a wall indicates that
   A. a flammable liquid was used to start the fire.
   B. there was a wide point of origin for the fire.
   C. the fire began on the ceiling.
   D. there was a flashover.

_____ 8. The process of carefully looking for evidence within the debris is referred to as
   A. layering.
   B. overhaul.
   C. evidence recovery.
   D. digging out.

_____ 9. Which type of evidence can be used to prove a theory, based on facts that were observed first-hand?
   A. Demonstrative evidence
   B. Trace evidence
   C. Circumstantial evidence
   D. Physical evidence

_____ 10. "Chain of custody" is a legal term that describes the process of maintaining continuous possession and control of evidence from
   A. the time it is discovered until it is presented in court.
   B. investigator to incident commander.
   C. investigator to law enforcement agency.
   D. discovery to isolation.

_____ 11. If a fire intensifies in a short period of time, it may indicate
   A. poor dispatch information.
   B. the use of an accelerant.
   C. extreme weather conditions.
   D. multiple points of origin.

_____ 12. An accelerant may have been used if the fire _____ when water is applied.
   A. spreads
   B. grows
   C. is extinguished
   D. rekindles

_____ 13. What might charring on the underside of a low horizontal surface, such as a tabletop, indicate?
   A. The fire's point of origin was on top of the surface.
   B. The fire was accidental.
   C. There was a pool of a flammable liquid.
   D. The fire started from a cigarette butt.

_____ 14. The fire department's authority over an incident ends when
   A. the property is formally released to the property owner.
   B. the property is under the investigator's supervision.
   C. any criminal or malicious intent regarding the fire's origin is ruled out.
   D. the property is secured and no hazards to public safety exist.

_____ 15. Which of the following terms is used to describe a device or mechanism that is used to start a fire?
   A. Arsonist device
   B. Incendiary device
   C. Accelerant
   D. Trailer

_____ 16. A pyromaniac is a pathological fire-setter who is often a(n)
   A. juvenile male.
   B. adolescent female.
   C. adult female.
   D. adult male.

_____ 17. Preadolescent male fire-starters are most often motivated to start fires by
   A. a need for attention.
   B. curiosity.
   C. revenge.
   D. excitement.

_____ 18. An arsonist who sets three or more fires at separate locations with no emotional cooling-off period between fires is called a
   A. spree arsonist.
   B. serial arsonist.
   C. mass arsonist.
   D. motivated arsonist.

_____ 19. An arsonist who sets three or more fires, with a cooling-off period between fires, is called a
   A. spree arsonist.
   B. serial arsonist.
   C. mass arsonist.
   D. motivated arsonist.

_____ 20. An arsonist who sets three or more fires at the same site or location during a limited period of time is called a
   A. spree arsonist.
   B. serial arsonist.
   C. mass arsonist.
   D. motivated arsonist.

Vocabulary

Define the following terms using the space provided.

1. Trailers:

   _____
   _____
   _____

2. Competent ignition source:

   _____
   _____
   _____

3. Depth of char:

   _____
   _____
   _____

4. Chain of custody:

   _____
   _____
   _____

Fill-in
Read each item carefully, and then complete the statement by filling in the missing word(s).

1. Insurance companies often investigate fires to determine the _____ of a claim.
2. A fire investigation can provide many _____, even if the specific cause and origin are never determined.
3. At the point of origin, an ignition source comes into contact with a(n) _____.
4. The fire investigation process usually begins with an examination of the building's _____.
5. Smoke residue and _____ patterns can be helpful in identifying the point of origin.
6. Anything that can be used to validate a theory is _____ evidence.
7. Evidence should not be _____ or altered from its original state in any way.
8. Until the fire is under control, fire fighters must concentrate on fighting the fire and _____ investigating the cause.
9. What a fire fighter _____ during an incident could be significant in an investigation of an incident.
10. The _____ of any victims found in the building should be noted.
11. An arsonist may place _____ to hinder the efforts of fire fighters.
12. _____ removed from any victim should be preserved as evidence.

True/False

If you believe the statement to be more true than false, write the letter "T" in the space provided.
If you believe the statement to be more false than true, write the letter "F."

1. _____ A cause-and-origin investigation determines where, why, and how a fire originated.
2. _____ A fire can be caused by an act or by an omission.
3. _____ The most important reason for investigating accidental fires is to identify the causes.
4. _____ Charring is usually deepest on the edges of the object.
5. _____ Evidence is most often found during the size-up phase of a fire.
6. _____ To avoid contaminating evidence, fire investigators always wash their tools between taking samples.
7. _____ Only one person should be responsible for collecting and taking custody of all evidence at a fire scene.
8. _____ For most fire fighters, "The fire is under investigation" is the best reply to any questions concerning the cause of the fire.
9. _____ The appearance and behavior of people at the scene of a fire can provide valuable clues.
10. _____ Arsonists often open the shades and windows of structures they burn.
11. _____ The color of the smoke often indicates what is burning.
12. _____ To assist with evidence collection, burned materials should be thrown into a pile.

Short Answer

Complete this section with short written answers using the space provided.

1. List the three steps to take if a fire investigator is not available and the premises needs to be maintained under the control of the fire department until the investigation can take place.

   _____
   _____
   _____

2. Describe the characteristics of fires set by pyromaniacs.

   _____
   _____
   _____

3. List the six common arson motives listed in NFPA 921, *Guide to Fire and Explosion Investigations*.

   _____
   _____
   _____
   _____
   _____
   _____

Word Fun

The following crossword puzzle is an activity provided to reinforce correct spelling and understanding of terminology associated with firefighting. Use the clues provided to complete the puzzle.

CLUES

Across
1. Evidence that is reported first-hand.
3. A pathological fire-setter.
4. Materials used to initiate or increase the spread of fire.
5. Items that can be examined in a laboratory, and presented in court to prove or demonstrate a point.
7. The exact location where a heat source and a fuel come in contact with each other and a fire begins.
11. Fire cause classification that includes fires with a proven cause that does not involve a deliberate human act.
12. The malicious burning of one's own or another's property with a criminal intent.
13. A series of fires set by the same offender, with a cooling-off period between fires.
14. Cause classification that includes fires for which the cause has not or cannot be proven.

Down
2. A device or mechanism used to start a fire or explosion.
6. A term used to describe evidence that may have been altered from its original state.
8. Combustible material used to spread fire from one point or area to other points or areas.
9. Involves an offender who sets three or more fires at the same site or location during a limited period of time.
10. A person who deliberately sets a fire to destroy property with criminal intent.

Fundamentals of Fire Fighter Skills

Fire Alarms

The following real case scenarios will give you an opportunity to explore the concerns associated with fire cause determination. Read each scenario, and then answer each question in detail.

1. You are on the scene of a suspected arson at a waterfront restaurant. You are assigned to work with one of the arson investigators in your department to help collect and process possible evidence. How do you proceed?

    _____
    _____
    _____
    _____
    _____
    _____

2. Your engine has responded to a structure fire in a church. A serial arsonist has been setting fire to churches in your community. This fire has some of the same characteristics as the other church fires. You and your partner confined the fire to the basement of the church. Now you and your partner have been assigned to help the fire investigator overhaul the fire scene. How should you proceed?

    _____
    _____
    _____
    _____
    _____

Answer Key

Chapter 1: The History and Orientation of the Fire Service

Matching

1. G (page 17)
2. E (page 17)
3. C (page 18)
4. H (page 18)
5. F (page 18)
6. D (page 19)
7. I (page 20)
8. A (page 18)
9. B (page 18)
10. J (page 20)

Multiple Choice

1. B (page 10)
2. C (page 16)
3. A (page 12)
4. C (page 14)
5. D (page 14)
6. A (page 15)
7. C (page 9)
8. A (page 9)
9. A (page 8)
10. B (page 8)
11. D (page 6)
12. C (page 6)
13. A (page 6)
14. D (page 6)
15. C (page 8)
16. C (page 9)
17. C (page 9)
18. C (page 14)
19. D (page 16)
20. A (page 16)

Vocabulary

1. **Incident commander (IC):** The incident commander is the individual who is responsible for the management of all incident operations. This position focuses on the overall strategy of the incident and is often assumed by the battalion/district chief. (page 4)
2. **Training officer:** The training officer is responsible for updating the training of current fire fighters and for training new fire fighters. He or she must be aware of the most current techniques of firefighting and EMS. (page 4)
3. **Company officer:** Usually a lieutenant or captain in charge of a team of fire fighters, both on the scene and at the station. The company officer is responsible for firefighting strategy, safety of personnel, and the overall activities of the personnel, and the overall activities of the fire fighters or their apparatus. (page 4)
4. **Safety officer:** The safety officer watches the overall operation for unsafe practices. He or she has the authority to halt any firefighting activity. (page 4)
5. **Emergency Medical Technician–Paramedic:** An EMT–Paramedic has completed the highest level of training in EMS. These personnel have extensive training in advanced life support, including IV therapy, administering drugs, cardiac monitoring, inserting advanced airways, manual defibrillation, and other advanced assessment and treatment skills. (page 5)

Fill-in

1. the city banned thatched roofs, wood chimneys (page 10)
2. Benjamin Franklin (page 10)
3. The Romans (page 14)
4. fire hydrants (page 14)
5. public call boxes (page 14)
6. Standard operating procedures (SOPs) (page 6)
7. geographic (page 8)
8. Emergency Medical Services (page 5)
9. incident command system (page 5)

Answer Key

True/False

1. F (page 10)
2. T (page 12)
3. T (page 16)
4. F (page 9)
5. T (page 8)
6. F (page 4)
7. F (page 4)
8. T (page 18)
9. T (page 10)
10. F (page 8)

Short Answer

1. (1) Unity of command, (2) Span of control, (3) Division of labor, (4) Discipline (page 9)
2. *Engine company*
   - An engine company is responsible for securing a water source, deploying handlines, conducting search-and-rescue operations, and putting water on the fire.

   Truck company
   - A truck company specializes in forcible entry, ventilation, roof operations, search-and-rescue operations above the fire, and deployment of ground ladders. They are also called ladder companies.

   Rescue company
   - A rescue company usually is responsible for rescuing victims from fires, confined spaces, trenches, and high-angle situations.

   Brush company
   - A brush company is dispatched to woodland and brush fires that larger engines cannot reach.

   Hazardous materials company
   - A hazardous materials company responds to and controls scenes involving spilled or leaking hazardous materials.

   Emergency Medical Services (EMS) company
   - An EMS company responds to and assists in transporting medical and trauma patients to medical facilities for further treatment. EMS personnel often have medications, defibrillators, and other equipment that can stabilize a critical patient during transport. (page 6)

3. *Fire fighter*
   - The fire fighter may be assigned any task from placing hose lines to extinguishing fires. Generally, the fire fighter is not responsible for any command functions and does not supervise other personnel, except on a temporary basis when promoted to an acting officer.

   Driver/operator
   - Often called an engineer, the driver is responsible for getting the fire apparatus to the scene safely, setting up, and running the pump once it arrives on the scene.

   Company officer
   - This is usually a lieutenant or captain in charge of a team of fire fighters. This person is in charge of the company both on scene and at the station. The company officer is responsible for initial firefighting strategy, personnel safety, and the overall activities of the fire fighters on their apparatus. Once command is established, the company officer focuses on tactics.

   Safety officer
   - The safety officer responds to scenes and watches the overall operation for unsafe practices. He or she has the authority to stop any firefighting activity until it can be done safely and correctly.

   Training officer
   - The training officer is responsible for updating the training of current employees and for training new fire fighters.

   Incident commander
   - The incident commander is the individual responsible for the management of all incident operations.

   Fire marshal/fire inspector/fire investigator
   - Fire marshals inspect businesses and enforce public safety laws and fire codes. They may respond to fire scenes to help investigate the cause of a fire.

   Fire and life safety education specialist
   - This person educates the public about fire safety and injury prevention, and presents juvenile fire safety programs.

9-1-1 dispatcher/telecommunicator
- From the communications center, the dispatcher takes the calls from the public, sends appropriate units to the scene, assists callers with emergency medical information, and assists the incident commander with needed resources.

Fire apparatus maintenance personnel
- Apparatus mechanics repair, service, and keep fire and EMS vehicles ready to respond to emergencies.

Fire police
- Fire police are usually fire fighters who control traffic and secure the scene from public access. Many fire police are sworn peace officers as well as fire fighters.

Information management
- "Info Techs" are fire fighters or civilians who take care of a department's computer and networking systems.

Public information officer
- The public information officer serves as a liaison between the incident commander and the news media.

Fire protection engineer
- The fire protection engineer usually has an engineering degree, reviews plans, and works with building owners to ensure that their fire suppression and detection systems will meet code and function as needed.

Aircraft/crash rescue fire fighter
- Aircraft rescue fire fighters are based on military and civilian airports and receive specialized training in aircraft fires, extrication, and extinguishing agents.

Hazardous materials technician
- "Hazmat" technicians have training and certification in chemical identification, leak control, decontamination, and clean-up procedures.

Technical rescue technician
- A "tech rescue" technician is trained in special rescue techniques for incidents involving structural collapse, trench rescue, swiftwater rescue, confined-space rescue, high-angle rescue, and other unusual situations.

SCUBA dive rescue technician
- Many fire departments, especially those around waterways, lakes, or an ocean, use SCUBA technicians who are trained in rescue, recovery, and search procedures in both water and under-ice situations.

Emergency Medical Services (EMS) personnel
- EMS personnel administer prehospital care to people who are sick and injured. Prehospital calls account for the majority of responses in many departments, so fire fighters are often cross-trained with EMS personnel.

Emergency Medical Technician–Basic (EMT-Basic)
- Most EMS providers are EMT-Basics. They have training in basic emergency care skills, including oxygen therapy, bleeding control, cardiopulmonary resuscitation (CPR), automated external defibrillation, use of basic airway devices, and assisting patients with certain medications.

Emergency Medical Technician–Intermediate (EMT-Intermediate)
- EMT-Intermediates can perform more procedures than EMT-Basics, but they are not yet EMT-Paramedics. They have training in specific aspects of advanced life support, such as intravenous (IV) therapy, interpretation of cardiac rhythms, defibrillation, and airway intubation.

Emergency Medical Technician–Paramedic (EMT-Paramedic)
- An EMT-Paramedic is the highest level of training in EMS. EMT-Paramedics have extensive training in advanced life support, including IV therapy, administering drugs, cardiac monitoring, inserting advanced airways (endotracheal tubes), manual (rather than automatic) defibrillation, and other advanced assessment and treatment skills. (page 4)

4. Regulations are developed by various government or government-authorized organizations to implement a law that has been passed by a government body.

Policies are developed to provide definite guidelines for present and future actions. Fire department policies outline what is expected in stated conditions. Policies often require personnel to make judgments and to determine the best course of action within the stated policy.

Standard operating procedures (SOPs) provide specific information on the actions that should be taken to accomplish a certain task. SOPs provide a uniform way to deal with emergency situations. They are vital because they enable everyone in the department to function properly and know what is expected for each task. (page 6)

Answer Key

Word Fun

```
 1F I 2R E 3P L U G                    4F
    I    E    I          5C A P T A I N
    R    S    E      6F           R
    E    C    U      I      7T    E    8D
    P    U    T      R      R     F    O
    O    E    E   9R E G U L A T I O N 10S
    L    C    N      M      I     G    A
    I    O  11B A T T A L I O N C H I E F
    C    M    N      R      I     T    E
    E    P    T      K    12B A N K E D T
         A                  G     R    Y
         N            13F I R E H O O K
         Y                  F   14D O F F
                            F         F
                      15D I S C I P L I N E
                            C         I
                       16P O L I C I E S
                            R         E
                                      R
```

Fire Alarms

1. Every member of the fire service will interact with the public. People may visit the fire station, requesting a tour or asking questions on specific fire safety issues. Fire fighters should be prepared to assist these visitors and use this opportunity to provide them with additional fire safety information. Use every contact with the public to deliver positive public relations and an educational message. Inform and encourage the use of wearing a helmet and explain the benefits.

2. During this course of study, you will need to practice and work hard. Do your best. Five guidelines will help to keep you on target to become a proud and accomplished fire fighter:

    (1) *Be safe*. Safety should always be uppermost in your mind. Keep yourself safe. Keep your teammates safe. Keep the public you serve safe.

    (2) *Follow orders*. Your supervisors have more training and experience than you do. If you can be counted on to follow orders, you will become a dependable member of the team.

    (3) *Work as a team*. Fighting fires requires the coordinated efforts of each department member. Teamwork is essential to success.

    (4) *Think!* Lives will depend on the choices you make. Put your brain in gear. Think about what you are studying.

    (5) *Follow the Golden Rule*. Treat each person, patient, or victim as an important person or as a member of your family. Everyone is an important person or family member to someone and deserves your best efforts.

Chapter 2: Fire Fighter Safety

Matching

1. G (pages 67)
2. F (pages 52)
3. E (pages 40)
4. A (pages 43)
5. B (pages 33)
6. D (pages 42)
7. H (pages 67)
8. C (pages 37)
9. I (pages 26)
10. J (pages 49)

Multiple Choice

1. B (pages 31)
2. D (pages 42)
3. C (page 52)
4. A (page 48)
5. A (page 40)
6. A (page 37)
7. C (page 49)
8. D (page 26)
9. A (page 50)
10. B (page 64)
11. B (page 52)
12. A (page 49)
13. B (page 37)
14. C (page 52)
15. C (page 25)
16. A (page 48)
17. D (page 51)
18. B (page 40)
19. C (page 42)

Vocabulary

1. **Employee assistance program (EAP):** Program adopted by many departments for fire fighters to receive confidential help with problems such as substance abuse, stress, depression, or burn-out that can affect their work performance. (page 28)
2. **National Institute for Occupational Safety and Health (NIOSH):** A U.S. Federal agency responsible for research and development on occupational safety and health issues. (page 50)
3. **Personnel accountability system:** A method of tracking the identity, assignment, and location of fire fighters operating at an incident scene. (page 32)
4. **Safety officer:** A designated individual who oversees safety practices at an emergency scene and during training. Safety officers have the authority to stop any activity that is deemed unsafe. (page 27)
5. **Standard operating procedures (SOPs):** Written rules, policies, regulations, and procedures enforced to structure the normal operations of most fire departments. (page 26)
6. **Buddy system:** A system in which two fire fighters always work as a team for safety purposes. (page 31)
7. **Immediately dangerous to life and health (IDLH):** A situation in which an atmospheric concentration of any toxic, corrosive, or asphyxiant substance poses an immediate threat to life or could cause irreversible or delayed health effects. (page 47)

Fill-in

1. incomplete combustion (page 47)
2. oxygen deficient (page 48)
3. safety (page 27)
4. freelancing (page 26)
5. supplied air respirator (page 49)
6. National Fallen Firefighters Foundation, 25 percent (page 27)
7. Employee assistance programs (page 28)
8. traffic regulations (page 30)
9. Safe driving practices (page 30)

Answer Key

True/False

1. T (page 73)
2. F (page 32)
3. F (page 52)
4. T (page 48)
5. F (page 29)
6. T (page 31)
7. T (page 28)
8. F (page 26)
9. T (page 30)
10. T (page 32)

Short Answer

1. (1) Standards and procedures; (2) personnel; (3) training; (4) equipment (page 25)
2. (1) Not easy to don; (2) heavy; (3) difficult for the body to cool itself; (4) limits mobility; (5) decreases normal sensory ability (page 42)
3. (1) Provides thermal protection; (2) repels water; (3) provides impact protection; (4) protects against cuts and abrasions; (5) furnishes padding against injury; (6) increases your visibility; (7) provides respiratory protection (page 36)
4. (1) Smoke; (2) toxic gases; (3) oxygen deficiency (page 47)
5. To provide a forum for firefighting and EMS personnel to discuss the anxieties, stress, and emotions triggered by a difficult call (page 34)

Word Fun

Fire Alarms

1. PPE that has been badly soiled by exposure to smoke, other products of combustion, melted tar, petroleum products, or other contaminants needs to be cleaned as soon as possible. Cleaning instructions are listed on the tag attached to the garment. Follow the manufacturer's cleaning instructions. Failure to do so may reduce the effectiveness of the garment and create an unsafe situation for the wearer. Some fire departments have special washing machines that are approved for cleaning PPE. Other departments contract with an outside firm to clean and repair PPE. In either case, the manufacturer's instructions for cleaning and maintaining the garment must be followed.
2. Turn on the red-colored emergency bypass valve on your SCBA regulator. This releases a constant flow of breathing air into the face piece. The emergency bypass mode uses more air, but it enables fire fighters to exit a hazardous environment if the regulator stops operating. A fire fighter who must use the emergency bypass mode must leave the hazardous area IMMEDIATELY.

Skill Drills

Skill Drill 2-1: Donning Personal Protective Clothing

1. Place your equipment in a <u>**logical**</u> order for donning.
2. Place your protective <u>**hood**</u> over your head and down around your neck.
3. Put on boots and pull up bunker pants. Place the <u>**suspenders**</u> over your shoulders and secure the front of the pants.
4. Put on your <u>**turnout**</u> coat and close the front of the coat.
5. Place the <u>**helmet**</u> on your head and adjust the chin strap securely. Turn up your coat collar and secure it in front.
6. Put on your <u>**gloves**</u>.
7. Have your partner <u>**check**</u> your clothing.

Skill Drill 2-2: Doffing Personal Protective Clothing

1. Remove your gloves.

2. Open the collar of your turnout coat.

3. Release the helmet chin strap and remove your helmet.

4. Remove your turnout coat.

5. Remove your protective hood.

6. Remove your bunker pants and boots.

Skill Drill 2-3: Donning SCBA from an Apparatus Seat Mount

1. Don full PPE ensemble prior to **mounting** fire apparatus. Safely mount the apparatus and sit in the seat, placing your arms through the SCBA shoulder straps.
2. Fasten your seat belt. Partially tighten the **shoulder** straps. When the apparatus stops, release the seat belt and release the SCBA from its brackets. Exit the apparatus.
3. Attach the **waist** belt and cinch down.
4. **Adjust** shoulder straps until they are tight.
5. Open the main **cylinder** valve.
6. Loosen or remove your helmet and pull the **hood** back. Don the face piece and check for leaks. Replace the protective hood and helmet, and secure the chin strap.
7. If necessary, connect the **regulator** to the face piece.
8. Activate the air flow and **PASS** alarm.

Skill Drill 2-5: Donning SCBA Using the Over-the-Head Method

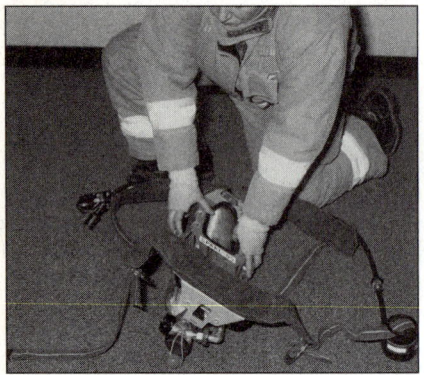

1. Open the case and lay out the SCBA with the cylinder valve away from you and the shoulder straps out to the sides.

2. Fully open the main cylinder valve.

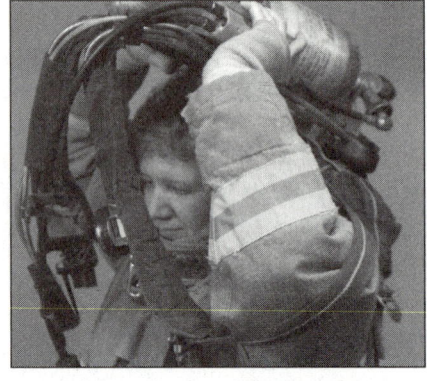

3. Bend down and grasp the SCBA backplate with both hands. Using your legs, lift the SCBA over your head. Rotate the SCBA 180 degrees, so the waist straps are pointed to the ground.

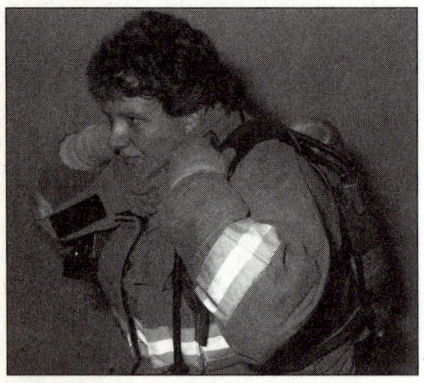

4. Slide the SCBA down your back while your arms slide into the shoulder straps. Tighten the shoulder straps and secure the waist belt.

5. Remove your helmet and pull the hood back. Don your face piece and check for an adequate seal. Pull your protective hood into position, replace your helmet, and secure the chin strap.

6. If necessary, connect the regulator to the face piece. Activate the air flow and PASS alarm.

Skill Drill 2-6: Donning SCBA Using the Coat Method

1. Open the case and lay out the SCBA with the cylinder valve away from you and the shoulder straps out to the sides. Fully open the main cylinder valve. Place your **dominant** hand on the opposite shoulder strap.
2. Lift the SCBA and **swing** it over your dominant shoulder.
3. Slide your other hand between the SCBA cylinder and the **corresponding** shoulder strap.
4. Tighten the **shoulder** straps.
5. Attach the **waist belt** and adjust tightness.
6. Remove your helmet and pull your hood back. Don the **face piece** and check for an adequate seal.
7. Pull the hood into position, replace the helmet, and secure the chin strap. If necessary, connect the regulator to the face piece. Activate the **air flow** and PASS alarm.

Skill Drill 2-8: Donning a Face Piece

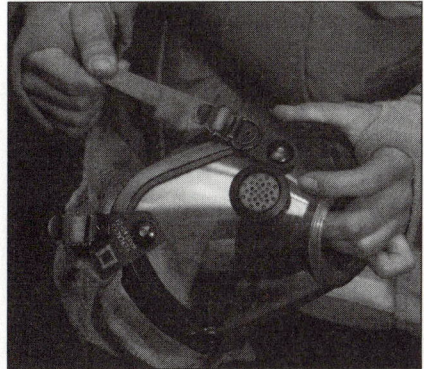

1. Fully extend the straps on the face piece.

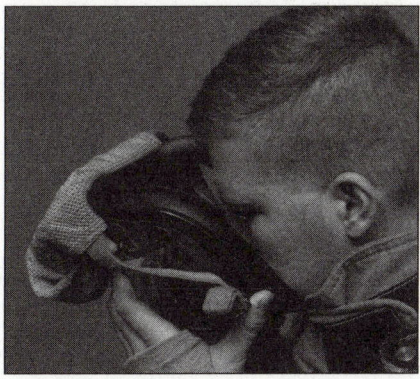

2. Place your chin in the chin pocket.

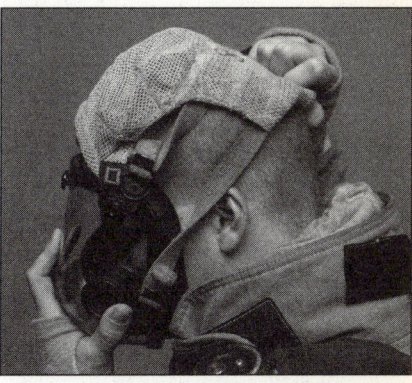

3. Fit the face piece to your face, bringing the straps or webbing over your head.

4. Tighten the lowest two straps.

5. If there are more straps, tighten the top straps last.

6. Check for a proper seal.

(continues)

Skill Drill 2-8: Donning a Face Piece (continued)

7. Pull your protective hood so it covers all bare skin. Don your helmet and secure the chin strap.

8. Install the regulator on your face piece or attach the low-pressure air supply hose to the regulator.

Skill Drill 2-13: Replacing an SCBA Cylinder

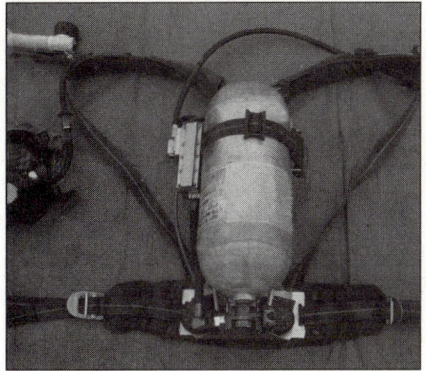

1. Place the SCBA on the floor or a bench.

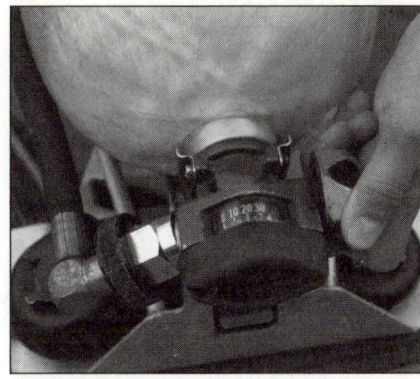

2. Turn off the cylinder valve.

3. Open the bypass valve to bleed off pressure.

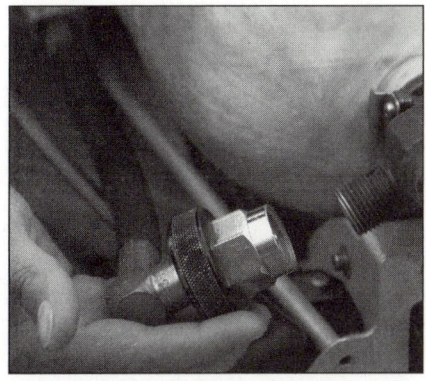

4. Disconnect the high-pressure supply hose.

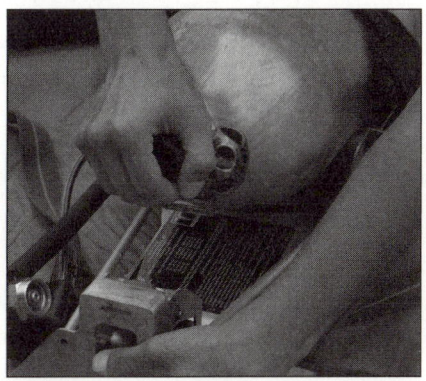

5. Release the cylinder from the backpack.

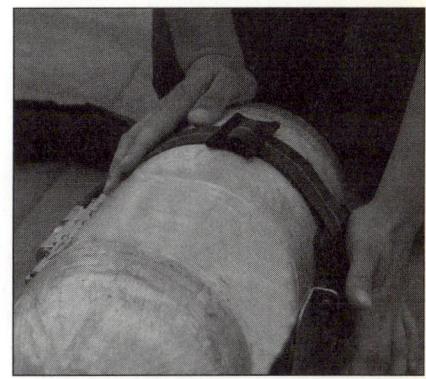

6. Slide a full cylinder into the backpack. Align the outlet to the supply hose. Lock the cylinder in place.

7. Check that the "O" ring is present and in good shape.

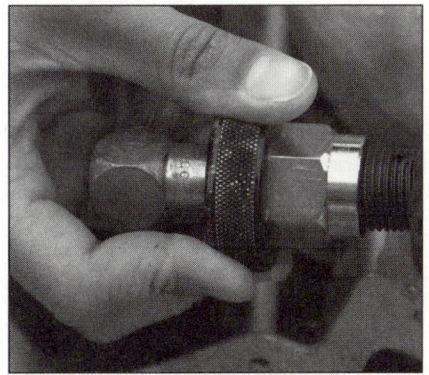

8. Connect the high-pressure hose to the air cylinder.

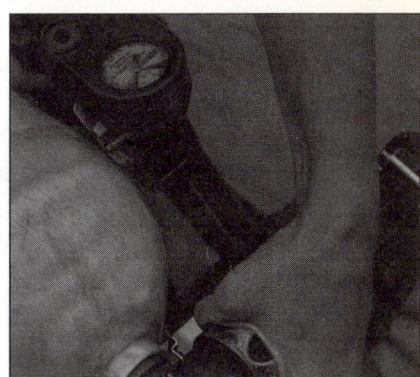

9. Open the cylinder valve. Check the gauge reading.

Chapter 3: Fire Service Communications

Matching

1. D (page 90)
2. E (page 92)
3. B (page 81)
4. A (page 96)
5. C (page 81)
6. F (page 96)
7. H (page 91)
8. I (page 92)
9. J (page 91)
10. G (page 85)

Multiple Choice

1. B (page 79)
2. C (page 81)
3. C (page 83)
4. A (page 83)
5. D (page 83)
6. B (page 83)
7. A (page 83)
8. A (page 86)
9. D (page 84)
10. D (page 86)
11. B (page 87)
12. D (page 90)
13. A (page 93)
14. B (page 93)
15. C (page 94)
16. C (page 96)
17. D (page 98)
18. C (page 94)
19. D (page 94)

Vocabulary

1. **TDD/TTY/text phone:** Special devices that display text rather than transmitting audio. (page 84)
2. **Ten-codes:** A system of coded messages that begin with the number 10. (page 94)
3. **Activity logging system:** A system that keeps a detailed record of every incident and activity that occurs. (page 81)
4. **Automatic location identification:** A system that queries a database to show the location of the telephone, the caller's name, and other details. (page 85)
5. **Evacuation signal:** A sequence of blasts or a siren that warns personnel to pull back to a safe location. (page 96)
6. **Run cards:** Documentation that lists units in the proper order of response, based on response distance or estimated response time. (page 86)
7. **Time marks:** Set intervals at which the communications center prompts the IC to report. (page 96)
8. **Computer-aided dispatch (CAD):** A system designed to assist a telecommunicator by performing specific functions more quickly and efficiently than they can be done manually. (page 81)

Fill-in

1. telecommunicator (page 80)
2. evacuation (page 96)
3. Mayday (page 96)
4. Time marks (page 96)
5. dispatch (page 99)
6. communication center (page 79)
7. simplex (page 91)
8. mobile (page 90)
9. duplex (page 91)
10. trunking (page 93)

True/False

1. T (page 82)
2. T (page 94)
3. T (page 83)
4. T (page 85)
5. T (page 84)
6. F (page 84)
7. F (page 80)
8. T (page 83)
9. T (page 94)
10. T (page 94)

Short Answer

1. (1) Dedicated 9-1-1 telephones; (2) public telephones; (3) direct-line phones to other agencies; (4) equipment to receive alarms from public or private fire alarm systems; (5) computers and/or hard-copy files and maps to locate addresses and select units to dispatch; (6) equipment for alerting and dispatching units to emergency calls; (7) two-way radio systems; (8) recording devices to record phone calls and radio traffic; (9) backup electrical generators (page 80)
2. (1) Call receipt; (2) location validation; (3) classification and prioritization; (4) unit selection; (5) dispatch (page 82)
3. (1) Receiving calls for emergency incidents and dispatching fire department units; (2) supporting the operations of fire department units delivering emergency services; (3) coordinating fire department operations with other agencies; (4) keeping track of the status of each fire department unit at all times; (5) monitoring the level of coverage and managing the deployment of available units; (6) notifying designated individuals and agencies of particular events and situations; (7) maintaining records of all emergency-related activities; (8) maintaining information required for dispatch purposes (page 82)
4. From a legal standpoint, records and reports are vital parts of the emergency process. Information must be complete, clear, and concise because these records can become admissible evidence in a court case. Improper or inadequate documentation can have long-term negative consequences. Fire reports are considered as public records under the Freedom of Information Act, so they may be viewed by an attorney, an insurance company, the news media, or the public. If a fatality occurs, incomplete or inaccurate reports may be used to prove that the fire department was negligent. The department, Fire Chief, and others may be held accountable. (page 97)

Word Fun

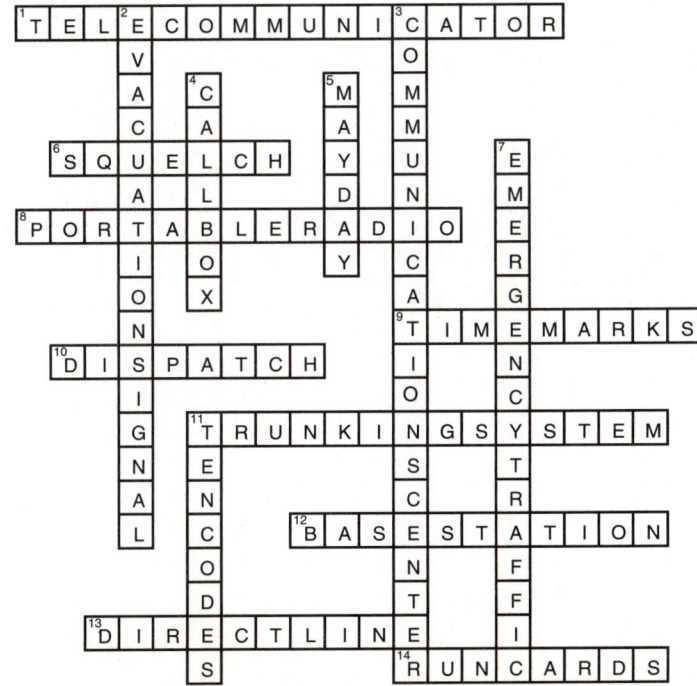

Fire Alarms

1. The most important emergency traffic is a fire fighter's call for help. Most departments use "mayday" to indicate that a fire fighter is lost, missing, or requires immediate assistance. If a mayday call is heard on the radio, all other radio traffic should stop immediately. The fire fighter making the mayday call should describe the situation, location, and help needed. Fire fighters should study and practice the procedure for responding to a mayday call.
2. Your department's SOPs should outline exactly what steps you should take in this situation, such as whether you should take the information yourself or connect the caller directly to the communications center. If you take the information from the caller, be sure to get as much information as possible, including type of incident/situation, location of the incident, cross streets or identifying landmarks, indication of scene safety, caller's name, caller's location (if different from the incident location), and caller's call-back number. If your station or your unit will be responding to the call, always advise the communications center immediately before responding.

Skill Drills

Skill Drill 3-3: Using a Radio
1. Before transmitting, determine that the **channel** is clear of any other traffic. Depress the "push-to-talk" (PTT) button and wait at least **two** seconds before speaking.
   Speak **across** the microphone at a(n) **45-degree** angle and hold the microphone 1 to 2 inches from the mouth.
   Speak clearly and keep the message brief. Do not release the **PTT** button until you have finished speaking.

Skill Drill 3-5: Operating and Answering the Fire Station Telephone and Intercom Systems
1. Determine immediately if the caller has an emergency. If there is an emergency, follow your department's **SOPs**.
2. If you take the information from the caller, focus on obtaining **vital** information. If your station or your unit will be responding to the call, advise the **communications** center immediately. Always be prepared to take accurate information or messages for emergency, nonemergency, and **personal** calls. Never leave someone on hold for a long time. Always let the caller **hang up** first.

Chapter 4: Incident Command System

Matching

1. F (page 113)
2. I (page 112)
3. H (page 112)
4. B (page 112)
5. J (page 108)
6. C (page 4)
7. A (page 108)
8. D (page 119)
9. G (page 119)
10. E (page 109)

Multiple Choice

1. B (page 108)
2. C (page 105)
3. A (page 108)
4. D (page 109)
5. C (page 107)
6. D (page 107)
7. A (page 108)
8. B (page 108)
9. C (page 108)
10. A (page 108)
11. B (page 109)
12. A (page 109)
13. D (page 109)
14. A (page 110)
15. B (page 110)
16. A (page 110)
17. B (page 111)
18. C (page 111)
19. D (page 111)
20. C (page 112)
21. A (page 112)
22. B (page 113)
23. B (page 114)
24. A (page 119)

Labeling

1. The ICS organization chart (page 109)

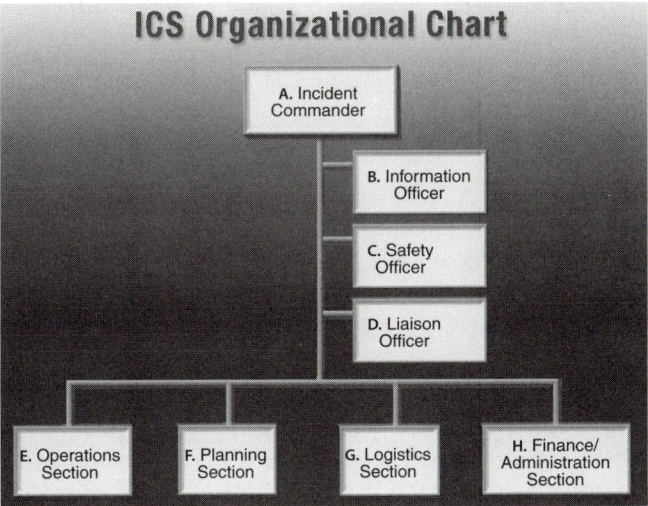

Figure 4-5

2. Major functional components of the ICS (page 110)

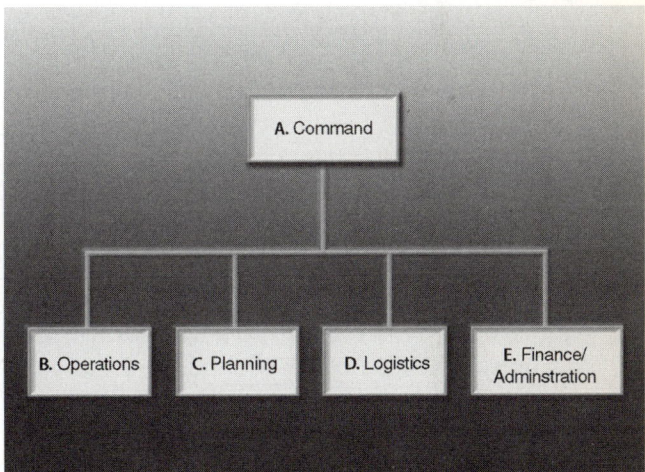

Figure 4-9

Vocabulary

1. **Command staff:** Individuals on the command staff perform functions that report directly to Command and cannot be delegated to other major sections of the organization. (page 109)
2. **Incident command system:** A command system that provides a standard approach, structure, and operational procedure to organize and manage any operation. (page 104)
3. **Incident action plan:** Oral or written plans containing general objectives reflecting the overall strategy for managing the incident. (page 106)
4. **Unified command:** A unified command brings representatives of different agencies together to work on one plan and ensures that all actions are fully coordinated. (page 106)
5. **Division:** Companies and/or crews working in the same geographic area. (page 112)
6. **National Incident Management System:** The standards and guidelines for incident management defined at a national level. (page 104)
7. **Resource management:** The use of a standard system of assigning and keeping track of the resources involved in the incident. (page 108)

Fill-in

1. approach, structure, operational (page 104)
2. communications (page 104)
3. Command, authority (page 108)
4. organization (page 115)
5. designated areas (page 108)
6. always (page 108)
7. aide (page 109)
8. general staff (page 110)
9. terminology (page 112)
10. effective span of control (page 112)

True/False

1. T (page 104)
2. T (page 105)
3. T (page 107)
4. F (page 108)
5. F (page 108)
6. T (page 106)
7. F (page 111)
8. F (page 111)
9. T (page 111)
10. T (page 112)

Short Answer

1. (1) Recognized jurisdictional authority and responsibility; (2) applicable to all risk and hazard situations; (3) applicable to day-to-day operations as well as major incidents; (4) unity of command; (5) span of control; (6) modular organization; (7) common terminology; (8) integrated communications; (9) consolidated incident action plans; (10) designated incident facilities; (11) resource management (page 105)
2. Command, Operations, Planning, Logistics, Finance/Administration (page 108)
3. Operations, Planning, Logistics, Finance/Administration (page 110)
4. Command is established at every incident, from the time that the first unit arrives until the time that the last unit leaves. The identity of Command may change, but there is always one single function who is in charge of the incident and responsible for everything that happens. SOPs may dictate who will be Command at any time.

   Each fire fighter always reports to one supervisor. A fire fighter's supervisor will usually be a company officer. The company officer directly supervises a small group of fire fighters, such as an engine company or ladder company, who work together. At an incident scene, the company officer provides instructions and must always know where each fire fighter is and what he or she is doing. If the company officer assigns two fire fighters to work together away from the rest of the company, both fire fighters remain under the supervision of the company officer. The company officer could be an acting officer (a fire fighter temporarily designated as a "fill-in" officer) or a fire fighter could be assigned to work temporarily under the supervision of a different officer.

   The company officer reports to Command. If there is only one company on the scene, the company officer is Command, at least until someone else arrives and assumes that role. At a small incident, the company officer may report directly to Command. At a large incident, there may be several layers of supervision between a company officer and Command. (page 116)
5. (1) Tactical priorities; (2) action plans; (3) hazardous or potentially hazardous conditions; (4) accomplishments; (5) assessment of effectiveness of operations; (6) current status of resources—assigned or working, available for assignment, or en route; (7) additional resource requirements (page 119)

Word Fun

Fire Alarms

1. The officer who is relinquishing command needs to give the new IC a current situation status report that includes tactical priorities, action plans, hazardous or potentially hazardous conditions, accomplishments, assessment of effectiveness of operations, current status of resources, and additional resource requirements.
2. The IMS was developed so that each person only has one supervisor to eliminate the confusion that can result when you are given orders from more than one boss. It also ensures that every member on the emergency scene is accounted for. Notify the Division D supervisor that you are assigned to Division C.

Chapter 5: Fire Behavior

Matching

1. C (page 136)
2. G (page 132)
3. H (page 129)
4. J (page 132)
5. A (page 131)
6. E (page 132)
7. I (page 129)
8. D (page 130)
9. B (page 131)
10. F (page 132)

Answer Key

Multiple Choice

1. B (page 128)
2. D (page 128)
3. A (page 130)
4. C (page 148)
5. A (page 129)
6. D (page 131)
7. D (page 135)
8. A (page 142)
9. C (page 131)
10. C (page 142)
11. C (page 129)
12. B (page 131)
13. C (page 131)
14. A (page 132)
15. A (page 132)
16. D (page 141)
17. B (page 141)
18. B (page 134)
19. D (page 134)
20. C (page 140)
21. B (page 137)
22. B (page 136)

Labeling

1. The fire tetrahedron. (page 130)

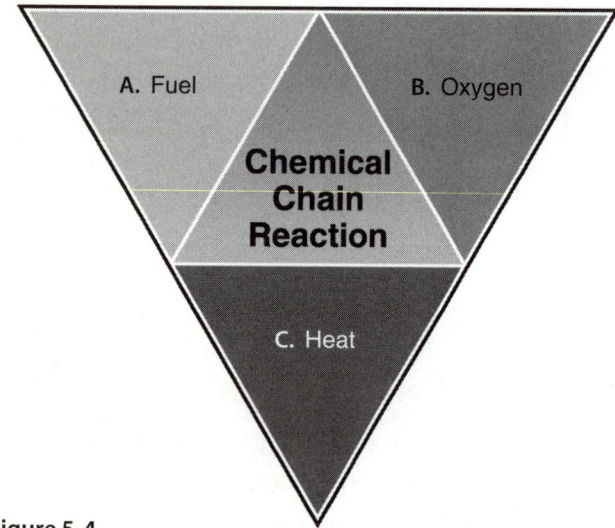

Figure 5-4

Vocabulary

1. **Lower flammable limit:** The minimum amount of gaseous fuel that must be present in a gas-air mixture for the mixture to be flammable. (page 142)
2. **BLEVE:** Boiling liquid, expanding vapor explosion; a deathly set of circumstances involving liquid and gaseous fuels. (page 142)
3. **Fire triangle:** The three basic conditions needed for a fire to occur—fuel, oxygen, and heat. (page 130)
4. **Ignition temperature:** The temperature at which a combustible material will ignite. (page 129)
5. **Thermal layering:** A property of gases such that gases rise as they are heated and form layers within a room. (page 140)
6. **Flashover:** An event in which the temperature in a room reaches a point where the combustible contents of the room ignite all at once. (page 137)
7. **Flash point:** The lowest temperature at which a liquid produces a flammable vapor. (page 141)

Fill-in

1. feet, inches, gallons, Fahrenheit, pounds per square inch (page 128)
2. meters, liters, Celsius, pascals (page 128)
3. solid, liquid, gas (page 129)
4. supply, foam (page 134)
5. exothermic (page 129)
6. monoxide (page 131)
7. volatility (page 141)
8. ignition (page 141)
9. volatile (page 136)
10. radiant heat (page 136)

True/False

1. F (page 130)
2. T (page 135)
3. T (page 129)
4. T (page 129)
5. T (page 131)
6. F (page 141)
7. T (page 138)
8. T (page 137)
9. T (page 140)
10. T (page 137)

Short Answer

1. Toxic gases; superheated, high temperatures (page 131)
2. (1) Cool the burning material; (2) exclude oxygen from the fire; (3) remove fuel from the fire; (4) interrupt the chemical reaction with a flame inhibitor (page 133)
3. (1) The fuel and air must be present at a concentration within a flammable range. (2) There must be an ignition source with enough energy to start ignition. (3) The ignition source and the fuel mixture must make contact for long enough to transfer the energy to the air-fuel mixture. (page 141)
4. (1) Ignition phase; (2) growth phase; (3) fully developed phase; (4) decay phase (page 135)
5. Temperatures in the room high enough to ignite the contents, adequate oxygen (page 137)
6. (1) Any confined fire with a large heat build-up; (2) little visible flame from the exterior of the building; (3) a "living fire" smoke puffing from the building that looks like it is breathing; (4) smoke that seems to be pressurized; (5) smoke-stained windows (an indication of a significant fire); (6) turbulent smoke; (7) ugly yellowish smoke, containing sulfur compounds (page 140)

Word Fun

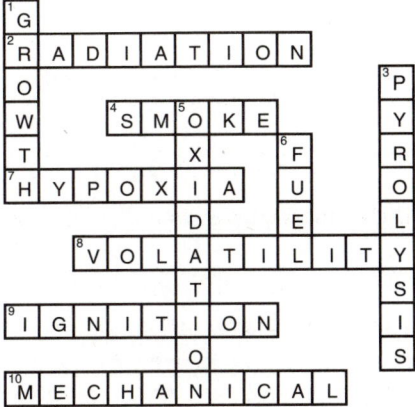

Fire Alarms

1. First, recognize that there may be a backdraft condition present. Do not open any openings. The best way to prevent a backdraft is to make a ventilation opening at a high level, so that hot gases can escape from the interior without allowing fresh air to enter.
2. Two actions are important. First, actions among fire fighters should be coordinated so that, whenever possible, the superheated gases are being vented from the fire room as you are attacking the fire. Second, it is important to use the proper fire stream for the situation. Opening a fog stream into a heated enclosed area will produce much more steam than a straight stream. Selecting a straight stream will allow more of the water to reach the seat of the fire where it can have the greatest effect. Use proper fire suppression techniques to avoid thermal imbalance.

Chapter 6: Building Construction

Matching

1. F (page 170)
2. I (page 163)
3. G (page 153)
4. C (page 155)
5. J (page 172)
6. H (page 169)
7. E (page 175)
8. A (page 176)
9. B (page 153)
10. D (page 152)

Multiple Choice

1. B (page 153)
2. A (page 153)
3. C (page 155)
4. D (page 175)
5. B (page 155)
6. B (page 156)
7. C (page 157)
8. D (page 157)
9. B (page 159)
10. D (page 157)
11. A (page 157)
12. B (page 159)
13. D (page 160)
14. C (page 163)
15. C (page 165)
16. B (page 165)
17. B (page 167)
18. A (page 168)
19. C (page 168)
20. D (page 170)
21. D (page 172)
22. A (page 172)

Vocabulary

1. **Dead load:** The weight of the building itself. (page 163)
2. **Thermoplastic materials:** Materials that melt and drip when exposed to high temperatures, even those as low as 500°F. (page 157)
3. **Interior finish:** The exposed interior surfaces of a building. (page 172)
4. **Load-bearing wall:** A wall that provides structural support by supporting a portion of the building's weight and its contents, transmitting that load down to the building's foundation. (page 169)
5. **Bowstring truss:** A truss in the shape of an archery bow, where the top chord represents the curved bow and the bottom chord represents the straight bow string. (page 168)
6. **Balloon-frame construction:** A method of construction in which the exterior walls are assembled with wood studs that run continuously from the basement to the roof. (page 162)

Fill-in

1. occupancy (page 152)
2. fire wall (page 170)
3. pyrolysis (page 156)
4. I (page 157)
5. contents of the building (page 158)
6. V (page 160)
7. Platform (page 162)
8. pitched (page 168)
9. bowstring truss roof (page 168)

True/False

1. T (page 154)
2. T (page 153)
3. T (page 157)
4. F (page 160)
5. F (page 163)
6. F (page 167)
7. T (page 168)
8. F (page 171)
9. T (page 172)
10. T (page 173)

Short Answer

1. Type I construction: Buildings with structural members made of noncombustible materials that have a specified fire resistance. Type II construction: Buildings with structural members made of noncombustible materials without fire resistance. Type III construction: Buildings with the exterior walls made of noncombustible or limited-combustible materials, but interior floors and walls made of combustible materials. Type IV construction: Buildings constructed with noncombustible or limited combustible exterior walls, and interior walls and floors made of large dimension combustible materials. Type V construction: Buildings with exterior walls, interior walls, floors, and roof structures made of wood. (pages 157–160)
2. (1) Combustibility; (2) thermal conductivity; (3) decrease in strength with increase in temperature; (4) rate of thermal expansion (page 153)
3. (1) Ignition source; (2) moisture content; (3) density; (4) preheating; (5) size and form (page 156)
4. (1) Foundations; (2) floors and ceilings; (3) roofs; (4) trusses; (5) walls; (6) doors and windows; (7) interior finishes (page 162)
5. Class A: Openings in fire walls and in walls that divide a single building into fire areas. Class B: Openings in enclosures of vertical communications through buildings and in two-hour-rated partitions providing horizontal fire separations. Class C: Openings in walls or partitions between rooms and corridors having a fire resistance rating of one hour or less. Class D: Openings in exterior walls subject to severe fire exposure from outside of the building. Class E: Openings in exterior walls subject to moderate or light fire exposure from outside of the building. (page 172)
6. A gusset is a metal plate used to tie chords and members of a truss together. They are embedded into the truss a depth of $3/8$ inch. Heating causes quick failure of the truss. (page 169)

Word Fun

Crossword answers:
- 2 Across: LAMINATEDGLASS
- 4 Across: WIREDGLASS
- 6 Across: PARTYWALLS
- 11 Across: THERMALCONDUCTIVITY
- 12 Across: OCCUPANCY
- 16 Across: FLATROOFS
- 17 Across: MASONRY
- 1 Down: FIREPARTITIONS
- 2 Down: LIVELOADS
- 3 Down: STEELJOISTS
- 5 Down: DEEDRESTRICTIONS (DEED...)
- 7 Down: FIRERESISTIVE (FIRE...)
- 8 Down: SPALLING (SPAYL...)
- 9 Down: PYROLYSIS
- 10 Down: PITSTS
- 13 Down: FIRES...
- 14 Down: RAFTERS
- 15 Down: GYPSUM

Fire Alarms

1. Stay out of the structure. Exterior walls also could collapse if a fire causes significant damage to the interior structure. Because the exterior walls, the floors, and the roof are all connected in a stable building, the collapse of the interior structure could make the freestanding masonry walls unstable and likely to collapse.

2. Nursing homes are occupied 24 hours a day by persons who will probably need assistance to evacuate. A building of Type III construction has two separate fire loads: the contents and the combustible building materials used. A nursing home will have a lot of mattresses and curtains that will burn readily. A fire involving both the contents and the structural components can quickly destroy the building. Type III construction presents several problems for fire fighters. For example, an electrical fire can begin inside the void spaces within the walls, floors, and roof assemblies and extend to the contents. The void spaces also allow a fire to extend vertically and horizontally, spreading from room to room and from floor to floor. Fire fighters will have to open the void spaces to fight the fire. An uncontrolled fire within the void spaces is likely to destroy the building.

Chapter 7: Portable Fire Extinguishers

Matching

1. E (page 208)
2. D (page 190)
3. F (page 209)
4. G (page 209)
5. A (page 192)
6. C (page 210)
7. H (page 180)
8. B (page 189)
9. J (page 210)
10. I (page 192)

Multiple Choice

1. D (page 184)
2. A (page 184)
3. A (page 187)
4. C (page 189)
5. B (page 194)
6. D (page 187)
7. B (page 196)
8. A (page 192)
9. D (page 197)
10. C (page 207)
11. C (page 196)
12. D (page 183)
13. B (page 181)
14. C (page 191)
15. A (page 184)
16. A (page 184)
17. B (page 185)
18. A (page 185)
19. C (page 189)
20. D (page 187)
21. C (page 183)
22. B (page 198)

Labeling

1. Basic parts of the portable fire extinguisher. (page 191)

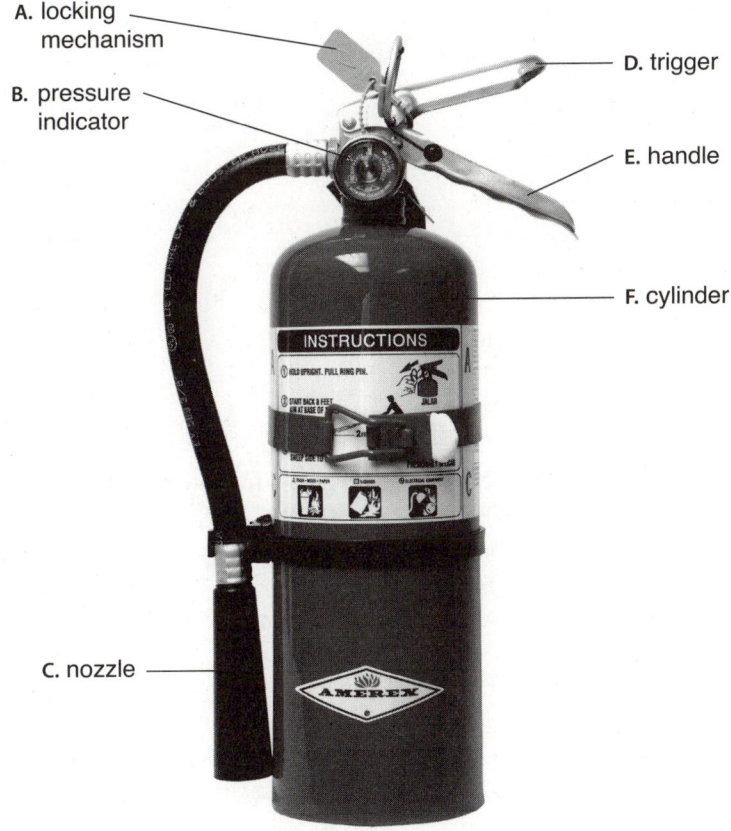

A. locking mechanism
B. pressure indicator
C. nozzle
D. trigger
E. handle
F. cylinder

Figure 7-19

Vocabulary

1. **Cartridge/cylinder fire extinguisher:** An extinguisher that relies on an external cartridge of pressurized gas, which is released only when the extinguisher is to be used. (page 191)
2. **Extra hazard locations:** Locations that contain more Class A combustibles and/or Class B flammables than do ordinary hazard locations. (page 186)
3. **Underwriters Laboratories, Inc. (UL):** The organization that developed the standards, classification, and rating system for portable fire extinguishers. (page 183)
4. **Rapid oxidation:** The scientific terminology for burning. (page 187)
5. **Multipurpose dry chemical extinguisher:** An extinguisher rated for Class A, B, and C fires. The chemicals in these extinguishers form a crust over Class A combustible fuels to prevent rekindling. (page 188)
6. **Class K fires:** Fires involving combustible cooking oils and fats. (page 183)
7. **Extinguishing agent:** The substance contained in a portable fire extinguisher that puts out a fire. (page 187)
8. **Polar solvent:** A water-soluble flammable liquid, such as an alcohol, acetone, ester, or ketone. (page 190)

Fill-in

1. nozzle (page 191)
2. clean agent (page 190)
3. A (page 182)
4. Carbon dioxide (page 196)
5. kindling (page 187)
6. B (page 184)
7. recharged (page 208)
8. Carbon dioxide (page 196)
9. Backpack (page 180)
10. basic (page 180)

True/False

1. T (page 180)
2. F (page 181)
3. F (page 184)
4. T (page 185)
5. T (page 190)
6. F (page 184)
7. F (page 198)
8. T (page 186)
9. F (page 208)
10. T (page 208)

Short Answer

1. (1) Cylinder; (2) handle; (3) nozzle; (4) trigger; (5) locking mechanism; (6) pressure indicator (page 191)
2. (1) Water; (2) dry chemicals; (3) carbon dioxide; (4) foam; (5) wet chemicals; (6) halogenated agents; (7) dry powder (page 188)
3. (1) Locate the fire extinguisher. (2) Select the proper classification of extinguisher. (3) Transport the extinguisher to the location of the fire. (4) Activate the extinguisher to release the extinguishing agent. (5) Apply the extinguishing agent to the fire for maximum effect. (6) Ensure your personal safety by having an exit route. (page 197)
4. (1) Pull the safety pin. (2) Aim the nozzle at the base of the flames. (3) Squeeze the trigger to discharge the agent. (4) Sweep the nozzle across the base of the flames. (page 198)
5. (1) The pressure gauge reading is outside the normal range. (2) The inspection tag is out of date. (3) The tamper seal is broken, especially in extinguishers with no pressure gauge. (4) The extinguisher does not appear to be full of extinguishing agent. (5) The hose and/or nozzle assembly is obstructed. (6) There are signs of leakage around the discharge valve or nozzle assembly. (page 208)

Word Fun

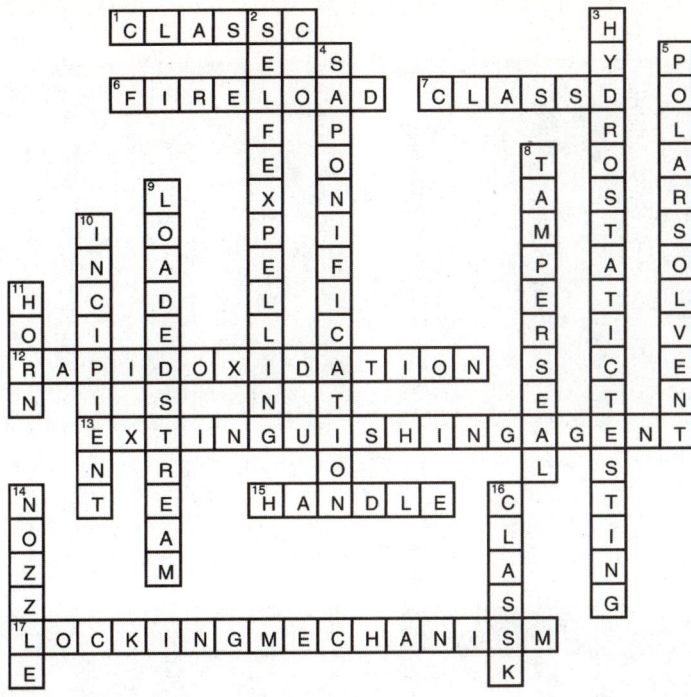

Fire Alarms

1. When inspecting fire extinguishers, you should be looking for some common indications that an extinguisher needs maintenance. Common indications include:
   - The pressure gauge reading is outside the normal range.
   - The inspection tag is out-of-date.
   - The tamper seal is broken.
   - Any indication that the extinguisher is not full of extinguishing agent.
   - The hose and/or nozzle assembly is obstructed.
   - There are signs of physical damage, corrosion, or rust.
   - Signs of leakage around the discharge valve or nozzle assembly can be seen.
2. Ensure the damper on the fireplace is open. Use a Class A type extinguisher, preferably a stored-pressure water type. Open the fireplace door and spray a small amount of water on the fire until it is extinguished.

Skill Drills

Skill Drill 7-2: Operating a Carbon Dioxide Extinguisher

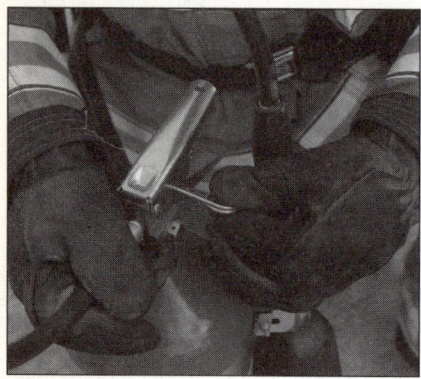

1. Pull the pin on the handle.

2. Remove the horn or nozzle from the secured position on the extinguisher and aim in the direction of your approach.

3. Give the trigger a quick squeeze to ensure that the extinguisher is operational and the agent discharges properly.

4. Approach the fire with an exit to your back. Never let the fire get between you and the exit.

5. Aim the nozzle of the fire extinguisher at the base of the fire and squeeze the trigger.

6. Sweep the extinguishing agent from side to side, continuing to aim at the base of the flames. Continue to use the extinguisher until the fire is out or the extinguisher is empty.

7. Back away from the fire. Overhaul the fire to ensure that it is completely extinguished.

Skill Drill 7-3: Attacking a Class A Fire with a Stored-Pressure Water-Type Fire Extinguisher

1. Begin to attack the fire from a **safe** distance.
2. Aim the stream **directly** at the **base** of the flames. **Sweep** the nozzle back and forth, moving **closer** as the fire goes out.
3. After the flames are out, position your finger **in front** of the nozzle to create a spray and soak the fuel.
4. Break apart the fuel with a stick and apply the **extinguishing agent** to any smoldering, smoking, or glowing surfaces.
5. Apply additional water spray to prevent the fire from **reigniting**.

Skill Drill 7-5: Attacking a Class B Flammable Liquid Fire with a Dry Chemical Fire Extinguisher

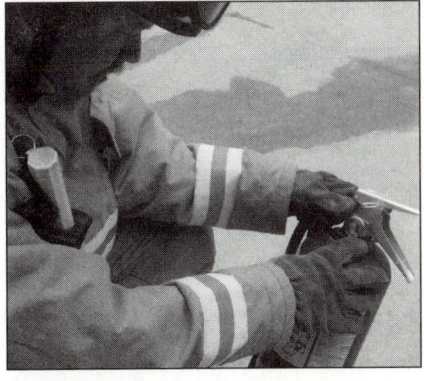

1. Check the pressure gauge to ensure that the extinguisher is properly charged.

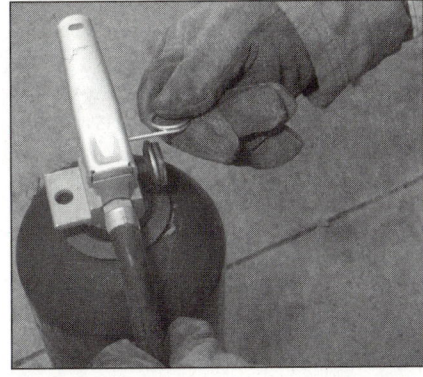

2. Pull the pin on the handle.

3. Begin fighting the fire from a safe distance.

4. Do not aim the initial discharge into the liquid at close range.

5. Discharge the stream at the base of the flame, starting at the near edge of the fire and working toward the back.

6. Sweep the nozzle back and forth across the surface of the flammable liquid.

7. Look for hot or smoldering objects that could reignite the liquid.

Skill Drill 7-7: Using a Wet Chemical Fire Extinguisher

1. Begin to apply the agent to the deep fat fryer from a **safe** distance.
2. Do not direct the agent stream **directly** into the burning liquid. **Lob** the extinguishing agent, which is expelled as a **fire spray**, lightly onto the burning surface to create a **foam** blanket.
3. Continue to apply the agent until the **foam blanket** has extinguished all flames.
4. Do not **disturb** the foam blanket even after all flames have died down. If reignition occurs, **repeat** these steps.

Chapter 8: Fire Fighter Tools and Equipment

Matching

1. D (page 230)
2. A (page 221)
3. C (page 222)
4. B (page 233)
5. G (page 234)
6. E (page 217)
7. I (page 234)
8. J (pages 218, 234)
9. H (page 233)
10. F (page 218)

Multiple Choice

1. C (page 214)
2. C (page 214)
3. D (page 215)
4. A (page 222)
5. D (page 224)
6. B (page 231)
7. D (page 231)
8. C (page 221)
9. A (page 228)
10. D (page 224)
11. D (page 228)
12. D (page 229)
13. B (page 230)
14. A (page 230)
15. B (page 230)
16. C (page 231)
17. D (page 231)
18. A (page 228)
19. B (page 228)
20. D (page 226)

Vocabulary

1. **Ceiling hook:** A tool consisting of a long wood or fiberglass pole and a metal point with a spur at right angles that can be used to probe ceilings and pull down plaster lath material. (page 218)
2. **Crowbar:** A straight bar made of steel or iron with a forked-like chisel on the working end. (page 219)
3. **Kelly tool:** A steel bar with two main features—a large pick and a large chisel or fork. (page 219)
4. **Hydrant wrench:** A tool used to open or close a hydrant by rotating the valve stem, and to remove the caps from the hydrant outlets. (page 217)
5. **Reciprocating saw:** A saw powered by either an electric motor or a battery motor that rapidly pulls the saw blade back and forth. (page 226)
6. **Gripping pliers:** A hand tool with a pincer-like working end that can also be used to bend wire or hold smaller objects. (page 217)
7. **Cutting torch:** A torch that produces an extremely high-temperature flame and is capable of heating steel until it melts, thereby cutting through the object. (page 226)
8. **Seat belt cutter:** A specialized cutting device that cuts through seat belts. (page 234)
9. **Pike pole:** A wood or fiberglass pole with a metal head attached to one end. (page 218)
10. **Spanner wrench:** A special wrench used to tighten or loosen hose couplings. (page 217)

11. **Claw bar:** A tool with a pointed claw-hook on one end and a forked- or flat-chisel pry on the other end that can be used for forcible entry. (page 219)
12. **Overhaul:** The phase in which you examine the fire scene carefully and ensure that all hidden fires are extinguished. (page 230)

Fill-in

1. Safely (page 214)
2. efficiently (page 212)
3. Cutting torches (page 226)
4. Pushing, pulling (page 217)
5. multiple-function (page 226)
6. Response/size-up (page 227)
7. tool staging (page 230)
8. instruction (page 231)
9. pick-head axe (page 222)
10. maul (page 221)

True/False

1. T (page 215)
2. T (page 214)
3. T (page 226)
4. T (page 228)
5. F (page 230)
6. T (page 215)
7. F (page 228)
8. F (page 228)
9. T (page 218)
10. T (page 229)

Short Answer

1. If you know which tools and equipment are needed for each phase of firefighting, you will be able to achieve the desired objective quickly and have the energy needed to complete the remaining tasks. (page 215)
2. Rotating tools apply a rotational force to make something turn. Examples: box-end wrench, gripping pliers, hydrant wrench, open-end wrench, pipe wrench, screwdriver, socket wrench, spanner wrench.
   Pushing/pulling tools extend the reach or increase the power on an object. Examples: ceiling hook, Clemens hook, drywall hook, K tool, multipurpose hook, pike pole, plaster hook, roofman's hook, San Francisco hook.
   Prying/spreading tools are used for creating or increasing the size of an opening. Examples: claw bar, crowbar, flat bar, Halligan tool, Hux bar, hydraulic spreader, Kelly tool, pry bar, rabbet tool.
   Striking tools are used to apply an impact force on an object. Examples: battering ram, chisel, flat-head axe, hammer, mallet, maul, pick-head axe, sledgehammer, spring-loaded center punch.
   Cutting tools' sharp edges are used for severing an object. Examples: axe, bolt cutter, chainsaw, cutting torch, hacksaw, handsaw, hydraulic shears, reciprocating saw, rotary saw, seat belt cutter.
   Multiple-function tools have a variety of uses, mainly to reduce the number of tools needed. Example: flat-head axe. (pages 215–226)
3. *Response/size-up:* This phase begins when the emergency call is received and continues as the units travel to the incident scene. The last part of this phase involves the initial observation and evaluation of factors used to determine which strategy and tactics will be employed.
   Forcible entry: This phase begins when entry to buildings, vehicles, aircraft, or other confined areas is locked or blocked, requiring fire fighters to use special techniques to gain access.
   Interior attack: During this phase, a team of fire fighters is assigned to enter a structure and attempt fire suppression.
   Search and rescue: This phase involves a search for any victims trapped by the fire and their rescue from the building.
   Rapid intervention: During this phase, a rapid intervention company/crew (RIC) provides immediate assistance to injured or trapped fire fighters.
   Ventilation: This step involves changing air within a compartment by natural or mechanical means.
   Overhaul: The final phase is to ensure that all hidden fires are extinguished after the main fire has been suppressed. (page 226)
4. Approved helmet, firefighting hood, eye protection, face shield, approved firefighting gloves, turnout coat, bunker pants, boots, self-contained breathing apparatus (SCBA), personal alert safety system (page 214)

5. (1) A prying tool, such as a Halligan tool; (2) a striking tool, such as a flat-head axe or a sledgehammer; (3) a cutting tool, such as an axe; (4) a pushing/pulling tool, such as a pike pole; (5) a strong hand light or portable light (page 228)
6. (1) Pushing tool (short pike pole); (2) prying tool (Halligan tool); (3) striking tool (sledgehammer or flat-head axe); (4) cutting tool (axe); (5) hand light (page 228)
7. (1) Thermal imaging device; (2) additional portable lighting; (3) life lines; (4) prying tools; (5) striking tools; (6) cutting tools, including a power saw; (7) SCBA and spare air cylinders (page 229)
8. (1) Positive-pressure fans; (2) negative-pressure (exhaust) fans; (3) pulling and pushing tools (long pike poles); (4) cutting tools (power saws and axes) (page 230)
9. (1) Pushing tools (pike poles of varying lengths); (2) prying tools (Halligan tool); (3) striking tools (sledgehammer, flat-head axe, hammer, mallet); (4) cutting tools (axes, power saws); (5) debris-removal tools (shovels, brooms, rakes, buckets, carryalls); (6) water-removal equipment (water vacuums); (7) ventilation equipment (electric, gas, or water-powered fans); (8) portable lighting; (9) thermal imaging device (page 230)
10. All tools and equipment must be properly maintained so that they will be ready for use when they are needed. Every tool and piece of equipment must be ready for use before you respond to an emergency incident. Preventive maintenance will help ensure that equipment will operate properly when it is needed. (page 231)

Word Fun

[Crossword puzzle with answers: 1-Across COUPLING; 3-Down PLASTERHOOK; IRONS; FORCIBLE; HALLIGAN; CLEM; HUX; MAUL; RABBETTOOL; BAN; RESPONSE; BOLTCUTTER; HACKSAW; ENTRY; KTOOL]

Fire Alarms

1. The special equipment that a rapid intervention crew should carry includes thermal imaging device, additional portable lighting, lifelines, prying tools, striking tools, cutting tools (including a power saw), SCBA, and spare air cylinders.
2. All debris should be removed and the tool should be clean and dry. All fuel tanks should be filled completely with fresh fuel. Any dull or damaged blades/chains should be replaced. Belts should be inspected to ensure they are tight and undamaged. All guards should be securely in place. All hydraulic hoses should be cleaned and inspected. All power cords should be inspected for damage. All hose fittings should be cleaned, inspected, and tested to ensure tight fit. The tools should be started to ensure that they operate properly. Tanks on water vacuums should be emptied, washed, cleaned, and dried. Hoses and nozzles on water vacuums should be cleaned and dried.

Chapter 9: Ropes and Knots

Matching

1. F (page 242)
2. E (page 248)
3. J (page 242)
4. D (page 250)
5. C (page 249)
6. B (page 241)
7. A (page 241)
8. I (page 243)
9. H (page 239)
10. G (page 249)

Multiple Choice

1. C (page 239)
2. B (page 238)
3. A (page 241)
4. D (page 248)
5. A (page 243)
6. A (page 241)
7. D (page 248)
8. C (page 249)
9. B (page 248)
10. A (page 260)
11. C (page 238)
12. A (page 239)
13. D (page 240)
14. D (page 240)
15. C (page 241)
16. B (page 243)
17. A (page 243)
18. C (page 246)
19. D (page 250)
20. B (page 264)

Labeling

1. Twisted and braided rope

A. Twisted Rope
B. Braided Rope

Figure 9-9 (A and B)

Vocabulary

1. **Braided rope:** Ropes constructed by weaving or intertwining strands—typically synthetic fibers—together in the same way that hair is braided. (page 242)
2. **Kernmantle rope:** A rope that consists of two parts—the kern and the mantle. (page 242)
3. **Running end:** The part of the rope used for lifting or hoisting. (page 248)
4. **Working end:** The part of the rope used for forming the knot. (page 248)
5. **Knot:** A prescribed way of fastening lengths of rope or webbing to objects or to each other. (page 248)
6. **Round turn:** A turn formed by making a loop and then bringing the two ends of the rope parallel to each other. (page 249)
7. **Rope bag:** A bag used to protect and store ropes. (page 248)
8. **Harness:** A piece of rescue or safety equipment made of webbing and worn by a person. (page 243)
9. **Shock load:** An instantaneous load that places a rope under extreme tension, such as when a falling load is suddenly stopped as the rope becomes taut. (pages 246, 270)
10. **Depressions:** Flat spots or lumps on the inside of the rope. (page 247)

Answer Key

Fill-in

1. carabiner (page 243)
2. harness (page 243)
3. Confined space (page 245)
4. rope record (page 247)
5. Knots (page 248)
6. reduce (page 248)
7. bight (page 249)
8. round turn (page 249)
9. safety knot (page 250)
10. Hitches (page 250)

True/False

1. T (pages 239, 246)
2. T (page 250)
3. F (page 239)
4. F (page 240)
5. T (page 239)
6. T (page 242)
7. F (page 243)
8. F (page 246)
9. T (page 248)
10. T (page 250)

Short Answer

1. (1) Lose their load-carrying ability over time; (2) subject to mildew; (3) absorb 50 percent of their weight in water; (4) degrade quickly (page 240)
2. (1) Thinner without sacrificing strength; (2) less absorbent than natural fiber ropes; (3) greater resistance to rotting and mildew; (4) longer-lasting than natural fiber ropes; (5) greater strength and added safety; (6) more fire-retardant than natural fiber ropes (page 241)
3. (1) Nylon—has a high melting temperature, has good abrasion resistance, strong and lightweight; (2) polyester—second most common synthetic fiber used for life safety ropes; (3) polypropylene—lightest of the synthetic fibers, does not absorb water, often used for water rescue (page 241)
4. (1) Twisted—also called laid rope; made of individual fibers twisted into strands; strands are twisted together to make the rope. (2) Braided—made by weaving or intertwining strands together, like braiding hair; all strands outside the rope are subject to abrasion; will stretch under a load, not prone to twisting; double-braided rope has an inner core covered by a protective braided sleeve, the inner core is protected from abrasion. (3) Kernmantle—consists of two parts, the kern and the mantle; kern is the center core of the rope and provides about 70 percent of the rope's strength; mantle or sheath is the braided covering that protects the core; both parts are made with synthetic fibers, but fibers can be different for the kern and the mantle; each fiber in the kern extends the entire length of the rope without knots or splices, so that the inner core is protected from abrasion. (pages 241–242)
5. (1) Care, (2) clean, (3) inspect, (4) store (page 246)
6. (1) Protect the rope from sharp abrasive surfaces; (2) protect the rope from heat, chemicals, and flames; (3) protect the rope from rubbing against another rope; (4) protect the rope from prolonged exposure to sunlight; (5) never step on a rope; (6) follow the manufacturer's recommendations (page 246)
7. (1) Wash the rope with mild soap and water. (2) Use a rope washer or machine if recommended by the rope's manufacturer. (3) Air dry the rope out of direct sunlight. (page 246)
8. (1) Safety knot (overhand knot): Secures the leftover working end of the rope to the standing part of the rope; used to finish knots. (2) Half hitch: Knot that wraps around an object; used with other knots. (3) Clove hitch: Used to attach a rope firmly to a round object; used to tie a hoisting rope around an axe or pike pole; can be tied anywhere in a rope. (4) Figure eight: Used to produce a family of other knots. (5) Figure eight on a bight: Creates a secure loop at the working end of a rope; loop can be used to attach the end of the rope to a fixed object or a piece of equipment, or to tie a life safety rope around a person. (6) Figure eight with a follow-through: Creates a secure loop at the end of the rope when the working end must be wrapped around an object or passed through an opening before the loop can be formed. (7) Bowline: Can be used to form a loop; frequently used to secure the end of a rope to an object or anchor point. (8) Bend (sheet or Becket bend): Used to join two ropes together; sheet bend or Becket bend can be used to join two ropes of unequal size. (pages 249–260)

Word Fun

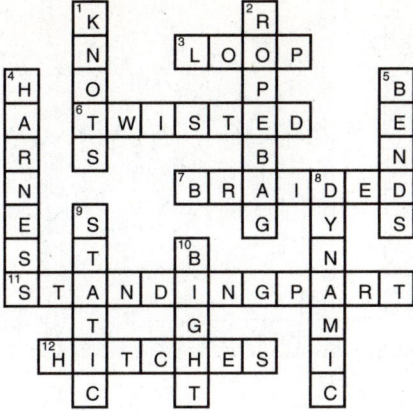

Fire Alarms

1. Many ropes made from synthetic fibers can be washed with a mild soap and water. A special rope washer can be attached to a garden hose. Some manufacturers recommend placing the rope in a mesh bag and washing it in a frontloading washing machine. Air-drying is usually recommended, but rope should not be dried in direct sunlight. The use of mechanical drying devices is not usually recommended. Life safety ropes must be inspected after each use, whether the rope was used for an emergency incident or in a training exercise. Inspect the rope visually, looking for cuts, frays, or other damage, as you run it through your fingers. Because you cannot see the inner core of a kernmantle rope, feel for any depressions. Examine the sheath for any discolorations, abrasions, or flat spots. If you have any doubt about whether the rope has been damaged, consult with your company officer.
2. Tie a figure eight knot in the rope about 39 inches (1 meter) from the working end of the rope. Loop the working end of the rope around the fan handle and back to the figure eight knot. Secure the rope by tying a figure eight with a follow-through. Thread the working end back through the first figure eight in the opposite direction. Tie a safety knot in the working end of the rope. Attach a tag line to the fan for better control. Prepare to hoist the fan.

Skill Drills

Skill Drill 9-5: Tying a Safety Knot

1. Take the loose end of the rope, beyond the knot, and form a loop around the **standing part** of the rope.
2. Pass the **loose** end of the rope through the loop.
3. Tighten the safety knot by **pulling** on both ends at the same time.

Skill Drill 9-6: Tying a Half Hitch

1. Grab the rope with your palm facing away from you.

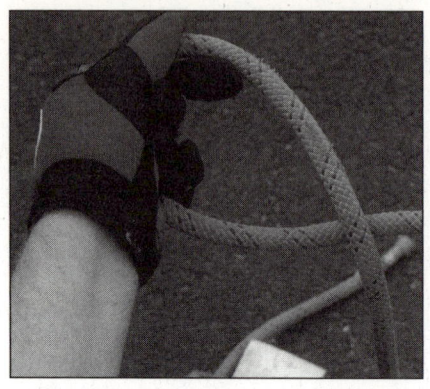

2. Rotate your hand so your palm is facing toward you. This will make a loop in the rope.

3. Pass the loop over the end of the object.

4. Finish the half-hitch knot by positioning it and pulling tight.

Skill Drill 9-7: Tying a Clove Hitch in the Open

1. Starting from **left** to **right** on the rope, grab the rope with crossed hands with the left positioned higher than the right.
2. Holding onto the rope, **uncross** your hands. This will create a loop in each hand.
3. Slide the **right** hand loop behind the **left** hand loop.
4. Slide both loops over the **object**.
5. Pull in opposite directions to **tighten** the clove hitch. Tie a safety knot in the working end of the rope.

Skill Drill 9-9: Tying a Figure Eight Knot

1. Form a **bight** in the rope.
2. Loop the **working end** of the rope completely around the **standing part** of the rope.
3. Thread the **working end** back through the bight.
4. Tighten the knot by pulling on both ends **simultaneously**.

Skill Drill 9-10: Tying a Figure Eight on a Bight

1. Form a bight and identify the end of the bight as the working end.

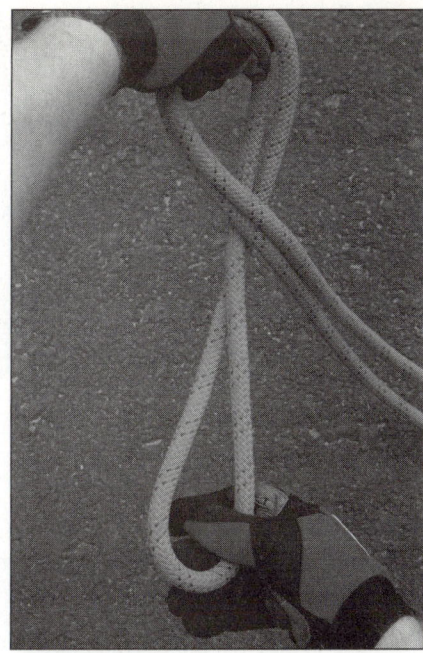

2. Holding both sides of the bight together, form a loop.

3. Feed the working end of the bight back through the loop.

4. Pull the knot tight.

5. Secure the loose end of the rope with a safety knot.

Skill Drill 9-12: Tying a Bowline

1. Make the desired sized **loop** and bring the working end back to the standing part.
2. Form another small loop in the **standing part** of the rope with the section close to the **working end** on top. Thread the working end up through this loop from the **bottom**.
3. Pass the working end **over** the loop, around and under the standing part, and back down through the same opening.
4. Tighten the knot by holding the working end and pulling the standing part of the rope **backward**.
5. Tie a **safety knot** in the working end of the rope.

Skill Drill 9-14: Hoisting an Axe

1. The team that needs the axe should lower a rope with enough extra rope to tie the required knot around the axe.

2. Tie a figure eight on a bight to make a small loop.

3. Place the loop over the axe handle near the head.

4. Pass the standing part of the rope around the head of the axe.

5. Place the standing part of the rope parallel to the axe handle.

6. Tie one or two half hitches along the axe handle.

7. Prepare to raise the axe.

Skill Drill 9-15: Hoisting a Pike Pole

1. The team that needs the pike pole should lower a rope with enough extra rope available for the required knot and the **tag line**.
2. Tie a clove hitch in the open and slip it over the handle. Secure it near the **head**.
3. Place a **half hitch** around the handle below the clove hitch.
4. Place a second **half hitch** around the handle near the bottom of the pike pole.
5. Prepare to raise the **pike pole**.

Skill Drill 9-16: Hoisting a Ladder

1. The team that needs the ladder should lower a rope with enough extra rope to tie onto the ladder.

2. Tie a figure eight on a bight to make a loop 3 or 4 feet in diameter.

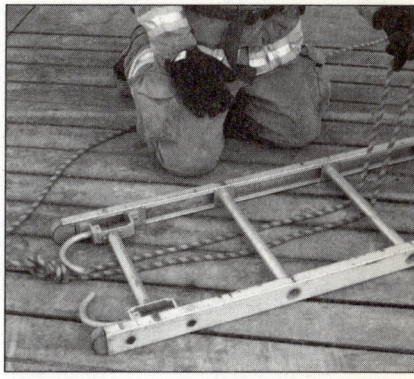

3. Pass the rope between the rungs of the ladder, three or four rungs from the top. Pull the loop under the rungs toward the top of the ladder.

4. Place the loop around the top of the ladder.

5. Remove the slack from the rope and allow the loop to slide down the ladder.

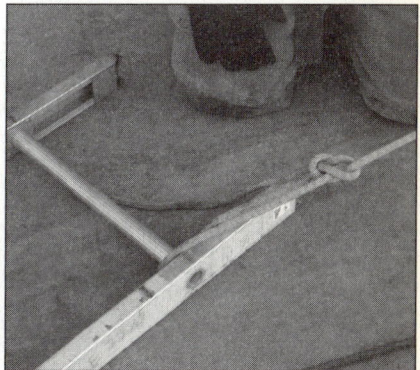

6. Attach a tag line from below to control the ladder as it is hoisted.

7. Prepare to raise the ladder.

Chapter 10: Response and Size-Up

Matching

1. C (page 287)
2. E (page 288)
3. F (page 287)
4. A (page 288)
5. B (page 274)
6. D (page 274)
7. J (page 278)
8. I (page 278)
9. H (page 282)
10. G (page 285)

Multiple Choice

1. C (page 275)
2. A (page 277)
3. B (page 278)
4. C (page 279)
5. C (page 280)
6. D (page 283)
7. A (page 282)
8. B (page 282)
9. A (page 285)
10. C (page 285)
11. D (page 285)
12. A (page 280)
13. B (page 286)
14. C (page 286)
15. A (page 286)
16. D (page 287)
17. B (page 283)
18. C (page 283)
19. A (page 276)
20. D (page 279)

Vocabulary

1. **Thermal imaging devices:** Electronic cameras that can detect sources of heat. They are valuable tools for finding fires in void spaces and containing the spread of fire. (pages 286, 288)
2. **Freelancing:** The dangerous practice of acting independently of command instructions. (pages 278, 288)
3. **Personal accountability tag (PAT):** An identification card used to track the location of a fire fighter on an emergency incident. (pages 279, 288)
4. **Personal alert safety system (PASS):** A device worn by a fire fighter that sounds an alarm if the fire fighter is motionless for a period of time. (pages 274, 288)
5. **Size-up:** The ongoing observation and evaluation of factors that are used to develop objectives, strategy, and tactics for fire suppression. (pages 274, 288)

Fill-in

1. reconnaissance (page 283)
2. offensive (page 286)
3. balloon (page 288)
4. Probabilities (page 283)
5. defensive (page 287)
6. overhaul (page 287)
7. size-up (page 274)
8. secured (page 275)
9. quarter (page 280)
10. systematic (page 282)

True/False

1. T (page 278)
2. T (page 277)
3. F (page 276)
4. T (page 278)
5. F (page 287)
6. T (page 277)
7. F (page 283)
8. F (page 280)
9. T (page 277)
10. F (page 277)

Short Answer

1. (1) Shutting off electricity eliminates potential ignition sources. (2) Shutting off gas decreases the potential for further leakage and explosions. (3) Shutting off water prevents electrical problems and helps to minimize additional damage to the structure and contents. (pages 279–280)
2. If there is no officer on the first-arriving unit, a fire fighter could be responsible for assuming command and conducting the preliminary size-up until an officer arrives. Individual fire fighters are often asked to obtain information or to report their observations for ongoing size-up. Fire fighters should routinely make observations during incidents to maintain their personal awareness of the situation and to develop their personal competence. (page 282)
3. Facts: Accurate information obtained from various sources such as preincident plans (building construction, layout, occupancy), trained specialists (engineers, utility representatives), and time of day, weather, and temperature. Probabilities: Events or outcomes that can be predicted or anticipated based on facts, observations, common sense, and previous experiences. (pages 282–285)
4. (1) Rescue any victims. (2) Protect exposures. (3) Confine the fire. (4) Extinguish the fire. (5) Salvage property and overhaul the fire. (page 286)

Word Fun

(Crossword solution)

Across: 2. RESPONSE, 3. SALVAGE, 9. THERMALIMAGINGDEVICES, 10. PERSONNELACCOUNTABILITYSYSTEM, 12. REKINDLE

Down: 1. SIZEUP, 2. RECONNAISSANCE, 4. PREINCIDENT, 5. DEFENSE, 6. OVERHAUL, 7. OFFENSIVE, 8. EXTENSION, 11. EXPOSURE

Fire Alarms

1. When the fire is too large or too dangerous to extinguish with an offensive attack, the IC will implement a defensive strategy. A defensive strategy is required when the IC determines that the risk to fire fighters' lives is excessive, as in situations where structural collapse is possible. The IC who adopts a defensive strategy has determined that there is no property left to save or that the potential for saving property does not justify the risk to fire fighters.
2. Consider approaching traffic, including other emergency vehicles, and other, less-obvious hazards. Always check for traffic before opening doors and dismounting from the apparatus. Watch out for traffic when working in the street. Follow departmental SOPs to close streets quickly and to block access to areas where operations are being conducted. Place traffic cones, flares, and other warning devices far enough away from the incident to slow approaching traffic and direct it away from the work area.

Skill Drills

Skill Drill 10-1: Mounting Apparatus

1. When mounting (climbing aboard) fire apparatus, always have at least one hand firmly grasping a **handhold** and at least one foot firmly placed on a foot surface. Maintain the one hand and one foot placement until you are seated.
2. Fasten your seat belt and then don any other required safety equipment for response, such as **hearing protection**. Eye and face protection are required for seating areas that are not fully enclosed.

Chapter 11: Forcible Entry

Matching

1. E (page 295)
2. I (page 297)
3. C (page 322)
4. A (page 310)
5. F (page 297)
6. D (page 296)
7. G (page 321)
8. B (page 310)
9. H (page 298)
10. J (page 294)

Multiple Choice

1. B (page 292)
2. C (page 292)
3. A (page 294)
4. D (page 294)
5. C (page 321)
6. B (page 295)
7. D (page 296)
8. A (page 296)
9. B (page 297)
10. C (page 298)
11. C (page 298)
12. A (page 301)
13. D (page 304)
14. B (page 304)
15. C (page 307)
16. A (page 310)
17. B (page 310)
18. A (page 310)
19. C (page 311)
20. D (page 313)
21. C (page 316)

Labeling

1. The parts of a door.

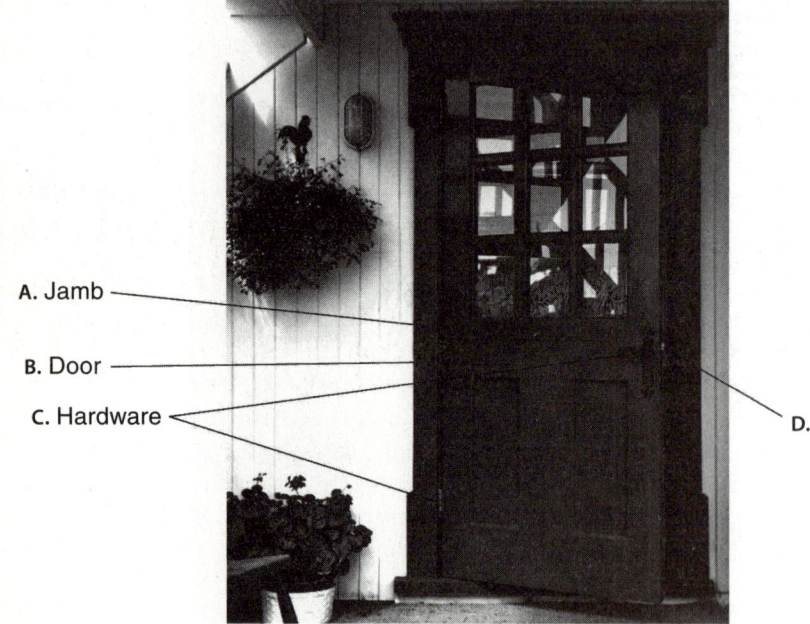

A. Jamb
B. Door
C. Hardware
D. Locking Device

Figure 11-8

2. The parts of a lock.

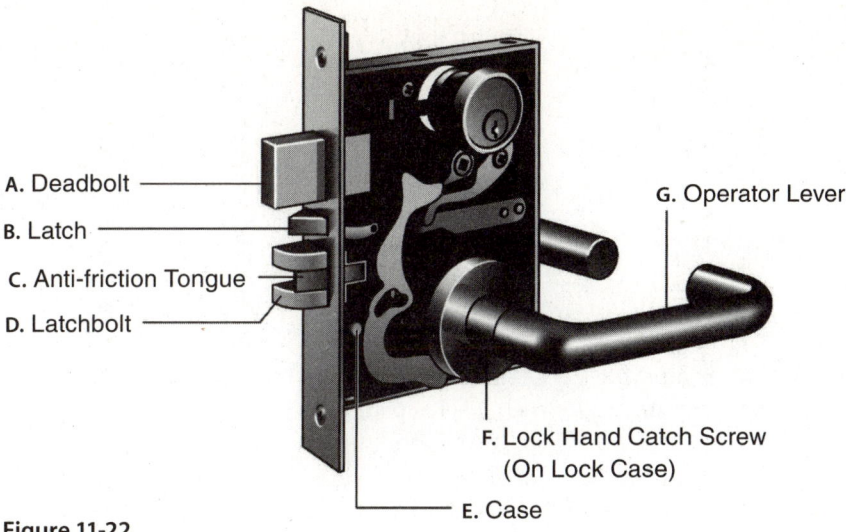

Figure 11-22

Vocabulary

1. **Projected windows:** Also called factory windows, usually found in older warehouse or commercial buildings. They project inward or outward on an upper hinge. (page 307)
2. **Jalousie windows:** Made of adjustable sections of tempered glass encased in a metal frame that overlap each other. Commonly found in mobile homes, and is operated by a hand crank. (page 305)
3. **Mortise locks:** Door locks with both a latch and a bolt built into the same mechanism; the two locking mechanisms operate independently of each other. Mortise locks are often found in hotel rooms. (page 313)
4. **Cylindrical locks:** The most common fixed lock in use today. The locks and handles are placed into predrilled holes in the doors. One side of the door will usually have a key-in-the-knob lock; the other side will have a keyway, a button, or some other type of locking/unlocking mechanism. (page 310)
5. **Casement windows:** Windows in a steel or wood frame that open away from the building via a crank mechanism. (page 307)
6. **K tool:** A tool that is designed to cut into a lock cylinder. (page 296)
7. **Jamb:** The part of a doorway that secures the door to the studs in a building. (page 322)
8. **Tempered glass:** A type of glass that is heat treated, making it four times stronger than regular glass. (page 305)
9. **Rabbet:** A type of door frame in which the stop for the door is cut into the frame. (page 295)

Fill-in

1. critical (page 292)
2. ready (page 294)
3. Bolt cutters (page 296)
4. Battery (page 296)
5. carbide (page 296)
6. rabbet tool (page 295)
7. Revolving (page 302)
8. security (page 302)
9. Windows (page 303)
10. Double-hung (page 305)
11. Awning (page 305)

True/False

1. T (page 292)
2. F (page 295)
3. T (page 295)
4. F (page 297)
5. T (page 297)
6. F (page 299)
7. T (page 302)
8. F (page 303)
9. T (page 304)
10. T (page 316)
11. F (page 316)

Short Answer

1. (1) Always wear the appropriate protective equipment. (2) Learn to recognize the materials used in building and lock construction and the appropriate tools and techniques for each. (3) Keep all tools clean, properly serviced according to the manufacturer's guidelines, and ready to use. (4) Do not leave tools lying on the ground or floor. (page 293)
2. (1) Do not carry a tool or piece of equipment that is too heavy or designed to be used by more than one person. (2) Always use your legs—not your back—when lifting heavy tools. (3) Keep all sharp edges and points away from your body at all times. Cover or shield them with a gloved hand to protect those around you. (4) Carry long tools with the head down toward the ground. Be aware of overhead obstructions and wires, especially when using pike poles. (page 293)
3. (1) Striking tools; (2) cutting tools; (3) prying tools; (4) through-the-lock tools (page 294)
4. (1) Door (the entryway); (2) jamb (the frame); (3) hardware (the handles, hinges, etc.); (4) locking device (page 297)

Word Fun

[Crossword puzzle solution]

Across: 1. DEADBOLT, 4. JAMB, 5. CLAW, 7. KTOOL, 8. TEMPERED, 10. HARDWARE, 13. LOADBEARING, 14. CASEMENTWINDOWS, 16. ANNEALED

Down: 1. DOOR, 2. LAMINATEDGLASS (LAMINATIONA), 3. SHACKLE, 6. BIKE, 9. PADSTAT, 11. RAB, 12. IRON, 15. ADZ

Fire Alarms

1. First try before you pry. Then, break one of the glass doors and clear the glass from the steel frame. Simply unlock the door by turning the lock. Block the door open by using a wedge. Enter the building and overhaul the fire.
2. The quickest way to force through a security roll-up door is to cut the door with a torch or saw. Make the cut in the shape of a triangle, with the point at the top. Pad the opening to prevent injury to those entering and exiting through the opening.

Skill Drills

Skill Drill 11-1: Forcing Entry into an Inward-Opening Door

1. Size up the door, looking for any safety hazards. Inspect the door for the location and number of locks and mechanisms.

2. Place the forked end of the Halligan tool into the door frame between the door jamb and the door stop. Insert the tool near the lock, with the beveled end of the tool against the door.

3. Once the Halligan tool is in position, have your partner, on your command, drive the tool farther into the gap between the rabbeted jamb or stop and the door. Make sure that the tool is not driven into the door jamb itself.

4. Once the tool is past the stop, but between the door and the jamb, push the Halligan tool toward the door to force it open. If more leverage is needed, your partner can slide the axe head between the bevel of the Halligan tool and the door. It may be necessary to push in on the door. Secure the door to prevent it from closing behind you.

Skill Drill 11-2: Forcing Entry into an Outward-Opening Door

1. **Size up** the door looking for any safety hazards. Place the **adz** end of the Halligan tool between the door and the frame either near the locking mechanism, or between the mechanism and a secondary lock.
2. Once the **Halligan tool** is in position, have your partner strike the **Halligan tool** on your command and drive the **adz** end farther into the gap.
3. Pry in a **downward** direction with the fork end of the tool and then force the door **outward**. Always secure the door to prevent it from closing behind you.

Skill Drill 11-7: Forcing Entry Using a K Tool

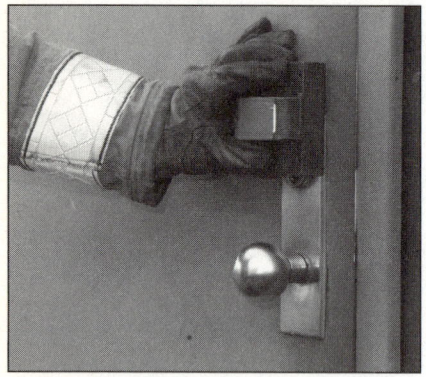 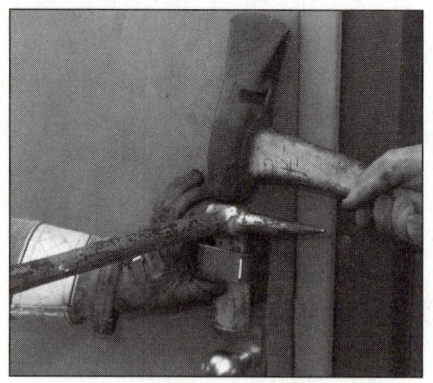 

1. Place the K tool over the face of the cylinder, noting the location of the keyway. Using a Halligan or similar pry tool, tap the K tool down into the cylinder.
2. Place the adz end of the Halligan tool into the slot on the K tool, and strike the end of the Halligan tool with a flat-head axe to drive the K tool farther into the lock cylinder.
3. Pry up on the Halligan tool to pull out the lock and expose the locking mechanism. Using the small tools that come with the K tool, turn the mechanism to open the lock.

Skill Drill 11-8: Forcing Entry Using an A Tool

1. Place the **cutting edges** of the A tool at the top of the cylinder, between the cylinder and the door frame.
2. Using a flat-head axe or similar tool, drive the **A tool** into the **cylinder**. Pry **up** on the A tool to remove the **cylinder** from the door. Insert a **key tool** into the hole to manipulate the locking mechanism and open the door.

Chapter 12: Ladders

Matching

1. E (page 329)
2. D (page 327)
3. F (page 326)
4. I (page 329)
5. A (page 328)
6. B (page 329)
7. C (page 327)
8. G (page 328)
9. H (page 328)
10. J (page 329)

Multiple Choice

1. A (page 327)
2. C (page 327)
3. A (page 327)
4. B (page 328)
5. D (page 328)
6. D (page 328)
7. A (page 328)
8. C (page 329)
9. B (page 329)
10. C (page 329)
11. B (page 330)
12. A (page 330)
13. D (page 331)
14. A (page 331)
15. C (page 331)
16. B (page 334)
17. B (page 335)
18. D (page 337)
19. B (page 337)
20. A (page 337)
21. C (page 338)
22. A (page 341)
23. C (page 341)
24. D (page 347)
25. B (page 347)
26. C (page 352)
27. A (page 355)
28. D (page 338)

Labeling

1. Basic components of a straight ladder.

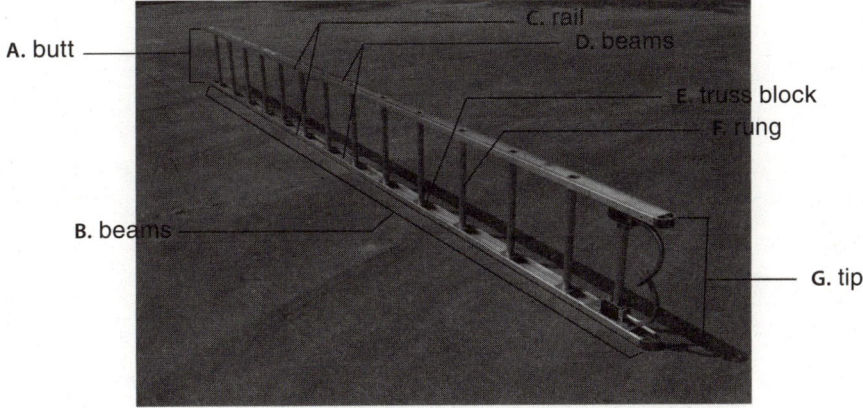

Figure 12-3

2. Components of an extension ladder.

Figure 12-5

Vocabulary

1. **Heat sensor label:** A label that identifies when the ladder has been exposed to specific heat conditions that could damage its structural integrity. (page 328)
2. **Protection plates:** Reinforcing pieces that are placed on a ladder at chaffing and contact points to prevent damage from friction or contact with other surfaces. (page 328)
3. **Halyard:** The rope or cable used to extend or hoist the fly sections of an extension ladder. (page 329)
4. **Ladder belt:** A piece of equipment specifically designed to secure a fire fighter to a ladder or elevated surface. (page 357)
5. **Pulley:** A small grooved wheel that is used to change the direction of the halyard pull. A downward pull on the halyard creates an upward force on the fly sections, extending the ladder. (page 329)
6. **Pawls:** The mechanical locking devices that are used to secure the extended fly sections of an extension ladder. They are sometimes called dogs, ladder locks, or rung locks. (page 329)
7. **Bed section:** The widest section of an extension ladder. It serves as the base; all other sections are raised from the bed section. (page 328)
8. **Guides:** Strips of metal or wood that guide a fly section as it is being extended. Channels or slots in the bed or fly section may also serve as guides. (page 329)
9. **Roof hooks:** Spring-loaded, retractable, curved metal pieces that are attached to the tip of a roof ladder. These hooks are used to secure the tip of the ladder to the peak of a pitched roof. (page 328)
10. **Tie rod:** Metal bar that runs from one beam of the ladder to the other and keeps the beams from separating. Tie rods are typically found in wood ladders. (page 328)

Fill-in

1. portable (page 326)
2. equipment (page 327)
3. rungs (page 328)
4. extension (page 328)
5. Aerial (page 329)
6. straight (page 330)
7. combination (page 331)
8. manufacturer's (page 332)
9. Rescue (page 338)
10. leg lock (page 338)
11. webbing (page 355)
12. three (page 356)

True/False

1. T (page 326)
2. T (page 326)
3. F (page 327)
4. F (page 328)
5. F (page 328)
6. F (page 330)
7. F (page 329)
8. F (page 359)
9. T (page 338)
10. F (page 341)
11. T (page 352)
12. F (page 352)

Short Answer

1. *Trussed beam:* Has a top rail and a bottom rail, which are joined by a series of smaller pieces called truss blocks. The rungs are attached to the truss blocks. Trussed beams are usually constructed of aluminum or wood.

   I-beam: Has thick sections at the top and the bottom, which are connected by a thinner section. The rungs are attached to the thinner section of the beam. This type of beam is usually made from fiberglass.

   Solid beam: Has a simple rectangular cross-section. Many wooden ladders have solid beams. Rectangular aluminum beams, which are usually hollow or C shaped, are also classified as solid beams. (page 327)
2. (1) Clean and lubricate the dogs, following the manufacturer's instructions. (2) Clean and lubricate the slides on extension ladders in accordance with the manufacturer's recommendations. (3) Replace worn halyards and wire rope on extension ladders when they fray or kink. (4) Clean and lubricate hooks. Remove rust and other contaminants, and lubricate the folding roof hook assemblies on roof ladders to keep them operational. (5) Check the heat sensor labels. Replace the sensors when they reach their expiration date. Remove a ladder that has been exposed to high

temperatures from service for testing. (6) Maintain the finish on fiberglass and wooden ladders in accordance with the manufacturer's recommendations. (7) Ensure that portable ladders are not painted except for the top and bottom 18 inches of each section, because paint can hide structural defects in the ladder. The tip and butt are painted for purposes of identification and visibility. (page 333)

3. (1) General safety; (2) lifting and moving ladders; (3) placement of ground ladders; (4) working on a ladder; (5) rescue operations; (6) ladder damage (page 335)

4. To work from a ladder, follow the steps in **Skill Drill 12-19**. (1) Climb to the desired work height and step up to the next higher rung. (2) Note the side of the ladder where the work will be performed. Extend the leg on the opposite side between the rungs. (3) Once the leg is between the rungs, bend the knee and bring that foot back under the rung and through to the climbing side of the ladder. (4) Secure the foot against the next lower rung or the beam of the ladder. Use the thigh for support and step down one rung with the opposite foot. (5) You are now free to lean out to the side of the ladder and work with two hands on the tool. (page 357)

Word Fun

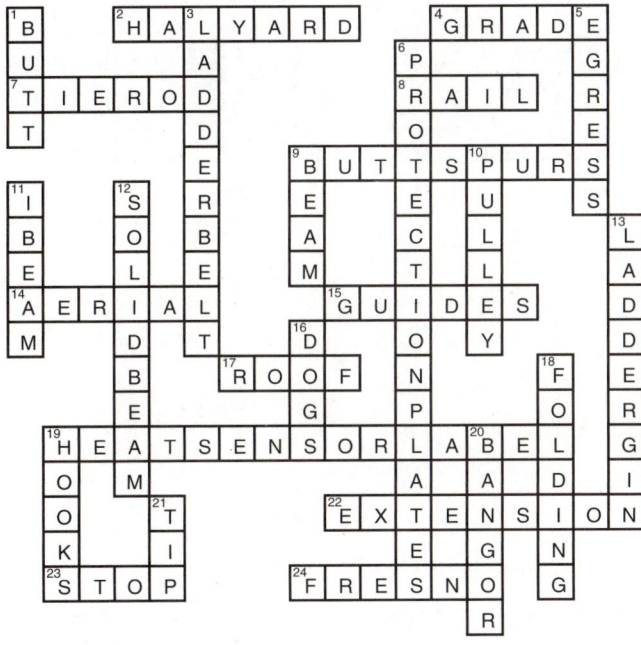

Fire Alarms

1. Think safety first. A person who is in extreme danger may not wait to be rescued. Jumpers risk their own lives and may endanger the fire fighters trying to rescue them. Several fire fighters have been seriously injured by persons who jumped before a rescue could be completed. A trapped person might try to jump onto the tip of an approaching ladder, or to reach out for anything or anyone nearby. You might be pulled or pushed off the ladder by the person you are trying to rescue.

    During rescue operations, the tip of the ladder should be immediately below the windowsill. This prevents the ladder from obstructing the window opening while a trapped occupant is removed.

2. Remove the roof ladder from the apparatus. Choose an appropriate ladder carry for the obstacles that you will have to maneuver around. Deploy the roof ladder up the extension ladder. Remember to use a leg lock or ladder belt to secure yourself to the ladder. Place the roof ladder on the roof, ensuring the hooks are pointing down.

Skill Drills

Skill Drill 12-2: Performing a One-Fire Fighter Carry

1. Locate the <u>center</u> of the ladder.
2. Place one arm through two rungs, just to one side of the <u>middle</u> rung.
3. Bring the top <u>beam</u> to rest on the fire fighter's shoulder.

Skill Drill 12-3: Performing a Two-Fire Fighter Shoulder Carry

1. Both fire fighters approach the ladder from the same side, facing the butt. One fire fighter stands near the butt, and the other stands near the tip.
2. Each fire fighter places an arm between two rungs and lifts the ladder onto the shoulder.
3. The butt spurs are covered with a gloved hand while the ladder is transported.

Skill Drill 12-5: Performing a Two-Fire Fighter Suitcase Carry

1. Both fire fighters face the <u>butt</u> of the ladder, at opposite ends.
2. The fire fighters grasp the upper <u>beam</u> of the ladder.
3. Pick up the ladder using good <u>lifting</u> techniques.

Skill Drill 12-11: Performing a One-Fire Fighter Rung Raise for Ladders Less Than 14 Feet Long

1. Carry the ladder to the structure. Check for overhead hazards before raising the ladder.
2. Place the ladder flat on the ground, with the heel positioned where it will be when the ladder is raised. By the tip, raise the ladder to hip level. Walk hand-over-hand down the rungs until the ladder is vertical.
3. Heel the ladder and lean it into place.

Skill Drill 12-13: Tying the Halyard

1. Wrap the excess halyard around two rungs and pull it tight over the upper rung.

2. Tie a clove hitch around the upper rung.

3. Pull the knot tight and add a safety knot.

Skill Drill 12-18: Climbing a Ladder while Carrying a Tool

1. Hold the tool in one hand and place it against the <u>beam</u>.
2. Slide the tool up the beam, while sliding the opposite <u>hand</u> up the other beam.

Skill Drill 12-19: Working from a Ladder

1. The fire fighter climbs the ladder to the desired work height and then one rung higher.

2. The fire fighter notes the side where the work will be performed. The opposite leg is extended between the rungs.

3. The knee is bent around the rung and the foot is brought back under the rung.

(continues)

Skill Drill 12-19: Working from a Ladder (continued)

4. The foot is secured around the lower rung or the beam. The fire fighter moves the other leg down one rung.

5. The fire fighter is now free to work with both hands.

Chapter 13: Search and Rescue

Matching

1. B (page 366)
2. A (page 366)
3. C (page 370)
4. D (page 374)
5. F (page 398)
6. E (page 371)
7. J (page 369)
8. I (page 371)
9. H (page 375)
10. G (page 378)

Multiple Choice

1. B (page 366)
2. A (page 367)
3. D (page 369)
4. A (page 369)
5. D (page 369)
6. C (page 370)
7. B (page 369)
8. A (page 372)
9. B (page 370)
10. C (page 378)
11. A (page 379)
12. D (page 380)
13. A (page 383)
14. B (page 385)
15. C (page 390)
16. C (page 390)
17. A (page 390)
18. D (page 395)
19. B (page 386)
20. A (page 385)

Chapter 13: Search and Rescue

Vocabulary

1. **Two-in/two-out rule:** The NFPA requirements state that a team of at least two fire fighters must enter together, and at least two other fire fighters must remain outside the danger area, ready to rescue the fire fighters who are inside the building. (page 378)
2. **Exit assist:** The simplest rescue if the victim is responsive and able to walk without assistance or with very little assistance. (page 379)
3. **Primary search:** A quick attempt to locate any potential victims who are in danger. (page 369)
4. **Shelter-in-place:** When the occupants are sheltered and kept in their present location instead of trying to remove them from a fire building. (page 379)

Fill-in

1. search (page 366)
2. lives (page 366)
3. Ventilation (page 366)
4. size-up (page 367)
5. greater (page 367)
6. controlling, extinguishing (page 369)
7. crawl (page 370)
8. secondary (page 370)
9. marked (page 371)
10. risk (page 377)

True/False

1. F (page 366)
2. T (page 368)
3. F (page 369)
4. T (page 367)
5. T (page 368)
6. F (page 371)
7. T (page 378)
8. T (page 378)
9. T (page 378)
10. T (page 379)

Short Answer

1. (1) Occupancy; (2) size of the building; (3) construction of the building; (4) time of day and day of week; (5) number of occupants; (6) degree of risk to the occupants presented by the fire; (7) ability of occupants to exit on their own (page 368)
2. (1) Corridor layouts; (2) exit locations; (3) stairway locations; (4) apartment layouts; (5) number of bedrooms in apartments; (6) locations of handicapped residents' apartments; (7) special-function rooms or areas (page 369)
3. (1) Identify the shape of a human body; (2) show furniture, walls, doorways, and windows; (3) navigate through the interior of a smoke-filled building; (4) locate a fire in a smoke-filled building or behind walls or ceilings; (5) locate the fire source and the direction of fire spread from the exterior; (6) scanning a door before opening it can indicate whether the room is safe to enter (page 372)
4. (1) Work from a single plan. (2) Maintain radio contact with the IC, both through the chain of command and via portable radios. (3) Monitor fire conditions during the search. (4) Coordinate ventilation operations with search and rescue activities. (5) Adhere to the personal accountability system. (6) Stay with a partner. (page 378)
5. (1) PPE; (2) portable radio; (3) hand light or flashlight; (4) forcible entry (exit) tools; (5) hose lines; (6) thermal imaging devices; (7) ladders; (8) long ropes; (8) a piece of tubular webbing or short rope (16 to 24 feet) (page 377)
6. (1) Two-person extremity carry; (2) two-person seat carry; (3) two-person chair carry; (4) cradle-in-arms carry (page 380)

Fire Alarms

1. The search and rescue needs to be coordinated with suppression and ventilation. It would also be necessary to position a hose line to protect the entry and exit paths. A marking system should be used to indicate which rooms have been searched.
2. Make sure you have a single communicated plan. Maintain radio contact with the IC using the chain of command, monitor fire conditions during the search, coordinate ventilation with your search and rescue activities, maintain your accountability system, stay with your team, and if at all possible, carry a thermal imaging device to speed up the search and help locate exits and hazards.

Skill Drills

Skill Drill 13-5: Performing a Two-Person Extremity Carry

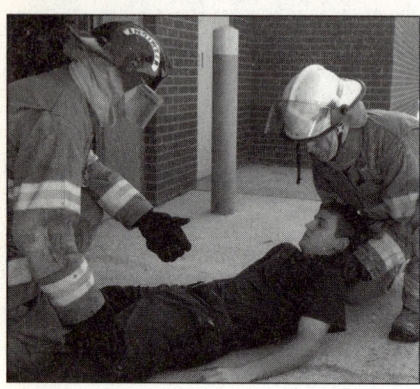

1. Two fire fighters help the victim to sit up.

2. The first fire fighter kneels behind the victim, reaches under the victim's arms, and grasps the victim's wrists.

3. The second fire fighter backs in between the victim's legs, reaches around, and grasps the victim behind the knees.

4. The first fire fighter gives the command to stand and carry the victim away, walking straight ahead. Both fire fighters must coordinate their movements.

Skill Drill 13-6: Performing a Two-Person Seat Carry

1. Kneel beside the victim near the victim's **hips**.
2. Raise the victim to a **sitting** position and link arms behind the victim's **back**.
3. Place your free arms under the victim's **knees** and link arms.
4. If possible, the victim puts his or her arms around your **necks** for additional support.

Skill Drill 13-9: Performing a Clothes Drag

1. **Crouch** behind the victim's **head**, and grab the shirt or jacket around the collar and shoulder area.
2. Lift with your legs until you are fully upright. Walk **backward**, dragging the victim to safety.

Skill Drill 13-12: Performing a Webbing Sling Drag

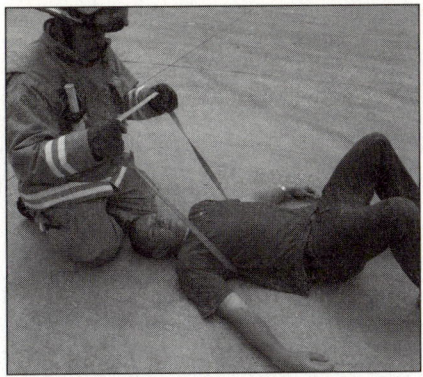 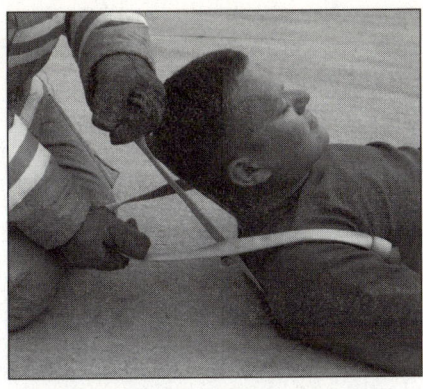 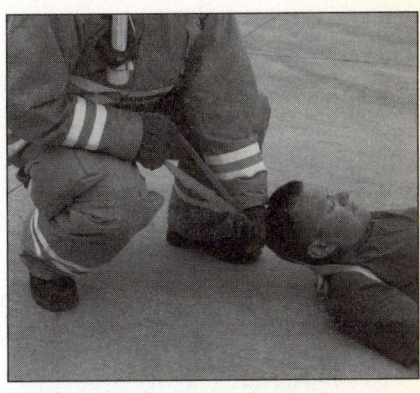

1. Place the victim in the center of the loop so the webbing is behind the victim's back.
2. Take the large loop over the victim and place it above the victim's head. Reach through, grab the webbing behind the victim's back, and pull through all the excess webbing. This creates a loop at the top of the victim's head and two loops around the victim's arms.
3. Adjust hand placement to protect the victim's head while dragging.

Skill Drill 13-13: Performing a Fire Fighter Drag

1. **Tie** the victim's **wrists** together with anything that is handy.
2. Get down on your **hands** and **knees** and straddle the victim.
3. Pass the victim's tied hands around your **neck**, straighten your arms, and drag the victim across the floor by **crawling** on your **hands** and **knees**.

Skill Drill 13-14: Performing a One-Person Emergency Drag from a Vehicle

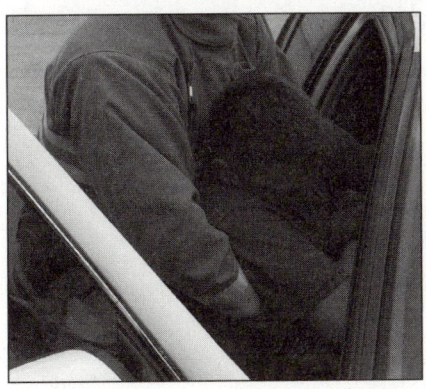 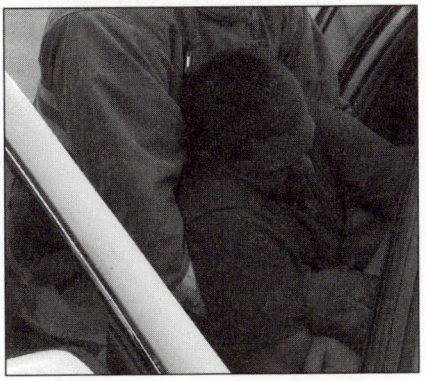 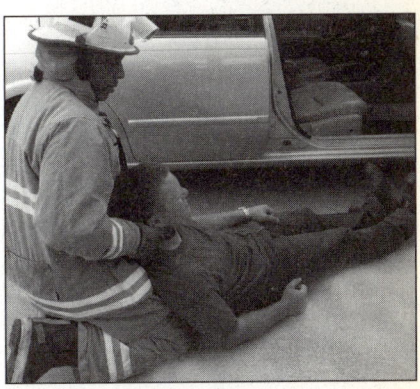

1. Grasp the victim under the arms and cradle the head between your arms.
2. Gently pull the victim out of the vehicle.
3. Lower the victim down into a horizontal position in a safe place.

Skill Drill 13-17: Rescuing an Unconscious Victim from a Window

1. The tip of the ladder is placed just **below** the windowsill.
2. One fire fighter **enters** to assist the victim. The second fire fighter climbs to the window.
3. The fire fighter waiting on the ladder places both hands on the **rungs**, with one leg **straight** and the other leg horizontal to the ground with the knee at an angle of **90 degrees**. The interior fire fighter will then pass the victim through the window and onto the ladder, keeping the victim's **back** toward the ladder.
4. The victim is lowered so that he or she **straddles** the fire fighter's leg. The fire fighter's **arms** should be positioned under the victim's arms holding onto the **rungs**.
5. Step down one rung at a time, **transferring** the victim's **weight** from one leg to the other. The victim's **arms** can also be secured around the fire fighter's neck.

Chapter 14: Ventilation

Matching

1. D (page 403)
2. F (page 423)
3. E (page 408)
4. I (page 422)
5. A (pages 165, 423)
6. J (page 406)
7. C (page 415)
8. H (page 424)
9. B (page 423)
10. G (pages 405, 439)

Multiple Choice

1. A (page 403)
2. A (page 403)
3. D (page 403)
4. C (page 403)
5. B (page 403)
6. C (page 404)
7. D (page 406)
8. A (page 406)
9. B (page 407)
10. C (page 410)
11. C (page 411)
12. A (page 412)
13. B (page 412)
14. A (page 406)
15. C (page 415)
16. D (page 418)
17. A (page 418)
18. B (page 420)
19. C (page 422)
20. B (page 423)
21. B (page 423)
22. D (page 423)
23. D (page 424)
24. C (page 427)
25. D (page 429)

Vocabulary

1. **Fire-resistive construction:** Buildings that have structural components of noncombustible materials with a specified fire resistance. Materials can include concrete, steel beams, and masonry block walls. Fire-resistive construction is also known as Type I building construction, as defined in NFPA 220, *Standard on Types of Building Construction*. (page 407)
2. **Ordinary construction:** Buildings whose exterior walls are made of noncombustible or limited-combustible materials, but whose interior floors and walls are made of combustible materials. Ordinary construction is also known as Type III building construction, as defined in NFPA 220, *Standard on Types of Building Construction*. (page 408)
3. **Stack effect:** A response to the differences in temperature inside and outside a building. A cold outer atmosphere and a heated interior will cause smoke to rise quickly through stairways, elevator shafts, and other vertical openings, filling the upper levels of the building. (page 434)

4. **Ventilation:** The process of removing smoke, heat, and toxic gases from a burning building and replacing them with cooler, cleaner, more oxygen-rich air. (page 403)
5. **Smoke inversion:** The condition in which smoke hangs low to the ground on a cool, damp day with very little wind. (page 411)
6. **Sounding:** The process of striking a roof with a tool to test the roof's stability. (page 420)
7. **Vertical ventilation:** The process of making openings in roofs or floors so that smoke, heat, and toxic gases can escape vertically from a structure. (page 411)
8. **Horizontal ventilation:** Using horizontal openings in the structure, such as windows and doors, to allow smoke, heat, and gases to escape horizontally from a building. (page 411)
9. **Gusset plates:** Connecting plates used in trusses, typically made of wood or lightweight metal. (page 423)

Fill-in

1. Horizontal (page 411)
2. Mushrooming (page 403)
3. high (page 418)
4. upwind (page 421)
5. bowstring (page 424)
6. Trench cut (page 432)
7. Wind, temperature, humidity (page 407)
8. Thermopane (page 406)
9. Hydraulic (page 418)
10. roof collapse (page 420)

True/False

1. F (page 411)
2. F (page 425)
3. T (page 427)
4. T (page 427)
5. T (page 434)
6. F (page 434)
7. F (page 408)
8. T (page 410)
9. F (page 404)
10. T (page 411)

Short Answer

1. The objective of any roof ventilation operation is simple: to provide the largest opening in the appropriate location, using the least amount of time and the safest technique. (page 424)
2. A secondary cut is used to limit the fire spread; a primary cut is located over the seat of the fire. (page 433)
3. Principles of heat transfer (convection, conduction, radiation); ventilation system (negative, positive, natural pressures); building construction; tactical priorities (life safety, fire containment, property conservation); the fire (size, stage of combustion, location, color, movement and amount of smoke) (pages 406–410)
4. (1) Any spongy feeling or indication that the roof is not as solid as when the venting operation began; (2) any visible indication of sagging roof supports; (3) any indication that the roof assembly is separating from the walls, such as the appearance of fire or smoke near the roof edges; (4) any structural failure of any portion of the building, even if it is some distance from the ventilation operation; (5) any sudden increase in the intensity of the fire from the roof opening (page 422)

Word Fun

Across:
4. HORIZONTAL
7. FLASHOVER
8. VERTICAL
9. CHURNING
10. BEARINGWALL
14. CONVECTION
15. EJECTORS
17. KERFCUT
18. NATURAL
19. BACKDRAFT

Down:
1. PRIMARY
2. MECHANICAL
3. HYDRAULIC
5. LECAF
6. STACKEFFECT
11. AFFLICTION
12. LOUVER
13. SOUNDING
16. CHS

Fire Alarms

1. Determine the location of the fire within the building and the direction of attack. Place the fan 4 to 10 feet in front of the opening to be used for attack. Provide an exhaust opening at or near the fire. This opening can be made before starting the fan or when the fan is started. Check for interior openings that could allow the products of combustion to be pushed into unwanted areas. Start the fan and check the cone of air produced. It should completely cover the opening. This can be checked by running a hand around the doorframe to feel the direction of air currents. Allow smoke to clear—usually 30 seconds to 1 minute depending on the size of the area to be ventilated and smoke conditions.

2. Sound the roof with a tool to locate the roof supports. Make two parallel cuts, perpendicular to the roof supports. Do not cut through the roof supports. Rock the saw over them to avoid damaging the integrity of the roof structure. Make cuts parallel to the supports and between pairs of supports in a rectangular pattern. Strike the nearest side of each section of the roofing material with an axe or maul, pushing it down on one side; use the support at the center of each panel as a fulcrum. This hole should be the same size as the opening made in the roof decking.

Skill Drills

Skill Drill 14-1: Breaking Glass with a Hand Tool

1. Position yourself to the **side** of the window.
2. With your back facing the wall, swing **backward** forcefully with the tip of the tool striking the top one-third of the glass.
3. Clear remaining glass from the opening with the **hand tool**.

Skill Drill 14-5: Delivering Positive-Pressure Ventilation

1. Place the fan in front of the opening to be used for the fire attack.

2. Provide an exhaust opening at or near the fire.

3. Start the fan and allow the smoke to clear.

Skill Drill 14-6: Sounding a Roof

1. Use a hand tool to check the roof before **stepping** onto it.
2. Use the tool to sound ahead and to both **sides** as you walk. Locate support members by sound and rebound. Check conditions around your work area periodically.
3. Sound the roof along your **exit** path.

Skill Drill 14-8: Making a Rectangular or Square Cut

1. Locate the roof supports by sounding. Make the first cut parallel to the roof support.

2. Make a triangle cut at the first corner.

3. Make two cuts perpendicular to the roof supports and then make the final cut parallel to another roof support.

(continues)

Skill Drill 14-8: Making a Rectangular or Square Cut (continued)

4. Pull out or push in the triangle cut.

5. Punch out the ceiling below. Be wary of a sudden updraft of hot gases or flames.

Skill Drill 14-9: Making a Louver Cut

1. Locate the roof supports by **sounding**.
2. Make two parallel cuts **perpendicular** to the roof supports.
3. Cut parallel to the supports and between pairs of supports in a(n) **rectangular** pattern.
4. Tilt the **panel** to a vertical position.

Skill Drill 14-10: Making a Triangular Cut

1. Locate the roof supports.

2. Make the first cut from just inside a support member in a diagonal direction toward the next support member.

3. Begin the second cut at the same location as the first, and make it in the opposite diagonal direction, forming a "V" shape.

4. Make the final cut along the support member so as to connect the first two cuts. Cutting from this location allows fire fighters the full support of the member directly below them while performing ventilation.

Chapter 15: Water Supply

Matching

1. E (page 458)
2. F (page 446)
3. B (page 458)
4. A (page 457)
5. D (page 450)
6. C (page 457)
7. J (page 449)
8. I (page 449)
9. H (page 450)
10. G (page 450)

Multiple Choice

1. A (page 449)
2. C (page 449)
3. B (page 449)
4. D (page 450)
5. A (page 450)
6. C (page 450)
7. C (page 450)
8. B (page 452)
9. B (page 453)
10. A (page 453)
11. C (page 456)
12. D (page 456)
13. A (page 457)
14. C (page 457)
15. B (page 457)
16. C (page 458)
17. B (page 458)
18. A (page 445)
19. D (page 448)
20. A (page 449)

Labeling

1. Dry hydrant placed near a static water source.

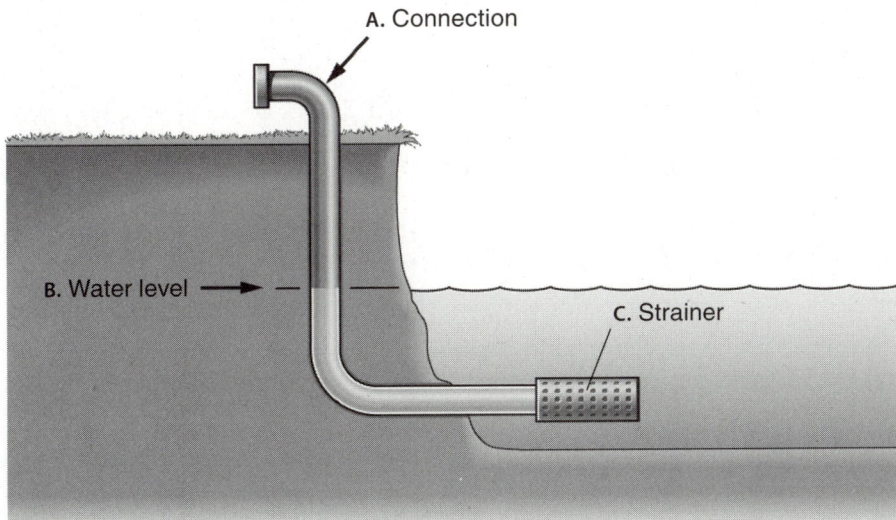

Figure 15-4

Vocabulary

1. **Gravity-feed system:** A water distribution system that depends on gravity to provide the required pressure. The system storage is usually located at a higher elevation than the end users. (page 449)
2. **Dry-barrel hydrant:** A type of hydrant used in areas subject to freezing weather. The valve that allows water to flow into the hydrant is located underground, and the barrel of the hydrant is normally dry. (page 452)
3. **Tanker shuttle:** A method of transporting water from a source to a fire scene using a number of mobile water supply apparatus. (page 460)
4. **Normal operating pressure:** The observed static pressure in a water distribution system during a period of normal demand. (page 457)
5. **Municipal water system:** A water distribution system that is designed to deliver potable water to end users for domestic, industrial, and fire protection purposes. (pages 449, 460)

Fill-in

1. water supply (page 444)
2. static water supplies (page 444)
3. reservoirs (page 449)
4. Control (page 450)
5. valves (page 452)
6. barrel (page 452)
7. fully, fully (page 453)
8. visibility (page 456)
9. volume (page 457)
10. potential (page 457)
11. gravity (page 457)
12. mobile (page 445)

True/False

1. T (page 444)
2. T (page 449)
3. F (page 449)
4. F (page 450)
5. T (page 452)
6. T (page 453)
7. T (page 455)
8. F (page 457)
9. T (page 457)
10. T (page 458)

Short Answer

1. (1) Municipal water systems furnish water under pressure through fire hydrants. (2) Rural areas may depend on static water sources such as lakes and streams. (page 444)
2. Wet-barrel hydrants are used in locations where temperatures do not drop below freezing. These hydrants always have water in the barrel and do not have to be drained after each use.

    Dry-barrel hydrants are used in climates where temperatures can be expected to fall below freezing. The valve that controls the flow of water into the barrel of the hydrant is located at the base, below the frost line, to keep the hydrant from freezing. (page 452)
3. The fire fighter making the connection opens a large outlet cap and then releases the valve just enough to ensure that water flows into the hydrant and flushes out any foreign matter. The fire fighter then closes the valve, connects the hose, and reopens the valve all the way. (page 453)
4. The first part of a hydrant inspection involves checking the exterior of the hydrant for signs of damage. Open the steamer port of a dry-barrel hydrant to ensure that the barrel is dry and free of debris. Make sure that all caps are present and that the outlet hose threads are in good working order.

    The second part of the inspection ensures that the hydrant works properly. Open the hydrant valve just enough to confirm that water flows out and flushes any debris out of the barrel. After flushing, shut down the hydrant. Leave the cap off dry-barrel hydrants to ensure that they drain properly. A properly draining hydrant will create suction against a hand placed over the outlet. When the hydrant is fully drained, replace the cap.

    If the threads on the discharge ports need cleaning, use a steel brush and a small triangular file to remove any burrs in the threads. Also check the gaskets in the caps to make sure they are not cracked, broken, or missing. Replace worn gaskets with new ones, which should be carried on each apparatus. Follow the manufacturer's recommendations for any parts that require lubrication. (page 456)

Word Fun

(Crossword answers)
- 1 Across: WATERSUPPLY
- 7 Across: RESIDUAL
- 9 Across: STATICWATER
- 10 Across: GRAVITYFEED
- 12 Across: STEAMERPORT
- 13 Across: PORTABLETANKS
- 2 Down: PRIMSMPVE (P R I M... — visible letters: P, R, I, M)
- 3 Down: DISTRIBUTORS
- 4 Down: FLOW
- 5 Down: PITOTGAUGE
- 6 Down: DUMPVALVE
- 8 Down: WETBARREL
- 9 Down: STAYCEDE
- 11 Down: DRYBARREL

Fire Alarms

1. Remove the tank from the tender; properly position the tank and expand the tank; assist the pump operator with hooking up the hard suction and strainer; and discharge your tanker water into the tank.
2. The first factors to check when inspecting hydrants are visibility and accessibility. Hydrants should always be visible from every direction, so they can be easily spotted. A hydrant should not be hidden by tall grass, brush, fences, debris, dumpsters, or any other obstruction. The second part of the inspection ensures that the hydrant works properly. Open the hydrant valve just enough to ensure that water flows out and flushes any debris out of the barrel. After flushing, shut down the hydrant. Leave the cap off dry-barrel hydrants to ensure that they drain properly. A properly draining hydrant will create suction against a hand placed over the outlet opening. When the hydrant is fully drained, replace the cap.

Skill Drills

Skill Drill 15-2: Operating a Fire Hydrant

1. Remove the cap from the **outlet** you will be using.
2. Quickly look inside the hydrant opening for **foreign** objects (dry-barrel hydrant only).
3. Check that the remaining **caps** are snugly attached (dry-barrel hydrant only).
4. Attach the hydrant wrench to the **stem** nut. Check for an arrow indicating the direction to turn to open.
5. Open the hydrant enough to verify the **flow** and flush the hydrant (dry-barrel hydrant only).
6. Shut off the **flow** of water (dry-barrel hydrant only).
7. Attach the hose or valve to the hydrant **outlet**.
8. When instructed, turn the **hydrant wrench** to fully open the valve.
9. Open the hydrant slowly to avoid a pressure **surge**.

Skill Drill 15-3: Shutting Down a Hydrant

1. Turn the wrench to slowly close the hydrant valve.

2. Drain the hose line. Slowly disconnect the hose from the hydrant outlet.

3. Leave one hydrant outlet open until the hydrant is fully drained (dry-barrel hydrant only).

4. Replace the hydrant cap.

Chapter 16: Fire Hose, Nozzles, Streams, and Foam

Matching

1. L (page 487)
2. O (page 480)
3. C (page 516)
4. G (page 526)
5. F (page 480)
6. M (page 487)
7. A (page 479)
8. K (page 466)
9. I (page 466)
10. B (page 480)
11. N (page 481)
12. J (page 465)
13. E (page 481)
14. H (page 522)
15. D (page 468)

Multiple Choice

1. C (page 465)
2. A (page 465)
3. B (page 465)
4. D (page 465)
5. A (page 465)
6. C (page 466)
7. C (page 467)
8. D (page 467)
9. B (page 468)
10. C (page 469)
11. D (page 472)
12. B (page 474)
13. A (page 479)
14. A (page 479)
15. D (page 481)
16. C (page 482)
17. C (page 492)
18. B (page 492)
19. A (page 503)
20. D (page 516)
21. B (page 517)
22. C (page 518)
23. A (page 521)
24. D (page 522)
25. C (page 524)

Labeling

1. Higbee indicators and Higbee cuts.

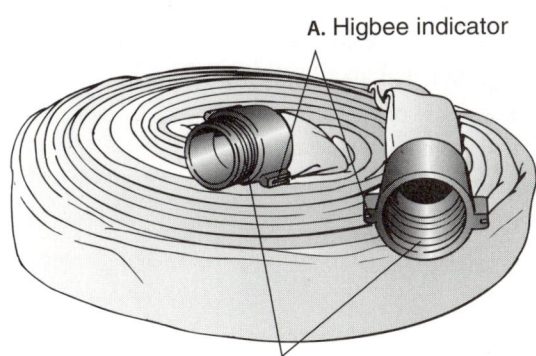

Figure 16-7

2. Ball valve, gate valve, and butterfly valve.

A. Butterfly valve   B. Gate valve   C. Ball valve

Figure 16-19

3. Pin lugs, recessed lugs, and rocker lugs.

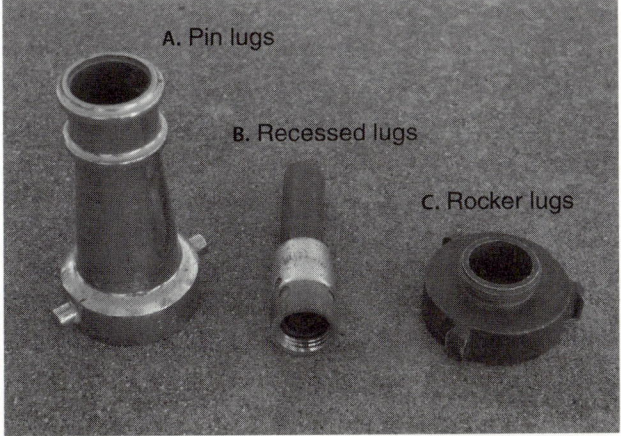

Figure 16-6

Vocabulary

1. **Fixed-gallonage fog nozzle:** A nozzle that delivers a preset flow at the rated discharge pressure. The nozzle could be designed to flow 30, 60, or 100 gpm. (page 518)
2. **Higbee indicators:** Sometimes called Higbee notches or Higbee cuts; show the position where the ends of the threads on a pair of couplings are properly aligned with each other. Using the Higbee indicators will help you to couple hose more quickly. When the indicators on the male and female couplings are aligned, the two couplings should connect quickly and easily. (page 468)
3. **Split hose lay:** A hose lay performed by two engine companies in situations where hose must be laid in two different directions to establish a water supply. This evolution could be used when the attack engine must approach a fire either along a dead-end street with no hydrant or down a long driveway. To perform a split hose lay, the attack engine drops the end of its supply hose at the corner of the street and performs a forward lay toward the fire. The supply engine stops at the intersection, pulls off enough hose to connect to the end of the supply line that is already there, and then performs a reverse lay to the hydrant or water source. When the two lines are connected together, the supply engine can provide water to the attack engine. (page 490)
4. **Water hammer:** An event in which flowing water stops suddenly; the velocity force of the moving water is transferred to everything it is in contact with. Water hammers can involve tremendous forces that damage equipment and cause injury. (page 466)
5. **Hard suction hose:** Hose designed to prevent collapse under vacuum conditions, ensuring that it can be used for drafting water from below the pump (e.g., lakes, rivers, wells, or sea water). (page 474)

Fill-in

1. Flow (page 465)
2. increase (page 465)
3. Water hammer (page 466)
4. attack hoses/lines (page 466)
5. Couplings (page 467)
6. rubber gasket (page 468)
7. ramps (page 476)
8. Mildew (page 476)
9. double female (page 480)
10. Valves (page 481)
11. supply, attack (page 486)
12. triple layer (page 499)
13. before (page 505)
14. after (page 510)

True/False

1. T (page 465)
2. T (page 466)
3. F (page 467)
4. T (page 466)
5. F (page 467)
6. F (page 467)
7. T (page 470)
8. T (page 476)
9. T (page 479)
10. F (page 516)
11. T (page 497)
12. T (page 502)
13. F (page 507)
14. T (page 512)

Short Answer

1. (1) Hose size, type, and manufacturer; (2) date when the hose was manufactured; (3) date when the hose was purchased; (4) dates when the hose was tested; (5) any repairs that have been made to the hose (page 479)
2. A split hose bed is divided into two or more sections for several purposes:
   - One compartment in a split hose bed can be loaded for forward lay (female coupling out) and the other side can be loaded for a reverse lay (male coupling out). This allows a line to be laid in either direction without adaptors.
   - Two parallel hose lines can be laid at the same time (called laying dual lines). Dual lines are beneficial if the situation requires more water than one hose line can supply.

- The split beds can be used to store hoses of different sizes. For example, one side of the hose bed could be loaded with 2 ½-inch hose that can be used as a supply line or as an attack line. The other side of the hose bed could be loaded with 5-inch hose for use as a supply line. This setup enables the use of the most appropriate-size hose for a given situation.
- All of the hose from both sides of the hose bed can be laid out as a single hose line. This is done by coupling the end of the hose in one bed to the beginning of the hose in the other bed. (page 490)

3. (1) Use a large area such as a parking lot for this procedure. (2) Disconnect any gate valves or nozzles from the hose before you begin. (3) Grasp the end of the hose, and pull it off the engine in a straight line. (4) When a coupling comes off the engine, disconnect the hose and pull off the next section of hose. (5) When all of the hose has been removed from the hose bed, use a broom to brush off any dirt or debris on both sides of the hose jacket. (6) Sweep out any debris or dirt from the hose bed. (7) Roll all of the hose into doughnut rolls. Place the male end out for supply lines, and the female end out for handlines. (8) Store hose rolls off the floor on a rack, in a cool dry area. (page 514)

Word Fun

[Crossword puzzle with answers including: CLASS B FOAM, HOSE ROLLER, SMALL DIAMETER HOSE, ATTACK LINE, HOSE LINER]

Fire Alarms

1. (a) A split hose lay. (b) To perform a split hose lay, the attack engine drops the end of its supply hose at the corner of the street and performs a forward lay toward the fire. The supply engine stops at the intersection, pulls off enough hose to connect to the end of the supply line that is already there, and then performs a reverse lay to the hydrant or water source. When the two lines are connected together, the supply engine can provide water to the attack engine. (page 490)

2. (1) Connect the female end of the first length of hose to the discharge outlet.

   (2) Flat load the hose even to the edges of the hose bed. At approximately the 30-feet mark, make a loop/ear (this will create a handle for advancing purposes). Continue the load until you reach the 60- to 70-feet mark. Make an additional loop/ear. Flat load the remainder of the 100-feet section, leaving extra hose to the side, in the opposite direction of deployment.

   (3) Assemble the remaining hose sections and attach the nozzle.

   (4) Place the nozzle on the hose bed.

(5) Load the remaining 100 feet of hose flat into the bed, alternating the folds from the left to right sides of the bed.

(6) Connect the last section loaded to the first section placed in the bed.

(7) Lay the remaining loose hose on top of the load. (page 499)

3. (1) Grasp the nozzle and the folds next to it.

(2) Pull the load approximately one-third out of the bed.

(3) Turn away from the hose bed and place the load on the shoulder. Walk away from the apparatus until all hose is clear of the hose bed.

(4) Continue walking away, pulling the remaining hose from the hose bed and then allowing the hose to deploy from the top of the load on the shoulder. (page 499)

Skill Drills

Skill Drill 16-4: Performing the One-Fire Fighter Knee-Press Method of Uncoupling a Fire Hose

1. Pick up the connection by the **female** coupling end.
2. Turn the connection **upright**, resting the male coupling on a firm surface.
3. Place a knee on the **female** coupling and press down on it with your body weight. Turn the female swivel counterclockwise and loosen the coupling.

Skill Drill 16-6: Uncoupling a Hose with Spanners

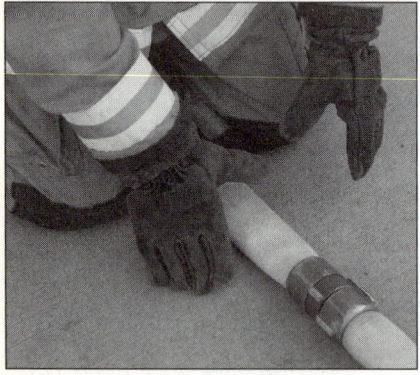

1. With the connection on the ground, straddle the connection above the female coupling.

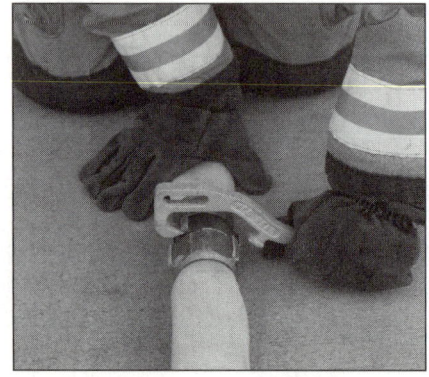

2. Place one spanner wrench on the female coupling with the handle of the wrench to the left.

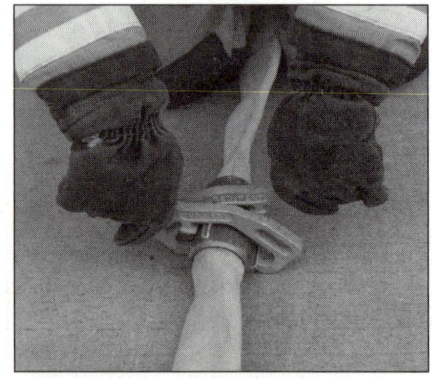

3. Place the second spanner wrench on the male coupling with the handle of the wrench to the right.

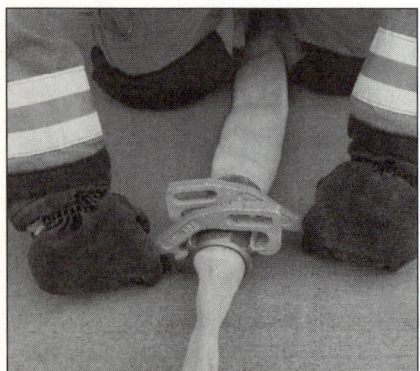

4. Push both spanner handles down toward the ground, loosening the connection.

Skill Drill 16-12: Performing a Single-Doughnut Roll

1. Place the hose **flat** and in a straight line.
2. Locate the **midpoint** of the hose.
3. From the midpoint, move **5 feet** toward the male coupling end. Start rolling the hose toward the female coupling.
4. At the end of the roll, wrap the excess hose of the **female** end over the male coupling to protect the threads.

Skill Drill 16-19: Performing a Flat Hose Load

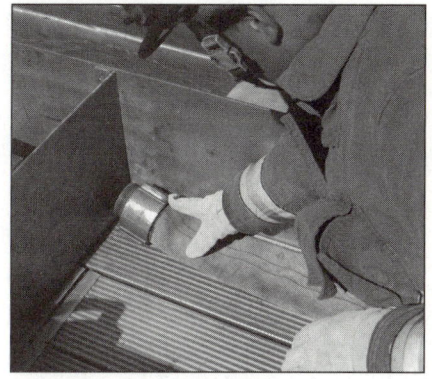
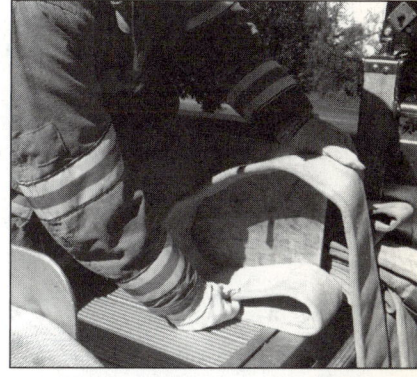

1. To set up the hose for a forward lay, place the male hose coupling in the hose bed first. To set up the hose for a reverse lay, place the female hose coupling in the hose bed first.
2. Start the hose lay with the coupling at the front end of the hose compartment.
3. Fold the hose back on itself at the rear of the hose bed.

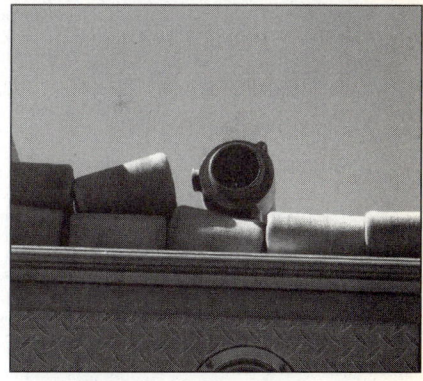

4. Run the hose back to the front end on top of the previous length of hose. Fold the hose back on itself so that the top of the hose is on the previous length.
5. While laying the hose back to the rear of the hose bed, angle the hose to the side of the previous fold.
6. Continue to lay the hose in neat folds until the whole hose bed is covered with a layer of hose. Continue to load the layers of hose until the required amount of hose is loaded.

Skill Drill 16-21: Performing an Accordion Hose Load

1. Lay the first length of hose in the hose bed on its **edge** against the side of the hose bed.
2. Double the hose back on itself at the rear of the hose bed. Leave the **female** end extended so that the two hose beds can be cross-connected.
3. Lay the hose next to the first length and bring it to the **front** of the hose bed. Fold the hose at the front of the hose bed so that the bend is even to the edge of the hose bed. Continue to lay folds of hose across the hose bed.
4. Alternate the length of the hose folds at each end to allow more room for the folded ends. When the bottom layer is completed, angle the hose **upward** to begin the second tier. Continue the second layer by repeating the steps used to complete the first layer.

Skill Drill 16-22: Attaching a Soft Suction Hose to a Fire Hydrant

1. The driver/operator positions the apparatus so that its inlet is the correct distance from the hydrant.

2. Remove the hose, any needed adaptors, and the hydrant wrench.

3. Attach the soft suction hose to the inlet of the engine if it is not already attached to the engine intake.

4. Unroll the hose.

5. Remove the large hydrant cap.

6. Attach the soft suction hose to the hydrant.

7. Ensure that there are no kinks or sharp bends in the hose that might restrict the flow of water.

8. Open the hydrant slowly when so indicated by the driver/operator. Check all connections for leaks. Tighten the couplings if necessary.

9. Place chafing blocks under the hose where it contacts the ground to prevent mechanical abrasion.

Skill Drill 16-24: Performing a Minuteman Hose Load

1. Connect the **female end** of the first length of hose to the discharge outlet.
2. Flat load the hose **even** to the edges of the hose bed. At approximately the 30-feet mark, make a loop/ear (this will create a handle for advancing purposes). Continue the load until you reach the 60- to 70-feet mark. Make an additional loop/ear. Flat load the remainder of the 100-feet section, leaving extra hose to the side, in the opposite direction of deployment.
3. Assemble the remaining hose sections and attach the **nozzle**. Place the nozzle on the **hose bed**.
4. Load the remaining 100 feet of hose flat into the bed, **alternating** the folds from the left to right sides of the bed.
5. **Connect** the last section loaded to the first section placed in the bed.
6. **Lay** the remaining loose hose on top of the load.

Skill Drill 16-25: Advancing a Minuteman Hose Load

1. Grasp the nozzle and the folds next to it.

2. Pull the load approximately one-third out of the bed.

3. Turn away from the hose bed and place the load on the shoulder. Walk away from the apparatus until all hose is clear of the hose bed.

4. Continue walking away, pulling the remaining hose from the hose bed and then allowing the hose to deploy from the top of the load on the shoulder.

Skill Drill 16-26: Performing a Preconnected Flat Load

1. Attach the **female end** of the hose to the preconnect discharge.
2. Begin laying the hose **flat** in the hose bed.
3. When about one-third of the hose is in the bed, make an 8-inch loop at the end of the hose bed. This loop will be used as a pulling handle. When two-thirds of the hose is loaded, make a second pulling loop that is about **twice** the size of the first loop.
4. Finish loading the hose, attach the nozzle, and place it **on top** of the hose bed. The preconnected flat load is now ready for use.

Skill Drill 16-27: Advancing a Preconnected Flat Hose Load

1. Place an arm through the larger lower loop. Grasp the smaller loop with the same hand.

2. Grasp the nozzle with the opposite hand.

3. Pull the load from the bed.

4. Walk away from the vehicle.

5. As the load deploys, drop the small loop. Extend the remaining hose to length.

Skill Drill 16-28: Performing a Triple-Layer Hose Load

1. Attach the female end of the hose to the preconnect **discharge**.
2. Connect the **sections** of hose together.
3. Extend the hose directly from the hose bed. Pick up the hose **two-thirds** of the distance from the discharge to the hose nozzle.
4. Carry the hose back to the apparatus, forming a **three-layer** loop.
5. Pick up the entire length of **folded** hose.
6. Lay the tripled folded hose in the hose bed in an **S-shape** with the nozzle on top.

Skill Drill 16-29: Advancing a Triple-Layer Hose Load

1. Grasp the nozzle and the top fold.

2. Turn away from the hose bed and place the hose on the shoulder.

3. Walk away from the vehicle until the entire load is out of the bed.

4. When the load is out of the bed, drop the fold.

5. Extend the nozzle the remaining distance.

Skill Drill 16-34: Advancing a Hose Line Up a Stairway

1. Use a __shoulder__ carry to advance up the stairs.
2. When ascending the stairway, lay the hose against the __outside__ of the stairs to reduce tripping hazards. Avoid sharp bends.
3. Arrange __excess__ hose so that it is available to fire fighters entering the fire floor.

Skill Drill 16-36: Advancing an Uncharged Hose Line Up a Ladder

1. Advance the hose line to the ladder.

2. Pick up the nozzle; place the hose across the chest, with the nozzle draped over the shoulder.

3. Climb up the ladder with the uncharged hose line. Once the first fire fighter reaches the first fly section of the ladder, a second fire fighter shoulders the hose to assist advancing the hose line up the ladder. To avoid overloading of the ladder, enforce a limit of one fire fighter per fly section.

4. The nozzle is placed over the top rung of the ladder and advanced into the fire area.

5. Additional hose can be fed up the ladder until sufficient hose is in position. The hose can be secured to the ladder with a hose strap to support its weight and keep it from becoming dislodged.

Skill Drill 16-42: Operating a Smooth-Bore Nozzle

1. Select the desired tip size and attach it to the nozzle **shut-off** valve.
2. Attain a stable **stance** (if standing).
3. Slowly open the **valve**, allowing water to flow.
4. Open the valve **completely** to achieve maximum effectiveness.
5. Direct the **stream** to the desired location.

Skill Drill 16-43: Operating a Fog-Stream Nozzle

1. Attain a stable stance (if standing).

2. Slowly open the valve, allowing water to flow.

3. Open the valve completely.

4. Select the desired water pattern, by rotating the bezel of the nozzle. Apply water where needed.

Chapter 17: Fire Fighter Survival

Matching

1. D (page 538)
2. C (page 542)
3. A (page 546)
4. E (page 538)
5. B (page 532)
6. J (page 536)
7. I (page 536)
8. G (page 533)
9. H (page 533)
10. F (page 533)

Answer Key

Multiple Choice

1. C (page 532)
2. A (page 533)
3. D (page 533)
4. B (page 534)
5. C (page 535)
6. A (page 536)
7. D (page 536)
8. C (page 538)
9. B (page 538)
10. B (page 542)
11. A (page 545)
12. C (page 549)
13. A (page 549)
14. A (page 532)
15. A (page 532)
16. D (page 533)
17. B (page 533)
18. D (page 534)
19. B (page 536)
20. C (page 538)

Vocabulary

1. **Air management:** The way in which an individual uses a limited air supply to ensure that it will last long enough to enter a hazardous area, accomplish needed tasks, and return safely. (page 542)
2. **Rapid intervention company/crew (RIC):** A company or crew that is assigned to stand by at the incident scene, fully dressed and equipped for action, and that is ready to deploy immediately when assigned to do so by the IC. (page 536)
3. **Critical incident stress management (CISM):** A system to help firefighting personnel deal with critical incident stress in a positive manner. Its aim is to promote long-term mental and emotional health after a critical incident. (page 547)
4. **Safe haven:** A temporary place of refuge in which fire fighters can await rescue. (page 542)
5. **Self-rescue:** A fire fighter's use of techniques and tools to remove himself or herself from a hazardous situation. (page 538)

Fill-in

1. incident commander (page 532)
2. continuous (page 533)
3. simple observation (page 533)
4. training (page 533)
5. integrity (page 534)
6. mayday (page 535)
7. oriented (page 538)
8. management (page 542)
9. rehabilitation (page 546)
10. not positive (page 547)

True/False

1. T (page 533)
2. T (page 533)
3. T (pages 534–535)
4. F (page 538)
5. F (page 538)
6. T (page 542)
7. F (page 542)
8. T (page 547)
9. F (page 547)
10. T (page 549)

Short Answer

1. (1) We will not risk our lives at all for persons or property that is already lost. (2) We will accept a limited level of risk, under measured and controlled conditions, to save property of value. (3) We will accept a higher level of risk only where there is a reasonable and realistic possibility of saving lives. (page 532)
2. Follow these steps:
   (1) Use your radio to call, "Mayday, mayday, mayday."
   (2) Give a LUNAR report:
   - Location
   - Unit number
   - Name

- Assignment
- Resources needed

(3) Activate your PASS device.

(4) Attempt self-rescue.

(5) If you are able to move, identify a safe haven where you can await rescue.

(6) Lie on your side in a fetal position with your PASS device pointing out so that it can be heard.

(7) Point your flashlight toward the ceiling.

(8) Slow your breathing as much as possible to conserve your air supply. (page 536)

3. (1) Anxiety; (2) denial/disbelief; (3) frustration/anger; (4) inability to function logically; (5) remorse; (6) grief; (7) reconciliation/acceptance (pages 548–549)

Word Fun

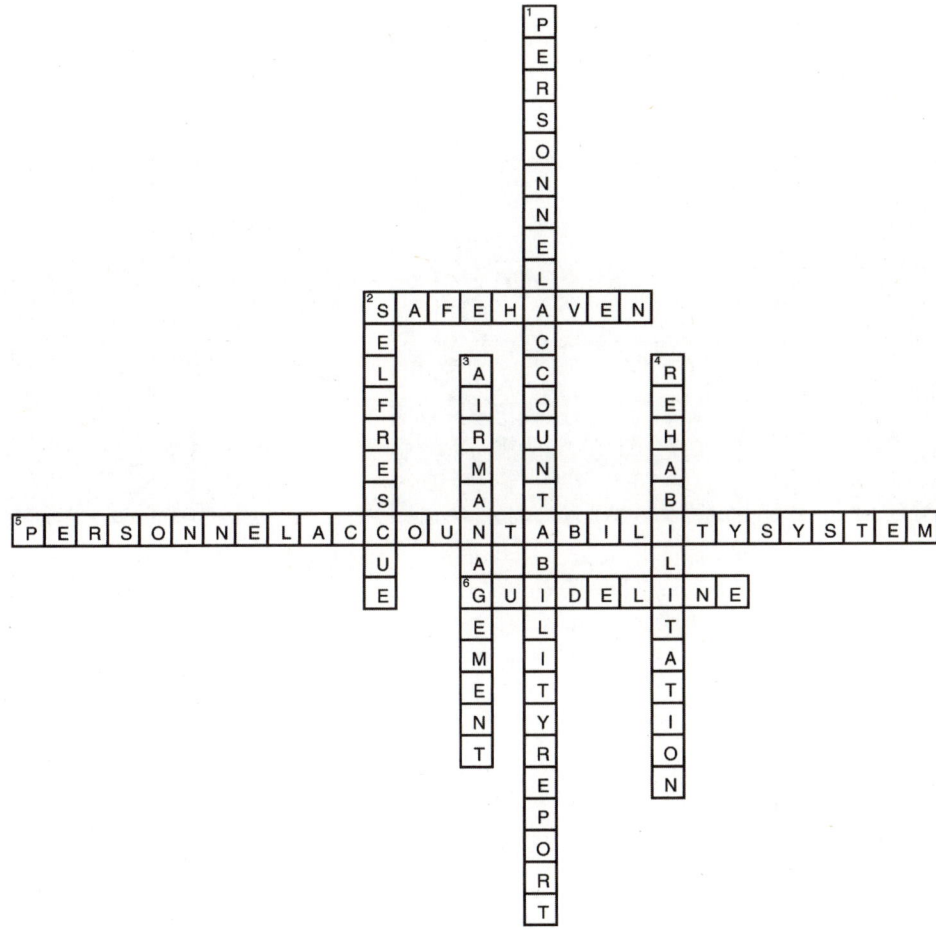

Fire Alarms

1. First, initiate a mayday over the portable radio. Manually activate the PASS device.

   Second, stay calm and control your breathing. Systematically locate a wall. Use a sweeping motion on the outside wall to locate an alternative exit. Identify the opening as a window, interior door, or external door. If the first opening identified is not adequate for an exit, continue to search.

   Third, maintain your orientation and stay low. Exit the room safely if possible. If unable to exit, assume the downed fire fighter position in a safe haven or find refuge. Keep command informed of your situation.

2. Initiate a mayday over the portable radio. Activate the PASS device. Stay calm and control your breathing. Change your position—back up or turn on your side to try to free yourself. Use the swimmer stroke to try to free yourself. Loosen the SCBA straps, remove one arm, and slide the air pack to the front of your body to try to free the SCBA. Cut the wires or cables causing the entanglement. Be aware of any possible electrocution risk. If you are unable to disentangle yourself, notify command of your situation. If you are able to exit, notify command that you are out of danger. (page 542)

Skill Drills

Skill Drill 17-1: Initiating a Mayday Call

1. Use your radio to call, "Mayday, mayday, mayday." State your name and **company number**. State your location. State the type of problem.
2. Activate your **PASS device**. Attempt self-rescue. If you are able to move, identify a safe haven where you can await rescue.
3. Lie on your side in a **fetal** position with your PASS device pointing out so that it can be heard.
4. Point your flashlight toward the **ceiling**. Slow your breathing as much as possible to conserve your air supply.

Skill Drill 17-2: Performing Self-Rescue

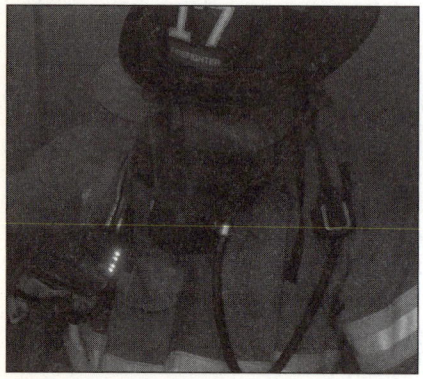  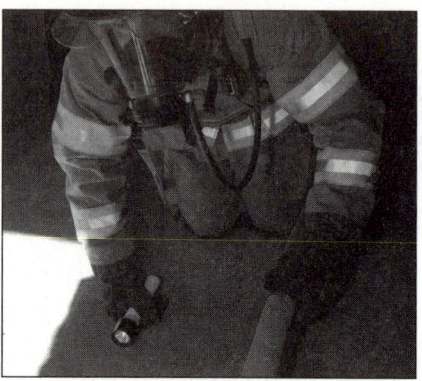

1. Initiate a mayday over the portable radio. Manually activate the PASS device. Stay calm and control your breathing.
2. Systematically search the room to locate a hose line. Follow the hose line to a hose coupling. Identify the male and female ends of the coupling. Move from the female coupling to the male coupling.
3. Follow the hose out. Exit the hazard area. Notify command of your location.

Skill Drill 17-3: Locating a Door or Window for Emergency Exit

1. Initiate a mayday over the portable radio. Manually activate the PASS device. Stay calm and control your **breathing**.
2. Systematically locate a **wall**.
3. Use a **sweeping** motion on the outside wall to locate an alternative exit. Identify the opening as a window, interior door, or external door.
4. If the first opening identified is not adequate for an exit, continue to search. Maintain your **orientation** and stay low. Exit the room safely if possible. If unable to exit, assume the downed fire fighter position in a safe haven or find refuge. Keep command informed of your situation.

Skill Drill 17-4: Opening a Wall to Escape

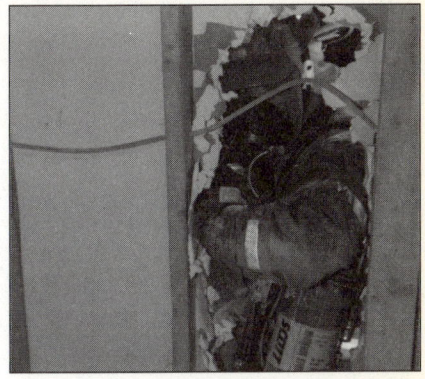

1. Identify deteriorating conditions that require exiting through a wall. Use the portable radio to initiate a mayday. Use a hand tool or your feet to open a hole in the wall between two studs. If using a hand tool, drive the hand tool completely through the wall to check for obstacles on the other side.

2. Enlarge the hole. Enter the hole head first to check the floor and fire conditions on the other side of the wall.

3. Loosen the SCBA waist strap and remove one shoulder strap. Maintain control of the regulator side of the SCBA, which will assist in donning of the air pack later. Sling the SCBA to one side to reduce your profile.

4. Escape through the opening in the wall. Adjust the SCBA straps to their normal position.

5. Assist others through the opening in the wall. Report your status to command.

Skill Drill 17-5: Escaping from an Entanglement

1. Initiate a <u>mayday</u> over the portable radio. Activate the PASS device. Stay calm and control your breathing.
2. Change your <u>position</u>—back up or turn on your side to try to free yourself.
3. Use the swimmer <u>stroke</u> to try to free yourself.
4. Loosen the <u>SCBA</u> straps, remove one arm, and slide the air pack to the front of your body to try to free the SCBA.
5. Cut the wires or cables causing the entanglement. Be aware of any possible <u>electrocution</u> risk. If you are unable to disentangle yourself, notify command of your situation. If you are able to exit, notify command that you are out of danger.

Chapter 18: Salvage and Overhaul

Matching

1. G (page 555)
2. D (page 557)
3. A (page 575)
4. F (page 564)
5. C (page 555)
6. B (page 555)
7. E (page 557)
8. J (page 564)
9. H (page 564)
10. I (page 577)

Multiple Choice

1. B (page 555)
2. C (page 559)
3. A (page 560)
4. D (page 560)
5. D (page 560)
6. B (page 565)
7. A (page 565)
8. C (page 566)
9. B (page 570)
10. D (page 561)
11. A (page 561)
12. B (page 561)
13. C (page 561)
14. D (page 564)
15. B (page 575)
16. A (page 575)
17. A (page 556)
18. B (page 556)
19. A (page 557)
20. C (page 558)

Vocabulary

1. **Floor runner:** A long section of protective material used to cover a section of flooring or carpet. Floor runners protect carpets or hardwood floors from water, debris, fire fighters' boots, and firefighting equipment. (page 570)
2. **Overhaul:** The process of examining all areas of the building and contents involved in a fire to ensure that the fire is completely extinguished. (page 570)
3. **Salvage cover:** Large square or rectangular sheets made of heavy canvas or plastic material that are spread over furniture and other items to protect them from water run-off and falling debris. (page 565)
4. **Balloon-frame construction:** A building in which the exterior walls are assembled with wood studs that run continuously from the basement to the roof. In a two-story building, the floor joists that support the first and second floors are nailed to these continuous studs. As a result, an open channel between each pair of studs extends from the foundation to the attic. (page 575)
5. **Sprinkler wedge:** A piece of wedge-shaped wood placed between the deflector and the orifice of a sprinkler head to stop the flow of water. (page 561)
6. **Sprinkler stop:** A mechanical device with a rubber stopper that can be inserted into a sprinkler head. (page 561)

Fill-in

1. Overhaul (page 555)
2. lower (page 555)
3. evidence (page 555)
4. secondary losses (page 559)
5. close (page 565)
6. replaced (page 563)
7. water (page 565)
8. safety officer (page 573)
9. Fire watch (page 575)
10. feel (page 576)
11. thermal imaging device (page 576)
12. carryalls (page 577)

True/False

1. T (page 555)
2. T (page 559)
3. T (page 560)
4. F (page 565)
5. T (page 570)
6. F (page 561)
7. T (page 575)
8. T (page 557)
9. T (page 577)
10. T (page 576)

Short Answer

1. Salvage covers: (1) treated canvas or plastic; (2) box cutter for cutting plastic; (3) floor runners; (4) wet/dry vacuums; (5) squeegees; (6) submersible pumps and hose; (7) sprinkler shut-off kit; (8) ventilation fans, power blowers; (9) small tool kit; (10) pike poles to construct water chutes (page 560)
2. (1) The structural safety of the building is often compromised. (2) Catastrophic building collapses have occurred during overhaul. (3) Heavy objects could lead to roof or ceiling collapse. (4) Debris could litter the area. (5) There could be holes in the floor. (6) Visibility is often limited, so fire fighters may have to depend on portable lighting. (7) Wet or icy surfaces make falls more likely. (8) Smoldering areas may burst into flames. (9) The air is probably not safe to breathe. (10) During overhaul operations, dangerous equipment including axes, pike poles, and power tools are used in close quarters. (page 571)
3. (1) Lightweight and/or truss construction; (2) cracked walls, out of alignment walls, sagging floors; (3) heavy mechanical equipment on the roof; (4) overhanging cornices or heavy signs; (5) accumulations of water (page 575)
4. (1) Pike poles and ceiling hooks—for pulling ceilings and removing gypsum wallboard. (2) Crowbars and Halligan-type tools—for removing baseboards and window or door casings. (3) Axes—for chopping through wood, such as floor boards and roofing materials. (4) Power tools such as battery-powered saws—for opening up walls and ceilings. (5) Pitchforks and shovels—for removing debris. (6) Rubbish hooks and rakes—for pulling things apart. (7) Thermal imaging cameras—for identifying hot spots. (page 577)

Word Fun

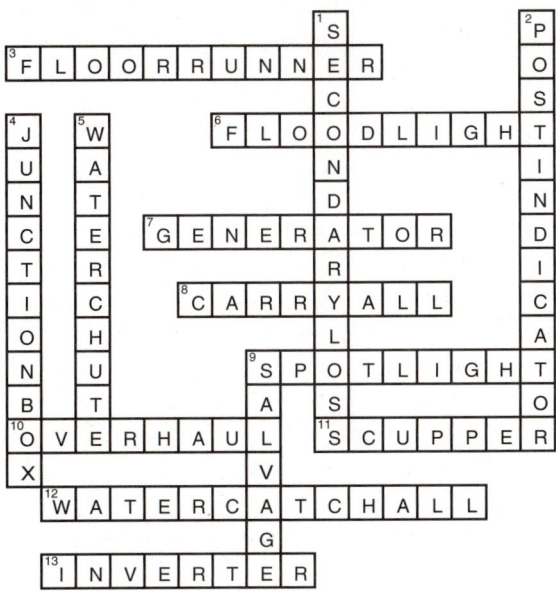

Fire Alarms

1. (a) The most efficient way to protect a room and contents is to move all the furniture to the center of the room, away from the walls, where water could damage the backs of the furniture. This reduces the total area that must be covered, enabling one or two fire fighters to cover the pile quickly and move on to the next room. Remove any pictures from the walls and place them with the furniture. Put smaller pictures and valuable objects in drawers or wherever they will be protected from breakage. If enough time is available, roll up any rugs and place them on the pile.

   (b) To construct a water chute:

   (1) Fully open a large salvage cover flat on the ground.

   (2) Roll the cover tightly from one edge toward the middle. If using pike poles, lay one pole on the edge and roll the cover around the handle. Roll the opposite edge tightly toward the middle in the same manner. Stop when the rolls are 1 to 3 feet apart.

   (3) Turn the cover upside down. Position the chute so that it collects the dripping water and channels it toward a drain or outside opening. The chute can be placed on the floor, with one end propped up by a chair or other object.

   (4) Use a stepladder or other tall object to support chutes constructed with pike poles. (page 564)

2. Salvage covers must be adequately maintained to preserve their shelf life. Salvage cover maintenance depends on the type of cover used. A canvas cover can usually be cleaned with a scrub brush and clean water. If the cover becomes particularly dirty, however, the user may have to use a mild detergent. Covers should be adequately rinsed if a detergent is used.

   Canvas covers must be properly dried before being returned to service. Effectively drying a canvas cover will reduce mildewing.

   Vinyl type covers are easily maintained by rinsing and do not mildew as easy as canvas covers.

   Once dried, salvage covers should be inspected for tears and holes. Any damage found can be patched by using duct tape or a sewn-on patch. (page 568)

3. To fold a salvage cover to prepare it for one fire fighter to deploy:

   (1) Spread the salvage cover flat on the ground. Stand at one end, facing your partner standing at the other end.

   (2) You and your partner each place one hand on the outer edge of the cover and the other hand one quarter of the way in from the edge.

   (3) Together, flip the outside edge in 3 inches from the middle of the cover, creating a fold at the quarter point.

   (4) Flip the outside fold in to the same point of the cover, creating a second fold. Repeat steps 2, 3, and 4, from the opposite side of the cover. The folded edges should meet at the middle of the cover, with the folds 6 inches apart.

   (5) Fold the two halves of the salvage cover together.

   (6) Starting from the middle of the cover, use a broom to brush the air out of the cover.

   (7) Move to the newly created narrow end of the salvage cover.

   (8) Fold the narrow end of the salvage cover 3 inches from the middle of the cover, creating a fold at the quarter point.

   (9) Flip the outside fold of the narrow end in to the same point of the cover, creating a second fold. The folded edge should meet at the middle of the cover, with the folds 6 inches apart.

   (10) Fold the two halves of the salvage cover together.

Skill Drills

Skill Drill 18-3: Using a Sprinkler Stop

1. Have a **sprinkler stop** in hand.
2. Place the **flat-coated** part of the sprinkler stop over the sprinkler head orifice and between the frame of the sprinkler head.
3. Push the **lever** to expand the sprinkler stop until it snaps into position.

Skill Drill 18-8: Constructing a Water Catch-All

1. Fully open a large salvage cover flat on the ground.

2. Roll two edges inward from the opposite sides, approximately 3 feet on each side.

3. Fold each of the four corners at a 90-degree angle, starting each fold approximately 3 feet in from the edge.

4. Roll the remaining two edges inward approximately 2 feet.

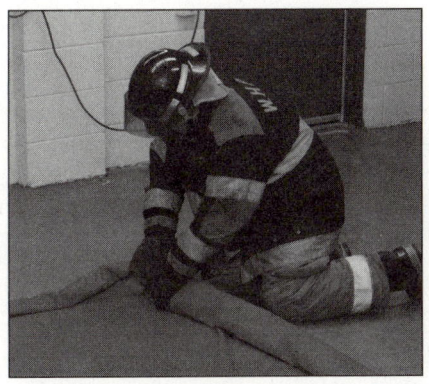

5. Lift the rolled edge over the corner flaps, and tuck it in under the flaps, to lock the corners in place.

Skill Drill 18-15: Pulling a Ceiling Using a Pike Pole

1. Select the appropriate-length pike pole based on the height of the ceiling. For most residential applications, a <u>**6**</u>-feet-long pole is sufficient; longer poles are needed for higher ceilings.
2. Determine which area of the ceiling will be opened. Typically, the most heavily damaged areas are opened <u>**first**</u>, followed by the surrounding areas. Position yourself to begin work with your back toward a door, so the debris you pull down will not block your access to the exit.
3. Using a strong, upward-thrusting motion, penetrate the ceiling with the <u>**tip**</u> of the pike pole. Face the hook side of the tip away from you.
4. Pull down and away from your <u>**body**</u>, so the ceiling material falls away from you.
5. Continue pulling down sections of the ceiling until the desired area is opened. Pull down any insulation, such as <u>**rolled fiberglass**</u>, found in the ceiling.

Skill Drill 18-16: Opening an Interior Wall

1. Determine which area of the wall will be opened up. The officer in charge usually makes this determination. Typically, the areas most heavily damaged by the fire are opened first, followed by the surrounding areas, working outward.

2. Use the axe blade to begin cutting near the top of the wall. Cut downward between wall studs. Survey the wall for electrical switches or receptacles, as they are evidence of electrical wires behind the wall.

3. After making two vertical cuts, use the pick end of the axe to pull the wall material away from the studs and open the wall. Work from top to bottom.

4. If necessary, remove items such as baseboards or window and door trim with a Halligan tool or axe. Continue opening sections of the wall until the desired area is open. Pull out any insulation, such as rolled fiberglass, found behind the wall.

Chapter 19: Fire Fighter Rehabilitation

Matching

1. F (page 584)
2. D (page 586)
3. E (page 591)
4. A (page 585)
5. C (page 589)
6. B (page 593)
7. J (page 590)
8. I (page 591)
9. H (page 594)
10. G (page 595)

Multiple Choice

1. A (page 586)
2. A (page 584)
3. D (page 588)
4. D (page 590)
5. A (page 591)
6. D (page 585)
7. A (page 593)
8. D (page 590)
9. C (page 594)
10. A (page 593)

Vocabulary

1. **Turnout gear:** A helmet, hood, coat, pants, boots, gloves, personal alert safety system, and self-contained breathing apparatus. Sometimes referred to as personal protective equipment (PPE). (page 596)
2. **Frostbite:** Damage to tissues as the result of exposure to cold; frozen or partially frozen body parts. (page 594)
3. **Critical incident stress management (CISM):** A system to help personnel deal with major critical incidents in a positive manner. Its aim is to promote long-term mental and emotional health after a critical incident. (page 595)
4. **Emergency incident rehabilitation:** A function on the emergency scene that cares for the well-being of the fire fighters. It includes physical assessment, revitalization, medical evaluation and treatment, and regular monitoring of vital signs. (page 584)
5. **Hypothermia:** A condition in which the internal body temperature falls below 95 °F, usually a result of prolonged exposure to cold or freezing temperatures. (page 590)

Fill-in

1. Rehabilitation (page 584)
2. intensity (page 585)
3. 1 to 1.5, 2 (page 593)
4. glucose (page 593)
5. Carbohydrates (page 593)
6. Sugar (page 593)
7. Frostbite (page 594)
8. Reassignment (page 595)
9. Bloating (page 591)
10. medical evaluation, treatment (page 594)

True/False

1. T (page 584)
2. F (page 591)
3. T (page 586)
4. T (page 586)
5. T (page 590)
6. F (page 591)
7. F (page 593)
8. T (page 593)
9. F (page 595)
10. T (page 595)

Short Answer

1. Without the opportunity to rest and recover, you may develop physical symptoms such as fatigue, headaches, or gastrointestinal problems. (page 584)

2. (1) The loud, jarring sound of the alarm. (2) Your personal protective equipment (PPE). (3) You may not have time to eat or get something to drink. (4) You may have to drive an emergency vehicle, haul hoses, position a ladder, and climb to the roof to cut a ventilation hole. All of these tasks require a significant amount of energy and concentration, and you must be able to move into action quickly with no time to warm up your muscles as athletes do before an event. (5) You may be called to a fire on the hottest day of the year, the coldest day of the year, and under all types of adverse circumstances. (6) You may feel an added emotional stress, which in turn affects your body. (7) Fire fighters often work in unfamiliar, smoke-filled environments. (page 585)

3. Dehydration reduces strength, endurance, and mental judgment. (page 586)

4. High-rise fires, wildland fires, hazardous materials incidents (pages 588–589)

5. (1) Medical monitoring; (2) revitalization; (3) medical monitoring and treatment; (4) transportation; (5) critical incident stress management; (6) reassignment. (page 590)

6. Drinks such as colas, coffee, and tea should be avoided because they contain caffeine; caffeine acts as a diuretic that causes the body to excrete more water. (page 591)

7. The sensation of thirst is not a reliable indicator of the amount of water the body has lost. Thirst develops only after the body is already dehydrated. (page 591)

8. During short-duration incidents, low-sugar, high-protein sports bars can be used to keep the glucose balance steady. During extended-duration incidents, a fire fighter should eat a more complete meal. The proper balance of carbohydrates, proteins, and fats will maintain energy levels throughout the emergency. To ensure peak performance, the meal should include complex carbohydrates such as whole-grain breads, whole-grain pasta, rice, and vegetables. It is also better to eat a series of smaller meals, rather than one or two large meals, because larger meals can increase glucose levels and slow down the body. (page 593)

9. Part of your responsibility is to know your own limits. No one else can know what you ate, whether you are lightheaded or dehydrated, whether you are feeling ill, or whether you need a breather. You are the only person who knows these things. Therefore, you may be the only person who knows when you need to request rehabilitation. (page 595)

Word Fun

```
 1            2
[D][E][H][Y][D][R][A][T][I][O][N]
              [E]
              [H]
              [A]
              [B]    3        4
              [I]   [G]      [F]
              [L]   [L]      [R]
              [I]   [U]      [O]
                    [C]      [S]
          5
         [E][L][E][C][T][R][O][L][Y][T][E][S]
              [A]   [S]      [B]
          6
         [H][Y][P][O][T][H][E][R][M][I][A]
                    [E]      [T]
                             [E]
```

Fire Alarms

1. Part of your responsibility is to know your own limits. No one else can know what you ate, whether you are lightheaded or dehydrated, whether you are feeling ill, or whether you need a breather. You are the only person who knows these things. Therefore, you may be the only person who knows when you need to request rehabilitation. It may be difficult to say, "I need a break," while your team members are still hard at work. Even so, it is better to take a break when you need it than to push yourself too far and have to be rescued by other members of your department. (pages 593–595)
2. Fire fighters responding to incidents during cold weather are subject to both hypothermia and frostbite (damage to tissues resulting from prolonged exposure to cold). In these cases, the rehabilitation center needs to be heated so that fire fighters can warm up before returning to the chilly environment. Fire fighters who are wet or severely chilled should be wrapped in warming blankets and moved into a well-heated area before they remove their turnout gear. As soon as possible, all wet clothing should be removed and replaced with warm, dry clothing. (page 594)

Chapter 20: Wildland and Ground Fires

Matching

1. C (page 605)
2. F (page 602)
3. E (page 601)
4. B (page 600)
5. D (page 601)
6. G (page 604)
7. A (page 604)
8. K (page 604)
9. J (page 604)
10. I (page 604)
11. H (page 604)

Multiple Choice

1. C (page 600)
2. A (page 601)
3. D (page 601)
4. B (page 602)
5. A (page 602)
6. D (page 604)
7. B (page 604)
8. C (page 604)
9. A (page 605)
10. B (page 606)
11. C (page 606)
12. D (page 606)
13. A (page 605)
14. B (page 607)
15. D (page 606)
16. B (page 605)

Vocabulary

1. **Fuel continuity:** The relative closeness of wildland fuels—a factor in a fire's ability to spread from one area of fuel to another. (page 602)
2. **Backpack pump extinguisher:** A portable fire extinguisher consisting of a 4-gallon to 8-gallon water tank that is worn on the user's back and features a hand-powered piston pump for discharging the water. (page 604)
3. **Heavy fuels:** Fuels of a large diameter, such as large brush, heavy timber, snags, stumps, branches, and dead timber on the ground. These fuels ignite and are consumed more slowly than light fuels. (page 601)
4. **Aerial fuels:** Fuels located more than 6 feet off the ground, usually part of or attached to trees. (page 601)
5. **Topography:** The features of the earth's surface; changes in land elevation and the position of natural and human-made features. (page 604)

Answer Key

Fill-in

1. Weather (page 602)
2. under, on, above (age 601)
3. quickly (page 601)
4. humidity (page 602)
5. Fixed (page 605)
6. Wildland, ground (page 605)
7. Backpack pump extinguishers (page 604)
8. area of origin (page 604)
9. spot (page 604)
10. oxygen (page 602)

True/False

1. T (page 600)
2. F (page 601)
3. T (pages 602, 604)
4. T (page 602)
5. T (page 608)
6. T (page 600)
7. F (page 601)
8. T (page 602)
9. T (page 604)
10. F (page 605)

Short Answer

1. (1) Driving on unimproved roads and steep terrain greatly increases the chance of fire apparatus rollovers.

   (2) When working in rough terrain, wildland and ground fire fighters are at increased risk for falls. Rough ground often contains holes that are difficult to see in smoky conditions. Steep terrain also increases the likelihood of falls.

   (3) Because wildland and ground firefighting involves working with sharp tools, it is important to prevent injuries caused by these tools.

   (4) Other hazards of fighting wildland and ground fires include burns and smoke inhalation. Because the personal protective equipment (PPE) worn by wildland and ground fire fighters provides less protection than the PPE worn by structural fire fighters, fire fighters must keep far enough from the heat of the fire to prevent burns.

   (5) Because much wildland and ground firefighting is done without self-contained breathing apparatus (SCBA), fire fighters must avoid inhaling poisonous gases and suspended smoke particles. Use SCBA in any conditions where it is needed.

   (6) When engaged in wildland and ground firefighting, be alert for the hazards posed by falling trees. During a fire, the lower parts of trees may burn away and weaken the support for the rest of the tree. Trees of all sizes can fall with little warning.

   (7) Be alert for the presence of electrical hazards. Electrical transmission lines and other electrical wires may be present in the location of a wildland fire. Wires that drop on vegetation may ignite a wildland and ground fire and pose an electrical hazard to fire fighters. Many of these safety hazards can be difficult to see at night and in smoky conditions. (page 608)

Word Fun

Fire Alarms

1. Driving on unimproved roads and steep terrain greatly increases the chance of fire apparatus rollovers. Given these dangers, drivers must thoroughly understand the operating characteristics of their fire apparatus and operate the apparatus within the safe limits for which it was designed. All fire fighters should keep their seat belts fastened whenever the apparatus is moving.
2. (1) Clear away a minimum area of 4 feet by 8 feet to bare earth, which is clear of any and all fuels (you can also select a 4 feet by 8 feet area that is already clear of fuels). (2) Discard all flammable items and any hand tools. (3) Remove the shelter and discard the plastic case. (4) Deploy and open the shelter. (5) Place one leg through the hold-down strap. (6) Place the upper part of the body inside the opposite ends of the hold-down strap. (7) Place the other leg into the shelter. Fall to the ground. (8) The shelter should now be deployed so that it covers the entire body and the inside flaps are tucked in and not exposed.

Chapter 21: Fire Suppression

Matching

1. B (page 625)
2. D (page 625)
3. A (page 626)
4. C (page 626)
5. J (page 624)
6. I (page 622)
7. H (page 621)
8. G (page 620)
9. F (page 620)
10. E (page 619)

Answer Key

Multiple Choice

1. D (page 617)
2. A (page 617)
3. C (page 617)
4. C (page 617)
5. A (page 618)
6. D (page 618)
7. A (page 618)
8. B (page 621)
9. C (page 620)
10. C (page 621)
11. A (page 624)
12. B (page 625)
13. D (page 626)
14. A (page 629)
15. D (page 630)
16. C (page 632)
17. B (page 632)
18. B (page 633)
19. C (page 633)
20. A (page 633)
21. C (page 637)

Vocabulary

1. **Indirect application of water:** The use of a solid object such as a wall or ceiling to break apart a stream of water, creating more surface area on the water droplets and thereby causing the water to absorb more heat. (page 620)
2. **Master stream device:** A large-capacity nozzle that can be supplied by two or more hose lines or fixed piping. It can flow in excess of 300 gallons per minute. Such devices include deck guns and portable ground monitors. (page 624)
3. **Boiling-liquid, expanding-vapor explosion (BLEVE):** If an LPG tank is exposed to heat from a fire, the temperature of the liquid inside the container will increase. The fire could then be fueled by propane escaping from the tank or from an external source. As the temperature of the product increases, the vapor pressure will also increase. The increasing pressure creates added stress on the container. If this pressure exceeds the strength of the cylinder, it could rupture catastrophically. An LPG cylinder can have the same explosive properties as dynamite. (page 635)
4. **Portable monitor:** A master stream appliance that is designed to be set up and then left to operate unattended. A portable monitor is used to flow large amounts of water onto a fire or exposed building from a ground-level position. (page 625)
5. **Straight stream:** A stream made by using an adjustable nozzle to provide a straight stream of water. (page 619)

Fill-in

1. combustion (page 617)
2. defensive (page 617)
3. defensive (page 617)
4. fog (page 619)
5. air (page 619)
6. interior (page 619)
7. direct (page 620)
8. large (page 622)
9. ignition (page 627)
10. basements (page 627)
11. oxygen (page 630)

True/False

1. T (page 617)
2. F (page 617)
3. T (page 618)
4. T (page 619)
5. T (page 619)
6. F (page 617)
7. T (page 624)
8. F (page 625)
9. T (page 627)
10. F (page 629)
11. F (page 632)
12. T (page 635)

Short Answer

1. (1) What are the risks versus the potential benefits? (2) Is it safe to send fire fighters into the building? (3) Are there any structural concerns? (4) Are there any lives at risk? (5) Does the size of the fire prohibit entry? (6) Are enough fire fighters on the scene to mount an interior attack? (Remember the two-in/two-out rule.) (7) Is there an adequate water supply? (8) Can proper ventilation be carried out to support offensive operations? (page 618)

Fire Alarms

1. Protect the scene and call for the appropriate utility company to come and turn off the power. Do not apply water to the transformer. Let it burn until the power can be shut off or extinguish with a dry chemical extinguisher if possible. Ensure that your apparatus is not placed under any power lines.
2. Exit the fire apparatus wearing full PPE, including SCBA. Select the proper hose line used to fight the fire depending on the fire's size, location, and type. Advance the hose line from the apparatus to the entry point of the structure. Activate your SCBA prior to entering the building. Signal the driver/operator that you are ready for water. Open the nozzle to purge air from the system and make sure water is flowing. Make sure ventilation is completed or in progress.

   Enter the structure and locate the seat of the fire. Apply water in either a straight or solid stream onto the base of the fire in short bursts. Watch for changing fire conditions; use only enough water to extinguish the fire. Locate and extinguish hot spots.

Skill Drills

Skill Drill 21-2: Performing a Direct Attack

1. Exit the fire apparatus wearing full PPE, including **SCBA**.
2. Select the proper hose line to fight the fire based on the fire's size, location, and **type**.
3. Advance the hose line from the apparatus to the entry point of the **structure**.
4. Don a face piece and activate the SCBA and **personal alert safety system (PASS)** device prior to entering the building.
5. Signal the operator/driver that you are ready for **water**.
6. Open the nozzle to **purge** air from the system and make sure water is flowing.
7. Make sure that **ventilation** is completed or in progress.
8. Enter into the structure and locate the **seat** of the fire.
9. Apply water in either a straight or solid stream onto the base of the fire until all visible **flame** has been extinguished.
10. Watch for **changes** in fire conditions.
11. Shut down the nozzle and **listen**.
12. Locate and **extinguish** hot spots.

Skill Drill 21-3: Performing an Indirect Attack

1. Exit the fire apparatus wearing full PPE, including SCBA.
2. Select the correct hose line to be used to attack the fire depending on the type of fire, its location, and its size.
3. Advance the hose line from the apparatus to the opening in the structure where the indirect attack will be made.
4. Don a face piece, and activate the SCBA and PASS device.
5. Notify the operator/driver that you are ready for water.
6. Open the nozzle and make sure that air is purged from the hose line and that water is flowing. If using a fog nozzle, ensure that it is set to the proper nozzle pattern for entry. Shut down the nozzle until you are in a position to apply water.
7. Advance with a charged hose line to the location where you will apply water.
8. Direct the water stream toward the upper levels of the room and ceiling into the heated area overhead, and move the stream back and forth. Flow water until the room begins to darken. Shut the nozzle off, and reassess the fire conditions.
9. Watch for changes and a reduction in the amount of fire. Once the fire is reduced, shut down the nozzle.
10. Confirm that ventilation has been completed.
11. Attack any remaining fire and hot spots until the fire is completely extinguished.

Skill Drill 21-4: Performing a Combination Attack

1. Don full **PPE** and SCBA. Select the correct hose line to accomplish the suppression task at hand.
2. Stretch the hose line to the **entry point** of the structure, and signal the operator/driver that you are ready to receive water.
3. Open the nozzle to get the air out and make sure that water is **flowing**.
4. Enter the structure, and locate the room or area where the fire **originated**.
5. Aim the nozzle at the **upper-left** corner of the fire and make either a "T," "O," or "Z" pattern with the nozzle. Start high and then work the pattern down to the fire level.
6. Use only enough water to **darken** down the fire without upsetting the thermal layering.
7. Once the fire has been reduced, find the remaining hot spots and complete fire extinguishment using a **direct** attack.

Skill Drill 21-5: Performing the One-Fire Fighter Method for Operating a Large Handline

1. Select the correct size of fire hose. Advance the hose into position. Signal that you are ready for water and open the nozzle to allow air to escape and to ensure water is flowing. Close the nozzle and then make a loop with the hose, ensuring that the nozzle is under the hose line that is coming from the fire apparatus.

2. Lash the hose sections together where they cross, or use your body weight to kneel or sit on the hose line at the point where the hose crosses itself.

3. Allow enough hose to extend past the section where the line crosses itself for maneuverability.

4. Open the nozzle and direct water onto the designated area.

Skill Drill 21-6: Performing the Two-Fire Fighter Method for Operating a Large Handline
1. <u>Stretch</u> the hose line from the fire apparatus into position.
2. <u>Signal</u> that you are ready for water and open the nozzle to allow air to escape and to ensure water is flowing. Advance the hose line as needed.
3. Before attacking the fire, the fire fighter on the nozzle should cradle the hose on his or her hip while grasping the nozzle with one hand and supporting the hose with the other hand. The second fire fighter should stay approximately <u>3</u> feet behind the fire fighter who is on the nozzle. The second fire fighter should grasp the hose with two hands and may use a knee to stabilize the hose against the ground if necessary.
4. Open the <u>nozzle</u> in a controlled fashion and direct water onto the fire or designated exposure.

Chapter 22: Preincident Planning

Matching

1. M (page 650)
2. J (page 656)
3. H (page 654)
4. A (page 649)
5. D (page 651)
6. C (page 656)
7. L (page 651)
8. K (page 647)
9. E (page 650)
10. I (page 644)
11. O (page 650)
12. B (page 644)
13. G (page 644)
14. N (page 652)
15. F (page 650)

Multiple Choice

1. D (page 642)
2. C (page 644)
3. A (page 645)
4. A (page 648)
5. D (page 650)
6. A (page 650)
7. D (page 651)
8. A (page 660)
9. C (page 656)
10. D (page 656)
11. B (page 649)

Vocabulary

1. **Horizontal ventilation:** The process of making openings so that the smoke, heat, and gases can escape horizontally from a building through openings such as doors and windows. (page 660)
2. **Dry hydrant:** A permanent piping system that provides access to draft water from a static source; sometimes called a drafting hydrant. (page 651)
3. **Fire alarm annunciator panel:** Part of the fire alarm system that indicates the source of an alarm within a building. (page 645)
4. **HVAC system:** A system to manage the internal environment that is often found in large buildings. (page 654)
5. **Ordinary construction:** Buildings where the exterior walls are made of noncombustible or limited-combustible materials, but the interior floors and walls are made of combustible materials. (page 648)
6. **Conflagration:** A large fire, often involving multiple structures. (page 644)
7. **Preincident plan:** A written document resulting from the gathering of general and detailed information to be used by public emergency response agencies and private industry for determining the response to reasonable anticipated emergency incidents at a specific facility. (page 643)

Fill-in

1. incident commander (IC) (page 642)
2. preincident planning (page 643)
3. computers (page 643)
4. target hazards (page 644)
5. survey (page 644)
6. annunciator panel (page 645)
7. response (page 646)
8. alternate (page 646)
9. Size-up (page 647)
10. concrete, steel beams, masonry block walls (page 648)
11. noncombustible, fire resistance protection (page 648)
12. V (page 648)
13. IV, heavy timber (page 648)
14. institutional (page 650)
15. retail stores, offices, industrial factories (page 650)
16. exposure (page 650)
17. sprinkler system (page 650)
18. Standpipe systems (page 650)
19. static water supply (page 651)
20. electricity, natural gas (page 652)
21. occupants, entrances, exits (page 653)
22. exterior, interior (page 653)
23. 75 (page 654)
24. defend-in-place (page 656)
25. horizontal (page 656)
26. material safety data sheets (MSDS) (page 658)

True/False

1. T (page 642)
2. F (page 658)
3. F (page 654)
4. F (page 652)
5. F (page 652)
6. T (page 650)
7. F (page 649)
8. F (page 644)
9. F (page 645)
10. T (page 648)
11. F (page 648)
12. T (page 647)
13. F (page 644)

Short Answer

1. (1) Gives you the tools and knowledge to become a much more effective fire fighter; (2) prevents you from going into situations "blind"; (3) alerts you to potential hazards and hidden dangers; (4) assists in better command decisions because the information is assembled before the emergency occurs (pages 643–644)
2. The preincident plan makes information available during an emergency incident that otherwise would not be readily evident or easily determined. (page 644)
3.
   - Building location
   - Apparatus access to exterior of the building
   - Access points to the interior of the building
   - Hydrant locations and/or alternative water supply
   - Size of the building (height, number of stories, length, width)
   - Exposures to the fire building and separation distances
   - Type of building construction
   - Building use
   - Type of occupancy (assembly, institutional, residential, commercial, industrial)
   - Floor plan
   - Life hazards
   - Building exit plan and exit locations
   - Stairway locations (note whether the stairways are enclosed or unenclosed)
   - Elevator locations and emergency controls
   - Built-in fire protection systems (sprinklers, sprinkler control valves, standpipes, standpipe connections)
   - Fire alarm systems and fire alarm annunciator panel (part of the fire alarm system that indicates the location of an alarm within the building) location
   - Utility shut-off locations
   - Ventilation locations

- Presence of hazardous materials
- Presence of unusual contents or hazards
- Type of incident expected
- Sources of potential damage
- Special resources required
- General firefighting concerns (page 645)

4.
   - Do several roads lead to the building, or just a few?
   - Where are the hydrants located?
   - Where are the fire department connections for automatic sprinkler and standpipe systems located?
   - Do any security barriers limit access to the site? Are there fire lanes to provide access to specific areas?
   - Are any barricades, gates, or other obstructions so narrow or low that they would prevent passage of apparatus?
   - Are there bridges or underground structures that will not support the weight of apparatus?
   - Do any gates require keys or a code to gain entry?
   - Will it be necessary to cut fences?
   - Does the site include natural barriers such as streams, lakes, or rivers that might limit access?
   - Does the topography limit access to any parts of the building?
   - Might the landscaping or the presence of snow prevent access to certain parts of the building? (page 647)

5.
   - Is there a lockbox containing keys to the building?
   - Do the keys work?
   - Where is the lockbox located? Is it clearly visible?
   - Is the lockbox within easy reach?
   - Are key codes needed to gain access to the building? Who has them?
   - Does the building have security guards? Is a guard always on duty?
   - Does the guard have access to all areas of the building?
   - Is a key holder available to respond to the alarm within a reasonable amount of time? How can this person be reached?
   - Where is the fire alarm annunciator panel located?
   - Is the fire alarm annunciator panel properly programmed so you can quickly determine the exact location of an alarm? (page 647)

6. (1) Bulk oil facilities and refineries; (2) high-rise buildings; (3) hospitals; (4) hotels and rooming houses; (5) large apartment buildings; (6) lumberyards; (7) manufacturing plants; (8) nursing homes and assisted-living facilities; (9) public assembly occupancies; (10) schools; (11) shopping centers; (12) storage structures for hazardous materials; (13) warehouses (page 644)

7. (1) Hospitals; (2) nursing facilities; (3) assisted-living facilities; (4) large apartment buildings; (5) hotels and rooming houses; (6) schools; (7) public assembly occupancies (page 644)

8. The goal of fire prevention is to identify hazards and minimize or correct them so that fires do not occur or have limited consequences. A preincident plan is intended to help to IC make informed decisions when an emergency incident occurs at a location. (page 643–644)

9. The property owner or a representative should be contacted before the preincident survey is conducted. Making this initial contact enables the fire department to schedule an acceptable time, to explain the purpose and the importance of the preincident survey, and to clarify that the information is needed to prepare fire fighters in the event that an emergency occurs at the location. The team members who conduct the survey should dress and conduct themselves in a manner appropriate to the department's mission. A representative of the property should accompany the survey team to answer questions and to provide access to various areas. Every effort should be made to obtain accurate, useful information.

   The preincident survey is conducted in a systematic fashion, following a uniform format. Begin with the outside of the building, gathering all of the necessary information about the building's geographic location, external features, and access points. Next, survey the inside of the building to collect information about every interior area. A good, systematic approach starts at the roof and works down through the building, covering every level of the structure, including the basement. If the property is large and complicated, it may be necessary to make more than one visit to ensure that all required information is obtained and recorded accurately. (page 645)

10. (a) Type I: Fire-resistive buildings where the structural members are made of noncombustible materials that have a specified fire resistance. Materials include concrete, steel beams, and masonry block walls, among others.

    (b) Type II: Noncombustible buildings where the structural members are made of noncombustible materials, but may not have fire resistance protection. Includes unprotected steel beams.

(c) Type III: Ordinary buildings where the exterior walls are made of noncombustible or limited-combustible materials, but the interior floors and walls are made of combustible materials.

(d) Type IV: Heavy timber buildings where the exterior walls are made of noncombustible or limited-combustible materials, but the interior walls and floors are made of combustible materials. The dimensions of the interior materials are greater than the dimensions of ordinary construction (typically with minimum dimensions of 8 feet by 8 feet).

(e) Type V: Wood-frame buildings where the exterior walls, interior walls, floors, and roof are made of combustible wood materials. (page 648)

11. (a) Public assembly: theatres, auditoriums, and churches; arenas and stadiums; convention centers and meeting halls; bars and restaurants

(b) Institutional: hospitals and nursing homes; schools

(c) Commercial: retail stores; industrial factories; warehouses; parking garages; offices (page 650)

12. (1) How much water will be needed to fight a fire in the structure? (2) Where are the nearest hydrants located? (3) How much water is available and at what pressure? (4) Where are hydrants that are on different water mains located? (5) How many feet of hose would be needed to deliver the water to the fire? (page 652)

13. (1) What is the nearest water supply? (2) Is the water readily available year-round? (3) Does it freeze during the winter or dry up during the summer? (4) How many tanker trucks will be needed to deliver the water? (5) How long will it take to establish a tanker shuttle? (page 652)

14. (1) Detailed floor plan; (2) location of sleeping areas; (3) information about any handicapped occupants; (4) escape routes and an outside rendezvous area; (5) fire hydrant or water source location (page 656)

15. (1) Chemical companies; (2) garden centers; (3) swimming pool supply stores; (4) hardware stores; (5) laboratories

Word Fun

Fire Alarms

1. • Schedule the survey in advance.
   • Make contact with a responsible person. Identify yourself by name, title, and department.
   • Present a neat and professional image.
   • Ensure that a representative accompanies you during the survey.
   • Take notes as needed, and start outside.
   • Note the building location.
   • Identify the building construction.

- Identify the building use and occupancy.
- Note any life hazards.
- Note the access points to the interior of the building.
- Note the utility shut-off locations.
- Assess the apparatus access to the building.
- Note hydrant locations and/or alternative water supplies.
- Note ventilation concerns.
- Record built-in fire detection and suppression systems.
- Sketch floor plans.
- Note the elevator and stairway locations.
- Review exit plans and exit locations.
- Identify any special hazards and hazardous materials.
- Note the building exposures.
- Anticipate the type of incident expected.
- Identify any special resources needed.
- Complete and file the preincident survey form.

2. (a) Bulk oil facilities and refineries; high-rise buildings; hospitals; hotels and rooming houses; large apartment buildings; lumberyards; manufacturing plants; nursing homes and assisted-living facilities; public assembly occupancies; schools; shopping centers; storage structures for hazardous materials; warehouses

3. (b) Hospitals; nursing facilities; assisted-living facilities; large apartment buildings; hotels and rooming houses; schools; public assembly occupancies

Chapter 23: Fire and Emergency Medical Care

Matching

1. H (page 672)
2. E (page 665)
3. F (page 672)
4. C (page 668)
5. A (page 666)
6. I (page 667)
7. D (page 672)
8. G (page 667)
9. B (page 667)
10. J (page 666)

Multiple Choice

1. A (page 664)
2. D (page 665)
3. A (page 666)
4. D (pages 668)
5. B (page 672)
6. D (page 666)
7. C (page 667)
8. A (page 665)
9. D (pages 670)
10. A (page 673)

Vocabulary

1. **EMT-Paramedic:** An EMT who has extensive training in advanced life support, including intravenous therapy, pharmacology, endotracheal intubation, and other advanced assessment and treatment skills. (page 667)
2. **Basic life support (BLS):** Noninvasive, emergency life-saving care that is used to treat airway obstruction, respiratory arrest, or cardiac arrest. (page 665)
3. **Advanced life support (ALS):** Advanced life-saving procedures, such as cardiac monitoring, administration of IV fluids and medications, and use of advanced airway adjuncts. (page 666)
4. **Combination EMS system:** A system in which the fire department provides medical first response, and another agency transports the patient to the hospital emergency room. (page 668)
5. **Fire department EMS system:** A system in which the fire department both provides medical first response and transports the patient to the hospital emergency room. (page 670)
6. **Medical director:** The physician providing direction for patient care activities in the prehospital setting. (page 672)

Fill-in

1. fire codes, public education (page 665)
2. emergency (page 665)
3. first responder (page 666)
4. EMT-Paramedic (page 667)
5. 80% (page 664)
6. basic, advanced (page 665)
7. save lives, property (page 664)
8. BLS (page 665)
9. ALS, standing orders, protocols, radio direction (page 666)
10. cross-train (page 670)

True/False

1. F (page 664)
2. F (page 667)
3. T (page 664)
4. T (page 672)
5. T (page 672)
6. T (page 668)
7. F (page 668)
8. F (page 668)
9. T (page 668)

Short Answer

1. The medical director serves as ongoing liaison between the medical community, hospitals, and the EMS providers. If treatment problems arise or different procedures should be considered, the medical director must be consulted for a decision and directions. The medical director also determines and approves the continuing education and training courses that are required to ensure that proper training standards are met. (page 672)
2. (1) The entry requirements for fire fighters help ensure the hiring of quality employees. (2) Most fire departments are required to regularly monitor the health and fitness of their employees and have mechanisms in place to provide for continuing education. (3) Fire departments have the infrastructure in place to support EMS operations, including radio systems, dispatch services, and fire stations located strategically throughout the community. (4) The fire department is prepared to respond quickly to EMS calls throughout the jurisdiction. (5) The fire department is familiar with the community and its people, which can be invaluable when responding to and treating patients. (page 668)
3. The first responder is expected to assess and provide basic care for the patient until EMT-Basics or EMT-Paramedics arrive to provide further care and transportation to a medical facility. The first responder is expected to stabilize a patient at the scene by providing immediate, life-saving care, such as controlling bleeding, establishing an airway, initiating CPR, and using an AED. (pages 666–667)
4. (1) Scene size-up; (2) securing the scene and scene safety; (3) patient assessment; (4) simple airway techniques; (5) CPR; (6) splinting; (7) bandaging; (8) administering oxygen; (9) lifting and moving ill and injured patients; (10) controlling external bleeding; (11) treating patients for shock; (12) ambulance operations; (13) performing cardiac defibrillation using an AED (page 667)
5. (1) Using electrocardiograms to evaluate the patient; (2) administering medications; (3) inserting endotracheal tubes into the airway; (4) electrically pacing the heart; (5) identifying causes and treatments of diseases (page 667)
6. Fire department EMS systems both provide the medical first response and transport patients. These medical and transport services may be provided at either BLS or ALS levels. As in a combination system, engine companies, truck companies, or special EMS units may be used for medical first response. Because all of the personnel work for the same agency, training is easily accomplished and efforts are coordinated under a united control. (page 670)
7. In a combination EMS system, the fire department provides medical first response and another agency operates the ambulances that transport the patients. (page 668)
8. Provide prompt, efficient, competent care for all members of your community, regardless of their age, socioeconomic background, or nature of injury. (page 670)

9. Direct or online medical control is provided by an EMS physician who can be reached by radio or telephone during a call. Local protocols define when EMS providers should give a radio report or obtain online medical direction. Once the crew has initiated any immediate, urgent care according to standing orders, the online medical control physician is contacted. The crew then provides a report that describes the patient's condition and any treatment provided. The physician either confirms or modifies this proposed treatment plan; he or she may also prescribe special orders that the EMS providers are to follow for that patient.

Indirect or offline medical controls are standing orders that direct the EMS providers to take specific actions when they encounter particular situations. (page 672)

Word Fun

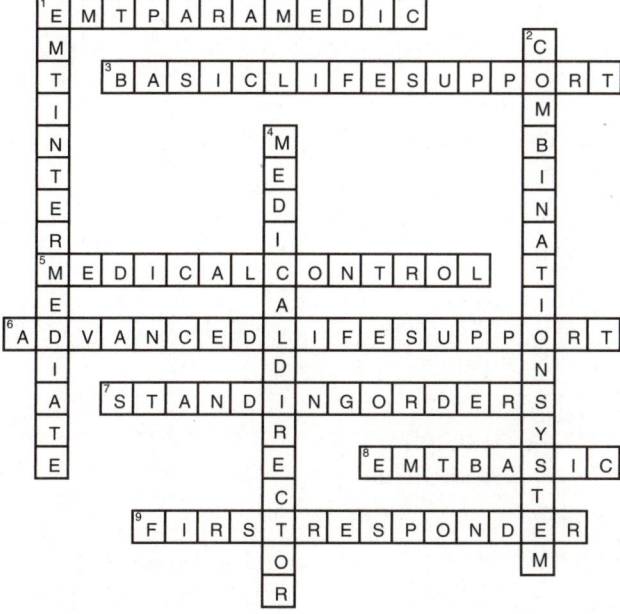

Fire Alarms

1. The EMT-Basic course involves more than 110 hours of training and teaches the following skills:
   - Scene size-up
   - Securing the scene and scene safety
   - Patient assessment
   - Simple airway techniques
   - CPR
   - Splinting
   - Bandaging
   - Administering oxygen
   - Lifting and moving ill and injured patients
   - Controlling external bleeding
   - Treating for shock
   - Ambulance operations
   - Performing cardiac defibrillation using an AED

2. There are several advantages to having EMS systems located within the fire department. For example, the entry requirements for fire fighters help ensure the hiring of quality employees. Most fire departments are required to regularly monitor the health and fitness of their employees, and most have mechanisms in place to provide for continuing education. Fire departments also have the infrastructure in place to support EMS operations, including radio systems, dispatch services, and fire stations located strategically throughout the community. As a result, the fire department is prepared to respond quickly to EMS calls throughout its jurisdiction. Furthermore, the fire department is familiar with the community and its people, which can be invaluable when responding to and treating patients. In short, many communities find that when they need to find a home for their EMS system, the fire department is a good fit.

Chapter 24: Emergency Medical Care

Matching

1. G (page 695)
2. C (page 697)
3. F (page 680)
4. K (page 711)
5. A (page 695)
6. L (page 680)
7. E (page 712)
8. B (page 720)
9. H (page 680)
10. D (page 713)
11. J (page 711)
12. I (page 695)

Multiple Choice

1. C (page 678)
2. A (page 679)
3. D (page 680)
4. B (page 680)
5. B (page 680)
6. D (page 680)
7. A (page 680)
8. C (page 680)
9. B (page 682)
10. B (page 683)
11. C (page 683)
12. B (page 684)
13. A (page 685)
14. A (page 687)
15. B (page 689)
16. D (page 689)
17. C (page 690)
18. C (page 690)
19. D (page 692)
20. A (page 695)
21. B (page 695)
22. D (page 695)
23. B (page 695)
24. C (page 699)
25. A (page 705)
26. C (page 705)
27. C (page 706)
28. B (page 708)
29. D (page 711)
30. A (page 716)

Labeling

1. The respiratory system.

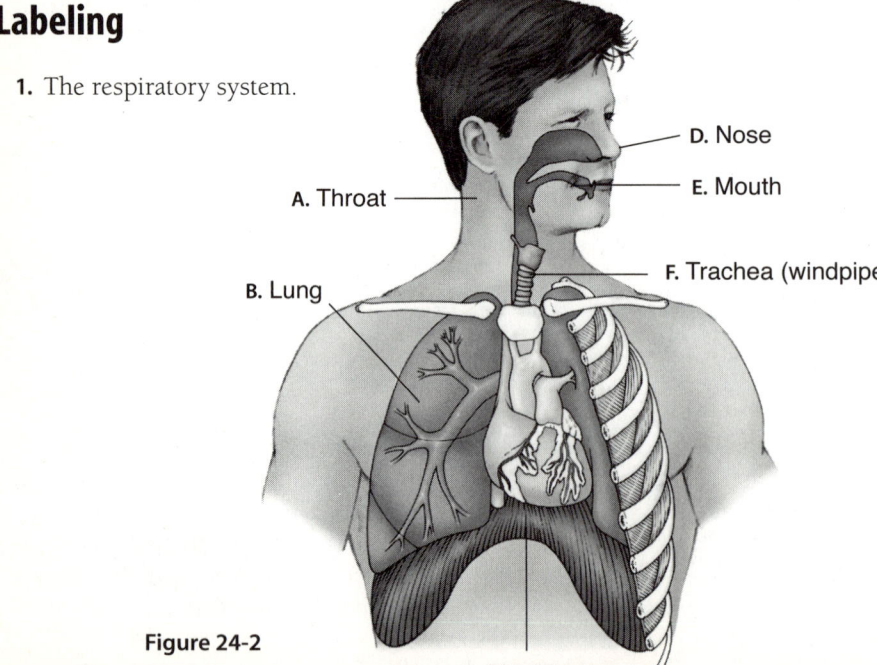

Figure 24-2

2. Anatomy of the respiratory system.

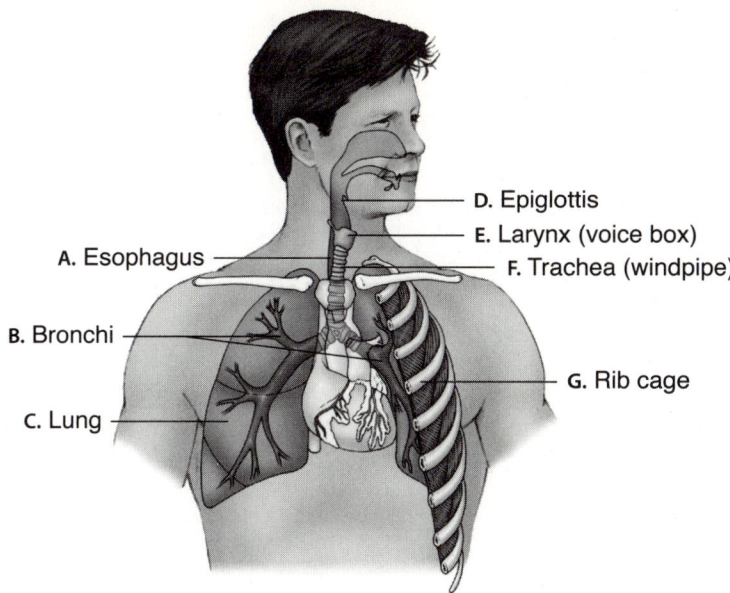

A. Esophagus
B. Bronchi
C. Lung
D. Epiglottis
E. Larynx (voice box)
F. Trachea (windpipe)
G. Rib cage

Figure 24-4

3. The CPR chain of survival.

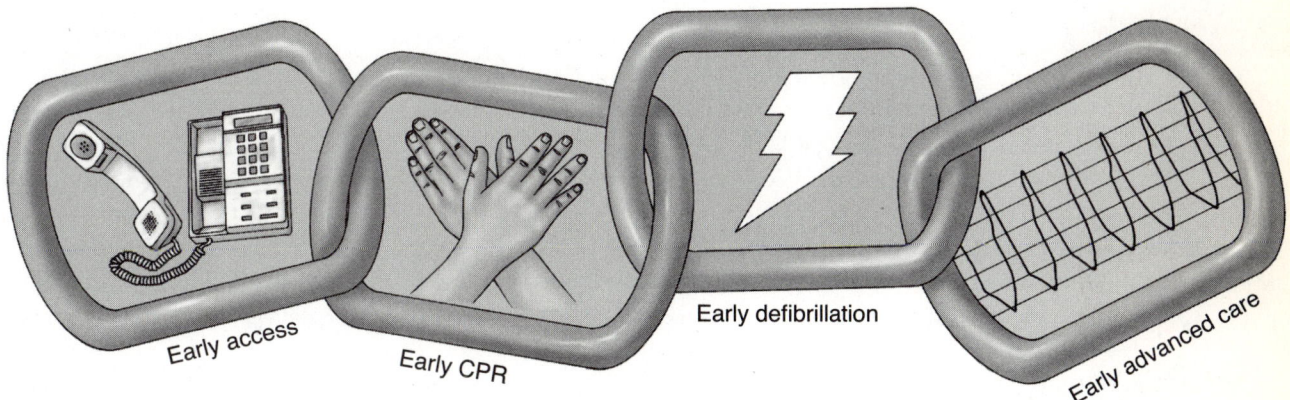

Early access   Early CPR   Early defibrillation   Early advanced care

Figure 24-14

4. Types of external bleeding.

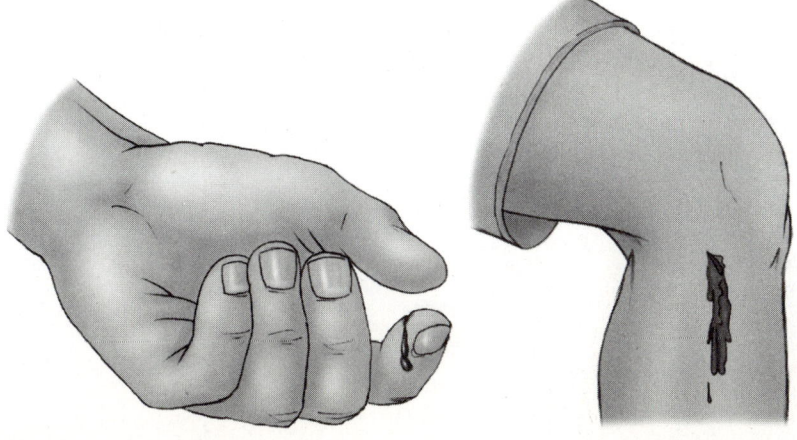 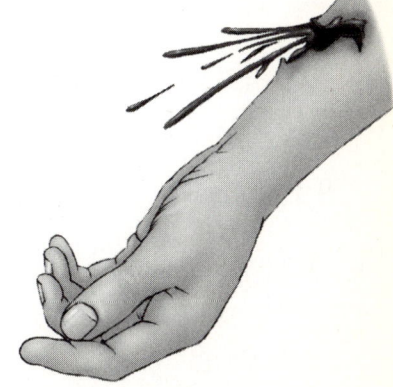

Figure 24-20   A. Capillary   B. Venous   C. Arterial

5. Location of pressure points.

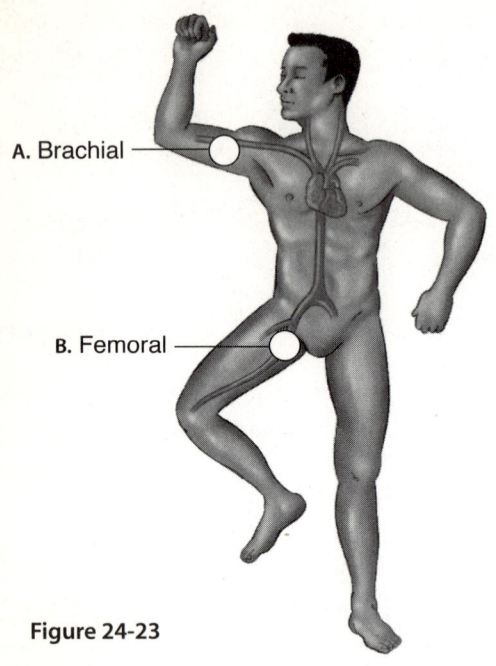

Figure 24-23

Vocabulary

1. **Cardiopulmonary resuscitation (CPR):** The artificial circulation of the blood and movement of air into and out of the lungs in a pulseless, nonbreathing victim. (page 695)
2. **Recovery position:** A position that helps keep the victim's airway open by allowing secretions to drain out of the mouth instead of into the trachea. Recovery position also uses gravity to prevent the victim's tongue and lower jaw from blocking the airway. (page 682)
3. **Dependent lividity:** The red or purple color that appears on those parts of the victim's body closest to the ground. It is caused by blood seeping into the tissues on the dependent, or lower, part of the person's body. (page 697)
4. **Shock:** A state of collapse of the cardiovascular system; the state of inadequate delivery of blood to the organs of the body. (page 705)
5. **Radial artery:** The major artery in the forearm. It is palpable at the wrist on the thumb side. (page 695)

Fill-in

1. HIV (page 678)
2. universal precautions (page 679)
3. oxygen, carbon dioxide (page 680)
4. Lungs (page 681)
5. responsiveness (page 681)
6. jaw-thrust (page 682)
7. hypothermia (page 683)
8. food (page 689)
9. gastric distention (page 692)
10. chest compressions (page 695)
11. 2 (page 699)
12. shock (page 705)

True/False

1. T (page 678)
2. F (page 679)
3. F (page 695)
4. T (page 680)
5. T (page 682)
6. F (page 685)
7. T (page 695)
8. T (page 695)
9. F (page 697)
10. F (page 700)
11. T (page 703)
12. T (page 704)

Short Answer

1. Always wear gloves when handling victims, and change gloves after contact with each victim. Wash your hands immediately after removing gloves. (Note that leather gloves are not considered to be safe, because leather is porous and traps fluids.)

   Always wear protective eyewear or a face shield when you anticipate that blood or other bodily fluids may splatter. Wear a gown or apron if you anticipate splashes of blood or other bodily fluids, such as those that occur with childbirth and major trauma.

   Wash your hands and other skin surfaces immediately and thoroughly if they become contaminated with blood and other bodily fluids. Change contaminated clothes and wash exposed skin thoroughly.

   Place used needles directly in a puncture-resistant container designed for sharps.

   Even though saliva has not been proven to transmit HIV, you should use a face shield, pocket mask, or other airway adjunct if the victim needs resuscitation. (page 679)

2. Respiratory arrest has many causes. By far the most common cause is heart attack, which claims more than 500,000 lives in the United States each year. Other major causes of respiratory arrest include the following conditions:
   - Mechanical blockage or obstruction caused by the tongue
   - Vomitus, particularly in a victim who has been weakened by an illness such as a stroke
   - Foreign objects such as teeth, dentures, balloons, marbles, pieces of food, or pieces of hard candy (especially in small children)
   - Illness or disease such as heart attack or severe stroke
   - Drug overdose
   - Poisoning
   - Severe loss of blood
   - Electrocution by electrical current or lightning (pages 683–684)

3. The human heart consists of four chambers, two on the right side and two on the left side. Each upper chamber is called an atrium. The right atrium receives blood from the veins of the body; the left atrium receives blood from the lungs. The bottom chambers are the ventricles. The right ventricle pumps blood to the lungs; the left ventricle pumps blood throughout the body. The most muscular chamber of the heart is the left ventricle, which needs the most power because it must force blood to all parts of the body. Together the four chambers of the heart work in a well-ordered sequence to pump blood to the lungs and to the rest of the body. One-way check valves in the heart and veins allow the blood to flow in only one direction through the circulatory system. The arteries carry blood away from the heart at high pressure; to withstand this pressure, they have very thick walls. The main artery carrying blood away from the heart is quite large (about 1 inch in diameter), but arteries become smaller as they branch out farther away from the heart. (pages 693, 695)

4. You should discontinue CPR only in the following circumstances:
   (1) Effective, spontaneous circulation and ventilation are restored.
   (2) Resuscitation efforts are transferred to another trained person who continues CPR.
   (3) A physician assumes responsibility for the victim.
   (4) The victim is transferred to properly trained EMS personnel.
   (5) Reliable criteria for death are recognized.
   (6) You are too exhausted to continue resuscitation, environmental hazards endanger your safety, or continued resuscitation would place the lives of others at risk. (page 697)

5. Signs and symptoms of shock are as follows:
   - Confusion, restlessness, or anxiety
   - Cold, clammy, sweaty, pale skin
   - Rapid breathing
   - Rapid, weak pulse
   - Increased capillary refill time
   - Nausea and vomiting
   - Weakness or fainting
   - Thirst (page 705)

6. To treat shock, follow these general steps:
   (1) Position the victim correctly.
   (2) Maintain the victim's ABCs.
   (3) Treat the cause of shock, if possible.

(4) Maintain the victim's body temperature by placing blankets under and over the victim.
(5) Make sure the victim does not eat or drink anything.
(6) Assist with other treatments (such as administering oxygen, if available).
(7) Arrange for immediate and prompt transportation to an appropriate medical facility. (page 706)

Word Fun

Across:
1. VENTRICLE
4. CAPILLARIES
7. BRACHIAL ARTERY
12. DRESSING
13. SHOCK
14. ABRASION
15. BRUISE

Down:
2. TRACHEA
3. PLASMA
5. PALATE
6. STERNUM
7. BLOOD VESSEL (BOGUELNS)
8. AVULSION
9. LUNG
10. ALVEOLI
11. AIRWAY

Fire Alarms

1. (a) Anaphylactic shock is caused by an extreme allergic reaction to a foreign substance (an allergen), such as venom from bee or insect stings, penicillin, or certain foods.

   (b) To treat shock, follow these general steps:
   (1) Position the victim correctly.
   (2) Maintain the victim's ABCs.
   (3) Treat the cause of shock, if possible.
   (4) Maintain the victim's body temperature by placing blankets under and over the victim.
   (5) Make sure the victim does not eat or drink anything.
   (6) Assist with other treatments (such as administering oxygen, if available).
   (7) Arrange for immediate and prompt transportation to an appropriate medical facility.

2. • CDC-recommended universal precautions include: Always wear gloves when handling victims, and change gloves after contact with each victim. Wash your hands immediately after removing gloves. (Note that leather gloves are not considered to be safe, because leather is porous and traps fluids.)
   • Always wear protective eyewear or a face shield when you anticipate that blood or other bodily fluids may splatter. Wear a gown or apron if you anticipate splashes of blood or other bodily fluids, such as those that occur with childbirth and major trauma.
   • Wash your hands and other skin surfaces immediately and thoroughly if they become contaminated with blood and other bodily fluids. Change contaminated clothes and wash exposed skin thoroughly.
   • Place used needles directly in a puncture-resistant container designed for sharps.
   • Even though saliva has not been proven to transmit HIV, you should use a face shield, pocket mask, or other airway adjunct if the victim needs resuscitation.

Skill Drills

Skill Drill 24-1: Performing the Head Tilt–Chin Lift Technique

1. Position yourself at the side of the **supine** victim.
2. Place your hand closest to the victim's head on the **forehead**.
3. With your other hand, place two fingers on the **underside** of the victim's chin.
4. Simultaneously apply backward and **downward** pressure to the victim's forehead and lift the jaw straight up. Do not depress the soft tissue below the chin.

Skill Drill 24-2: Performing the Jaw-Thrust Technique

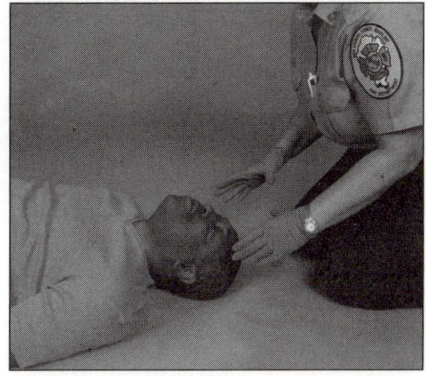 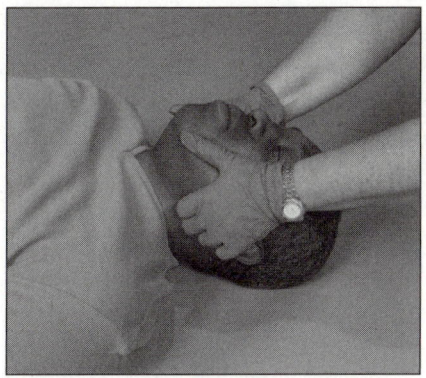 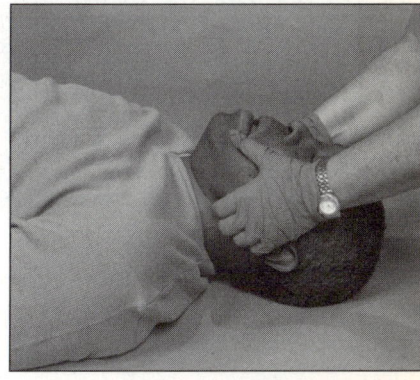

1. Position yourself at the top of the victim's head.
2. Place the meaty portion of the base of your thumbs on the arches of the jaw, and hook the tips of your index fingers under the angle of the mandible, in the indent below the ear.
3. While holding the victim's head still, displace the jaw forward and open the victim's mouth with your thumb tips.

Skill Drill 24-4: Performing Mouth-to-Mask Rescue Breathing

1. Open the victim's airway using the **head tilt–chin lift** technique.
2. If there is a suspected head injury, open the airway using the **jaw-thrust** technique.
3. Seal the **mask** against the victim's face.
4. **Breathe** through the mouthpiece.

Skill Drill 24-5: Performing Mouth-to-Barrier Device Rescue Breathing

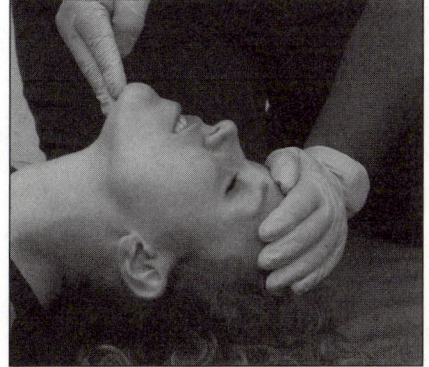 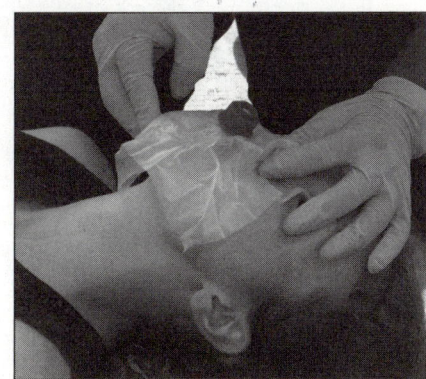 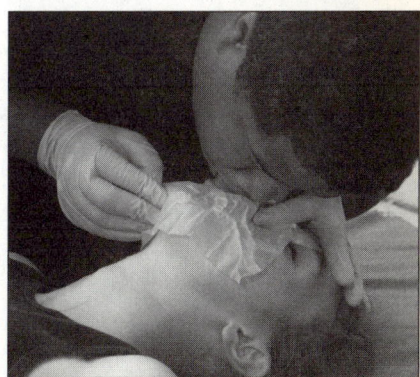

1. Open the victim's airway using the head tilt–chin lift technique.
2. Place the barrier device over the victim's mouth. Pinch the victim's nostrils together.
3. Perform rescue breathing.

Skill Drill 24-7: Performing the Abdominal-Thrust Maneuver

1. Stand behind the victim and deliver an abdominal thrust. Press into the victim's abdomen with a quick **inward** and upward thrust.
2. Repeat the abdominal thrusts until either the foreign body is expelled or the victim becomes **unconscious**.

Skill Drill 24-8: Treating Airway Obstruction in an Unconscious Adult

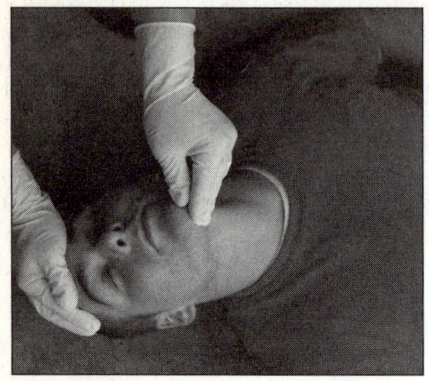
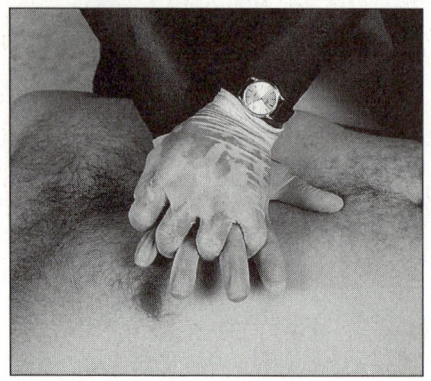
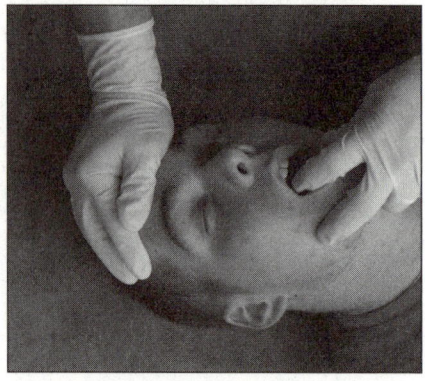

1. Open the victim's airway by using the head tilt–chin lift technique. If trauma to the spine is suspected, use the jaw-thrust technique. Attempt rescue breathing.

2. Reposition the victim's head to improve the likelihood of opening the airway. Attempt rescue breathing again. If it is unsuccessful, perform 30 chest compressions.

3. Open the victim's mouth so you can check for obstructions. Use your finger to sweep the mouth clear only if you see an object. Attempt rescue breathing again. Repeat until the airway is cleared.

Skill Drill 24-11: Performing Chest Compressions

1. Place the heel of one hand on the victim's **sternum** between the nipples.
2. Place the **heel** of your other hand over the first hand.
3. With your arms straight, lock your elbows, and position your shoulders directly over your **hands**. Depress the sternum 1½ to 2 inches using a direct downward movement.

Skill Drill 24-12: Performing One-Rescuer Adult CPR

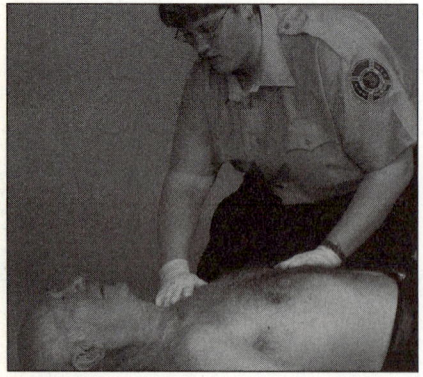
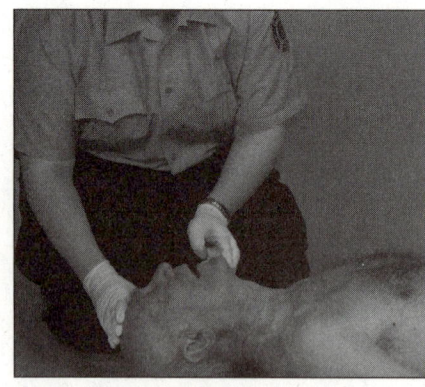
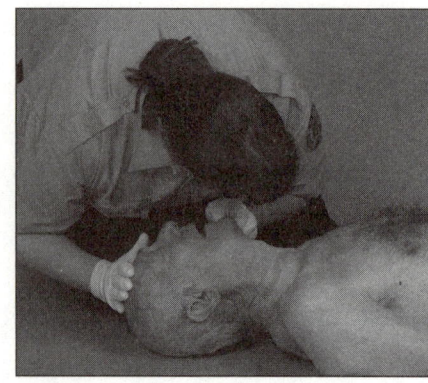

1. Establish unresponsiveness and call for help.

2. Open the airway.

3. Look, listen, and feel for breathing. If breathing is adequate, place the victim in the recovery position and monitor him or her.

(continues)

Skill Drill 24-12: Performing One-Rescuer Adult CPR (continued)

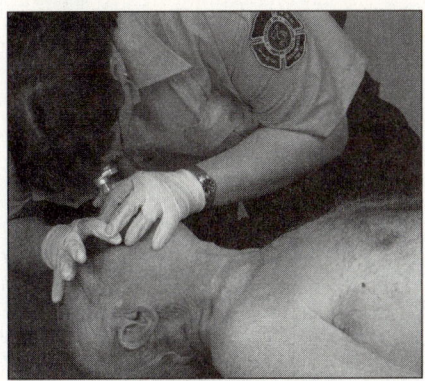

4. If the victim is not breathing, give two breaths of 1 second each.

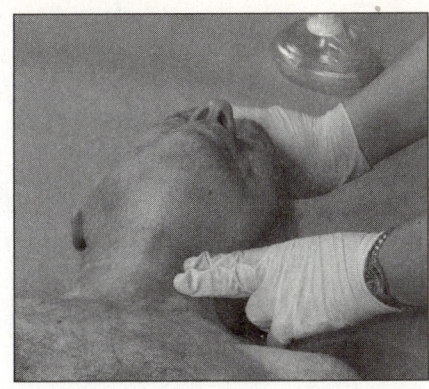

5. Check for a carotid pulse.

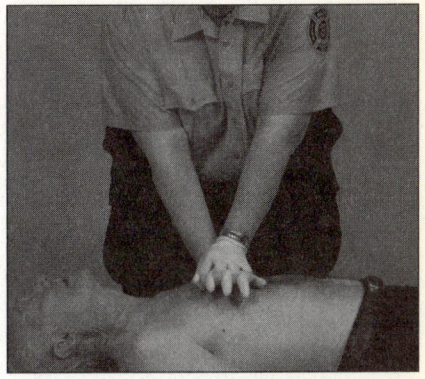

6. If no pulse is found, apply the AED. If no AED is available, place your hands in the proper position for chest compressions. Give 30 compressions at a rate of about 100 per minute. Open the airway and give two ventilations of 1 second each. Perform five cycles of chest compressions and ventilation. Stop CPR and check for return of the victim's pulse. Depending on the victim's condition, continue CPR, continue rescue breathing only, or place the victim in the recovery position and monitor his or her breathing and pulse.

Skill Drill 24-13: Performing Two-Rescuer Adult CPR

1. Establish the victim's unresponsiveness and take up your **positions**.
2. Open the **airway**.
3. Look, listen, and **feel** for breathing. If the victim's breathing is adequate, place the victim in the recovery position and monitor.
4. If the victim is not breathing, give **two** breaths of 1 second each.
5. Check for a **carotid** pulse. If no pulse is felt in 10 seconds, begin CPR.
6. If there is no pulse but an AED is available, apply it now. If no AED is available, begin chest compressions at a rate of **100** per minute (30 compressions for each 2 ventilations). After every five cycles, switch rescuer positions to minimize fatigue. Keep the switch time to 5 to 10 seconds. Depending on the victim's condition, continue CPR, continue ventilations only, or place the victim in recovery position.

Chapter 25: Vehicle Rescue and Extrication

Matching

1. E (page 730)
2. D (page 725)
3. B (page 725)
4. F (page 724)
5. A (page 729)
6. C (page 729)
7. J (page 730)
8. I (page 733)
9. H (page 735)
10. G (page 733)

Multiple Choice

1. C (page 724)
2. A (page 725)
3. C (page 729)
4. D (page 729)
5. B (page 730)
6. C (page 731)
7. A (page 733)
8. D (page 733)
9. B (page 733)
10. A (page 735)
11. C (page 736)
12. C (page 737)
13. D (page 738)
14. B (page 739)
15. A (page 739)
16. A (page 740)
17. C (page 741)
18. B (page 743)
19. A (page 740)
20. D (page 740)

Labeling

1. The anatomy of a vehicle.

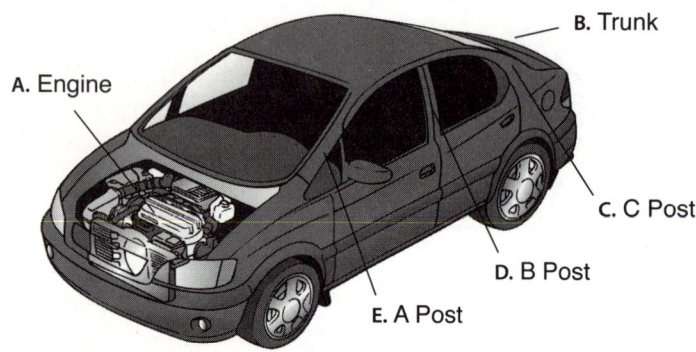

Figure 25-2

Vocabulary

1. **Unibody:** The frame construction most commonly used in vehicles. The base unit is made of formed sheet metal; structural components are then added to the base to form the passenger compartment. Subframes are attached to each end. This type of construction does away with the rail beams used in platform-framed vehicles. (page 725)
2. **Hybrid vehicle:** A vehicle that uses a battery-powered electric motor and a gasoline-powered engine. (page 725)
3. **Platform frame:** A type of vehicle frame resembling a ladder, which is made up of two parallel rails joined by a series of cross members. This kind of construction is typically used for luxury vehicles, sport utility vehicles, and all types of trucks. (page 725)
4. **Firewall:** The structure that divides the engine compartment from the passenger compartment. (page 725)
5. **Post:** One of the vertical support members of a vehicle that holds up the roof and forms the upright columns of the passenger compartment. (page 725)

Fill-in

1. BLEVE (page 724)
2. passenger (page 725)
3. "B" (page 725)
4. response (page 725)
5. hazards, scope (page 726)
6. law enforcement (page 726)
7. pressure (page 730)
8. door (page 733)
9. door (page 735)
10. "A" (page 741)

True/False

1. T (page 725)
2. T (page 726)
3. F (page 726)
4. T (page 728)
5. T (page 729)
6. T (page 730)
7. F (page 730)
8. F (page 735)
9. T (page 738)

Short Answer

1.
   - Never stack high-pressure lift air bags more than two high.
   - Do not use a rescue-lift air bag to pull a steering column.
   - Do not use a rescue-lift air bag as the sole means to stabilize a vehicle; cribbing must be the primary stabilizer.
   - Never operate the rescue-lift air bag system without having been properly trained and fully understanding how the system works.
   - Place a sheet of plywood on the ground under the air bag to protect it.
   - Do not use boards or plywood between or above rescue-lift air bags.
   - Clean rescue-lift air bags by following the recommendations of their manufacturer.
   - Test rescue-lift air bags regularly.
   - Never store a lift air bag near gasoline. (page 731)

2. (1) Stabilize or hold an object or vehicle: An example is stabilizing a vehicle with cribbing to keep them from moving. (2) Bend, distort, or displace: An example is bending a vehicle door back to get it out of the way. (3) Cut or sever: An example is cutting a roof. (4) Disassemble: An example is removing a vehicle door by unbolting the door hinges. (page 733)

3.
   - On most recently manufactured vehicles, the steering wheel contains the driver's-side air bag, which is a life-saving safety feature for the driver of the vehicle. Front-passenger air bags may also be present.
   - If the air bag deployed during the crash, it does not present a safety hazard for rescuers.
   - If the air bag did not deploy during the crash, it presents a hazard both for the occupant of the vehicle and for rescue personnel. An undeployed air bag could deploy if wires are cut or if it becomes activated during the rescue operation.
   - If the air bag did not deploy, disconnect the battery and allow the air bag capacitor to discharge. The time required to discharge the capacitor varies from one model of air bag to another.
   - Do not place a hard object such as a backboard between the victim and an undeployed air bag.
   - Do not attempt to cut the steering wheel if the air bag has not deployed.
   - For your safety, never get in front of an undeployed air bag. You could suffer serious injury if it activates unexpectedly.
   - Some vehicles contain side-mounted air bags or curtains that provide lateral protection for occupants. Check vehicles for the presence of these devices. (page 740)

Answer Key

Word Fun

Fire Alarms

1. (a) Cribbing, step chocks, wedges, and rescue-lift air bags
   (b) (1) Don PPE.
   (2) Lay out a tarp at the edge of the secure work area for staging tools and equipment, if indicated.
   (3) Enter the secure work area safely.
   (4) Assess the scene for hazards.
   (5) Chock both sides of one tire to prevent the vehicle from rolling.
   (6) Place one step chock at each corner of the vehicle, and deflate the tires or use wedges to stabilize the vehicle.

2. (a) Whenever possible, place emergency vehicles in a manner that will ensure safety and does not disrupt traffic any more than necessary. However, do not hesitate to request that the road be closed if necessary. Remember—safety first! Position large emergency vehicles so that they provide a barrier against motorists who fail to heed emergency warning lights. Many departments place apparatus at an angle to the crash. This position helps to push the apparatus to the side of a crash in the event that the emergency apparatus is struck from behind.

   Traffic cones or flares can be placed to direct motorists away from the crash. If needed, call for law enforcement to assist in traffic control.

   Fire fighters need to be readily visible at a crash scene. Personal protective equipment (PPE) should be bright to help ensure fire fighters' visibility during daylight hours; PPE that is used at night should be equipped with reflective material to increase visibility in the darkness. PPE must be worn at all motor vehicle crashes.

   (b) (1) Position the emergency vehicle so as to protect the scene. Size up the scene from inside this vehicle.
   (2) Transmit the initial report over the radio.
   (3) Establish command.
   (4) Perform a scene size-up outside the vehicle.
   (5) Check for overhead hazards.
   (6) Approach the crash scene vehicles and examine the space under the vehicles.
   (7) Perform a 360-degree walk-around of the entire scene.
   (8) Assess the hazards.
   (9) Determine the number of people involved.
   (10) Determine the severity of the victims' injuries.

Chapter 25: Vehicle Rescue and Extrication   507

(11) Determine the level of entanglement.

(12) Assess the resources available, and call for additional units if needed.

(13) Give an updated report.

(14) Establish a secure work area and stabilize the vehicles.

(15) Establish a staging resource area.

(16) Direct the placement of arriving apparatus.

(17) Assign personnel and tasks.

Skill Drills

Skill Drill 25-1: Performing a Scene Size-up at a Motor Vehicle Crash

1. Position the emergency vehicle so as to **protect** the scene. Size up the scene from inside this vehicle. Transmit the initial report over the radio. Establish command. Perform a scene size-up outside the vehicle. Check for overhead hazards.

2. Approach the crash scene vehicles and examine the space under the vehicles. Perform a **360**-degree walk-around of the entire scene. Assess the hazards. Determine the number of people involved. Determine the severity of the victims' injuries. Determine the level of entanglement. Assess the resources available, and call for additional units if needed. Give an updated report. Establish a secure work area. Establish a staging resource area. Direct the placement of arriving apparatus. Assign personnel and tasks.

Skill Drill 25-4: Breaking Tempered Glass

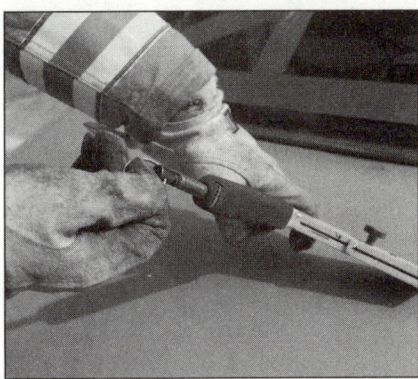

1. All fire fighters should be in proper PPE, including dust mask, safety glasses, or goggles. Attempt to lower windows as far as possible before breaking glass, and then select a spring-loaded center punch, a multipurpose tool, or axe.

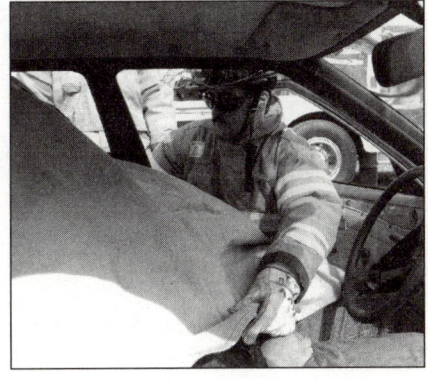

2. Ensure that the victim and other fire fighters are properly protected.

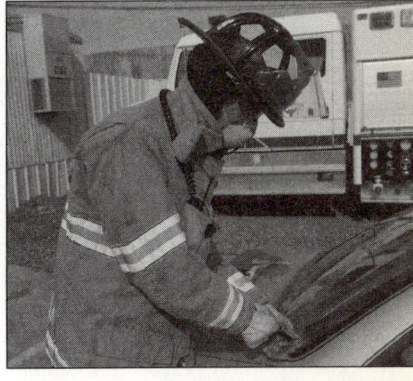

3. Warn personnel with the verbal command "Breaking glass," unless a stop/freeze call is made, while you continue with glass removal starting with a window farthest from the victim. Place the tool in the lower corner of the window and apply pressure until the spring is activated, or strike the lower corner with the axe.

4. Remove any loose glass around the window opening with a tool. Follow this procedure until all glass has been removed.

Skill Drill 25-8: Displacing the Dashboard of a Vehicle

1. Don PPE. Retrieve the required tools. Set up any power units. Couple hoses and cords if needed. Check the equipment for readiness. Enter and secure the work area. Assess the vehicle for hazards, including **passive restraint devices**.
2. Communicate with the victim and ensure that both the victim and rescuers are **protected** from hazards. Operate tools according to the manufacturer's recommendations.
3. Make relief cuts at the bottom of the "A" post. Use extra cribbing under the base end of the tool to keep the spreader from **slipping**. Guard against the tool kicking out.
4. Add **cribbing** as the dashboard is being lifted. Use good body mechanics when lifting. Maintain control of all tools at all times. Notify command that the dashboard has been displaced. Return the tools to the staging area upon completion of the tasks.

Skill Drill 25-9: Removing the Roof of a Vehicle

1. Don PPE. Retrieve the required tools. Set up any power units. Couple hoses and cords if needed. Check the equipment for readiness. Enter and secure the work area. Assess the vehicle for hazards, including passive restraint devices. Communicate with the victim; ensure that the victim and rescuers are protected from hazards. Remove any remaining glass.
2. Operate the required tools to remove the roof. Operate tools according to the manufacturer's instructions. Use good body mechanics.
3. Control the vehicle roof and protect against sharp edges at all times. Remove the roof to a safe location. Notify command when the roof is removed. Return the tools to the staging area upon completion of the tasks.

Chapter 26: Assisting Special Rescue Teams

Matching

1. C (page 767)
2. G (page 763)
3. E (page 753)
4. I (page 763)
5. F (page 753)
6. H (page 753)
7. A (page 762)
8. D (page 756)
9. B (page 761)
10. J (page 762)

Multiple Choice

1. B (page 750)
2. A (page 751)
3. D (page 752)
4. C (page 752)
5. C (page 753)
6. D (page 753)
7. B (page 754)
8. A (page 754)
9. B (page 754)
10. A (page 754)
11. B (page 754)
12. C (page 755)
13. D (page 755)
14. A (page 755)
15. B (page 755)
16. C (page 755)
17. A (page 756)
18. B (page 756)
19. C (page 757)
20. A (page 758)
21. D (page 759)
22. C (page 762)
23. D (page 763)
24. B (page 766)
25. C (page 768)

Vocabulary

1. **Hazardous materials:** Any materials or substances that pose an unreasonable risk of damage or injury to persons, property, or the environment if not properly controlled during handling, storage, manufacture, processing, packaging, use and disposal, or transportation. (page 766)
2. **Technical rescue incident:** A complex rescue incident involving vehicles or machinery, water or ice, rope techniques, a trench or excavation collapse, confined spaces, a structural collapse, wilderness search and rescue operation, or hazardous materials, and which requires specially trained personnel and special equipment. (page 750)
3. **Lockout/tagout system:** Methods of ensuring that electricity and other utilities have been shut down and switches are locked so that they cannot be switched on, so as to prevent flow of power or gases into the area where rescue is being conducted. (page 753)

Fill-in

1. awareness (page 750)
2. terminology (page 752)
3. proportional (page 753)
4. law enforcement (page 753)
5. packaging (page 754)
6. decontaminated (page 755)
7. stabilized (page 755)
8. asphyxiants (page 756)
9. incident command system (page 757)
10. training (page 758)
11. prior (page 761)
12. Structural (page 764)

True/False

1. T (page 751)
2. T (page 751)
3. F (page 751)
4. F (page 752)
5. F (page 753)
6. T (page 754)
7. T (page 755)
8. T (page 758)
9. T (page 759)
10. T (page 765)
11. T (page 765)

Short Answer

1. (1) Be safe. (2) Follow orders. (3) Work as a team. (4) Think. (5) Follow the Golden Rule of public service, which emphasizes the ethic of reciprocity: Treat others as you would like to be treated. (page 751)

2. • The fire officer's knowledge base and experience are greater than yours.
   • Orders come from superiors. Legitimate orders are those given by a fire officer or other designated person.
   • Follow rules and procedures. A fire fighter is required to follow rules, procedures, and guidelines regardless of his or her personal opinions.
   • You must do your own job.
   • Get the job done. In emergency situations, time is critical. Nevertheless, you must not act beyond your own skill and training level, and you must not violate any rules, procedures, standard operating guidelines, or orders of a superior in an effort to get the job done quickly. (page 751)

3. F: Failure to understand the environment, or underestimating it. A: Additional medical problems not considered. I: Inadequate rescue skills. L: Lack of teamwork or experience. U: Underestimating the logistics of the incident. R: Rescue versus recovery mode not considered. E: Equipment not mastered. (page 752)

4. (1) Preparation; (2) response; (3) arrival and size-up; (4) stabilization; (5) access; (6) disentanglement; (7) removal; (8) transport; (9) security of the scene and preparation for the next call; (10) postincident analysis (page 752)

5. • Scope and magnitude of the incident
   • Risk–benefit analysis
   • Number of known and potential victims
   • Hazards
   • Access to the scene
   • Environmental factors
   • Available and necessary resources
   • Establishment of control perimeters (page 756)

Word Fun

Fire Alarms

1. Respond safely to the emergency scene. Place the emergency vehicle in a safe position that protects the scene, and perform a size-up to assess for hazards. Secure the scene and call for needed assistance. Don the appropriate personal protective equipment (PPE). Use appropriate devices to establish a barrier, following the IC's orders.

2.
   - Personal protective equipment
   - Hydraulic tools (spreaders, cutters, rams)
   - Halligan tool
   - Cutting torch
   - Air chisels
   - Cribbing
   - Saber saw
   - Windshield cutter
   - Spring-loaded punch
   - Chains
   - Air bags (high and low pressure)
   - Extra SCBA tanks
   - Basic hand tools
   - Come along
   - Portable generator
   - Seat belt cutters
   - Hand lights and other scene lighting
   - Hose lines (protection)
   - Blanket

3. Safety is of paramount importance when approaching a trench or excavation collapse. Walking close to the edge of a collapse can trigger a secondary collapse, so stay away from the edge of the collapse and keep all workers and bystanders at a distance from the site. Vibration from equipment and machinery can also cause secondary collapses, so shut off all heavy equipment. Likewise, vibrations caused by nearby traffic can cause collapse, so it may be necessary to stop or divert traffic near the scene.

Chapter 27: Hazardous Materials: Overview

Matching

1. F (page 777)
2. A (page 777)
3. G (page 775)
4. C (page 781)
5. B (page 776)
6. E (page 776)
7. D (page 780)
8. J (page 778)
9. H (page 778)
10. I (page 778)

Multiple Choice

1. B (page 775)
2. B (page 776)
3. D (page 777)
4. D (page 778)
5. C (page 778)
6. A (page 778)
7. C (page 778)
8. A (page 777)
9. B (page 771)
10. C (page 777)

Vocabulary

1. **Local emergency planning committee (LEPC):** A group consisting of members of industry, transportation officials, the public at large, media, and fire and police agencies that gathers and disseminates information on hazardous materials stored in the community and ensures that there are adequate local resources to respond to a chemical event in the community. (page 778)
2. **Material safety data sheet (MSDS):** A form, provided by manufacturers and compounders (blenders) of chemicals, containing information about chemical composition, physical and chemical properties, health and safety hazards, emergency response, and waste disposal of the material. (page 778)
3. **Specialist level:** A level of fire fighter expertise at which the individual receives more specialized training than does a hazardous materials technician. Practically speaking, the two levels are not very different. Most of the training that specialist employees receive is either product or transportation mode specific. (page 777)
4. **HAZWOPER:** Hazardous Waste Operations and Emergency Response. This OSHA regulation governs hazardous materials waste sites and response training. Specifics can be found in 29 CFR 1910.120. Subsection (q) is specific to emergency response. (page 777)
5. **Operations level:** The level at which the responder should be able to recognize a potential hazardous materials incident, isolate the incident, deny entry to other responders and the public, and take defensive actions such as shutting off valves and protecting drains without having contact with the product. Operations-level responders act in a defensive fashion. (page 777)

Fill-in

1. Environmental Protection Agency (page 778)
2. material safety data sheet (page 778)
3. more (page 778)
4. Preplanning (page 780)
5. protective (page 777)
6. technicians (page 777)
7. NFPA 472 (page 777)
8. HAZWOPER (page 777)
9. Hazardous waste (page 776)
10. state-plan (page 776)

True/False

1. T (page 775)
2. F (page 778)
3. T (page 778)
4. T (page 780)
5. F (page 780)
6. T (page 780)
7. F (page 780)
8. T (page 780)
9. F (page 778)
10. T (page 778)

Short Answer

1. (1) Awareness-level personnel are persons who, in the course of their normal duties, could encounter an emergency involving hazardous materials/weapons of mass destruction (WMD). They are expected to recognize the presence of the hazardous materials/WMD, protect themselves, call for trained personnel, and secure the area. This level of training enables those that are first on scene of an incident to recognize a potential hazardous materials emergency, isolate the area, and call for assistance. Awareness-level trained persons are not seen as responders, but they do take protective actions.

    (2) Operations-level responders are persons who respond to hazardous materials/WMD incidents for the purpose of protecting nearby persons, the environment, or property from the effects of the release. Fire fighters in modern society are usually trained to the operations level because they should be able to recognize potential hazardous materials incidents, isolate and deny entry to other responders and the public, evacuate persons in danger, and take defensive actions such as shutting off valves and protecting drains without having contact with the product. Operations-level responders act in a defensive fashion.

    (3) Technician-level fire fighters are trained to enter heavily contaminated areas using the highest levels of personal protection. Hazardous materials technicians take offensive actions.

(4) Specialist-level fire fighters receive more specialized training than do hazardous materials technicians. Practically speaking, however, the two levels are not very different. The majority of the specialized training relates to a specific product such as chlorine or to a specific mode of transportation such as rail emergencies.

(5) The hazardous materials officer level of training is intended for those assuming command of a hazardous materials incident beyond the operations level. Individuals trained as hazardous materials officers receive operations-level training as well as additional training specific to commanding a hazardous materials incident. The hazardous materials officer is trained to act as a branch director or group supervisor for the hazardous materials component of the incident. (pages 777–778)

2. The Superfund Amendments and Reauthorization Act (SARA) was one of the first laws to affect how fire departments respond in a hazardous materials emergency. Finalized in 1986, SARA was the original driver for OSHA's HAZWOPER regulation. Indirectly, it indicated that workers handling hazardous wastes should have a minimum amount of training. Additionally, this law laid the foundation that ultimately allowed local fire departments and the community at large to obtain information on how and where hazardous materials were stored in their community. (page 778)

Word Fun

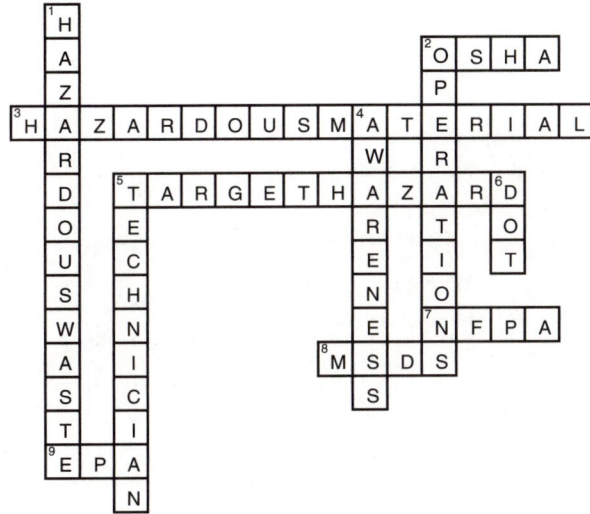

Fire Alarms

1. A hazardous material, as defined by the U.S. Department of Transportation (DOT), is a material that poses an unreasonable risk to the health and safety of operating emergency personnel, the public, and/or the environment if it is not properly controlled during handling, storage, manufacture, processing, packaging, use and disposal, or transportation.

2. The response begins with finding out about potential hazards in your area. Response agencies also should conduct preincident planning activities related to target hazards and other potential problem areas throughout the jurisdiction. Preplanning activities enable agencies to develop logical and appropriate response procedures for anticipated incidents. Jurisdictions that have no railways or maritime ports do not have to include training for those kinds of responses. Planning should focus on the real threats that exist in your community or adjacent communities you could be assisting. Once the threats have been identified, agencies must determine how they will respond. Some agencies establish parameters that guide their response to particular hazardous material incidents. Those parameters outline incident severity based on the nature of the chemical, the amount released, or the type of occupancy involved in the incident.

Chapter 28: Hazardous Materials: Properties and Effects

Matching

1. E (page 802)
2. F (page 789)
3. A (page 792)
4. H (page 798)
5. B (page 789)
6. D (page 790)
7. C (page 792)
8. I (page 790)
9. G (page 799)
10. J (page 788)

Multiple Choice

1. A (page 787)
2. C (page 787)
3. D (page 788)
4. C (page 788)
5. B (page 788)
6. C (page 788)
7. A (page 790)
8. B (page 790)
9. A (page 791)
10. C (page 792)
11. A (page 792)
12. D (page 792)
13. B (page 792)
14. C (page 792)
15. A (page 793)
16. A (page 793)
17. D (page 794)
18. B (page 797)
19. C (page 797)
20. B (page 798)
21. B (page 800)
22. A (page 800)

Labeling

1. Vapor density.

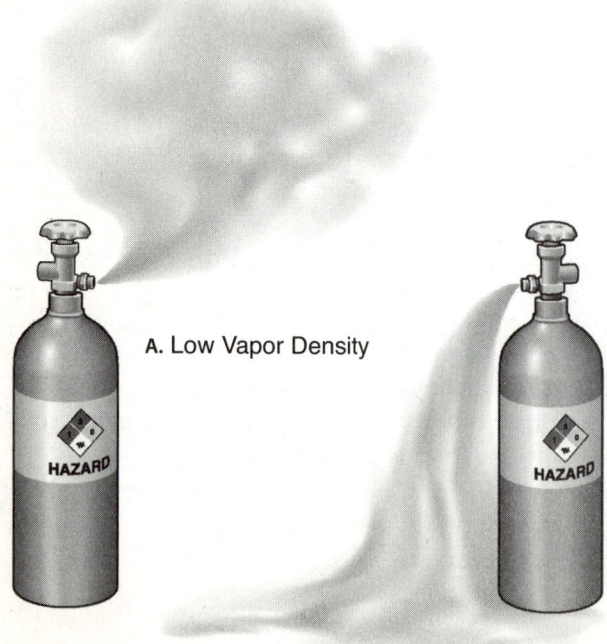

A. Low Vapor Density
B. High Vapor Density

Figure 28-3

2. Radiation.

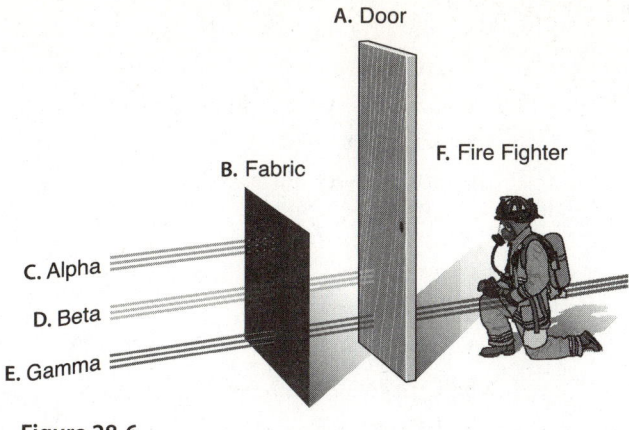

Figure 28-6

3. Four ways a chemical substance can enter the body.

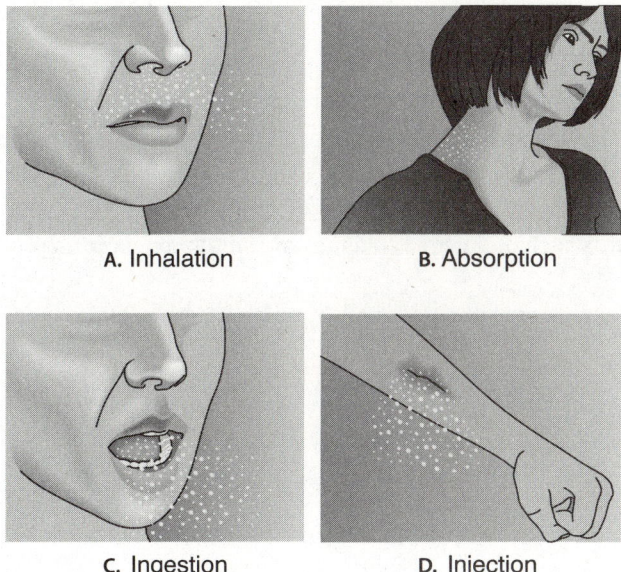

Figure 28-8

Vocabulary

1. **Flammable vapor:** Any substance that exists in the gaseous state at normal atmospheric temperature and pressure and is capable of being ignited and burned when mixed with the proper proportions of air, oxygen, or other oxidizers. (page 788)
2. **HEPA filter:** High-efficiency particulate air filter capable of catching particles down to 0.3-micron size—much smaller than a typical dust or alpha radiation particle. (page 793)
3. **Radiation:** The combined process of emission, transmission, and absorption of energy traveling by electromagnetic wave propagation between a region of higher temperature and a region of lower temperature. (page 793)
4. **Contamination:** The process of transferring a hazardous material from its source to people, animals, the environment, or equipment, all of which may act as carriers for the material. (page 794)
5. **Weapons of mass destruction (WMD):** Weapons whose use is intended to cause mass casualties, damage, and chaos. (page 794)

Fill-in

1. predict (787)
2. physical (page 788)
3. 14.7 (page 788)
4. −43 °F (page 789)
5. Flash point (page 802)
6. ignition (page 789)
7. float (page 791)
8. elements (page 793)
9. Beta (page 794)
10. convulsants (page 797)

True/False

1. T (page 788)
2. F (page 788)
3. T (page 789)
4. T (page 789)
5. T (page 791)
6. F (page 793)
7. F (page 793)
8. T (page 794)
9. T (page 796)
10. T (page 800)

Short Answer

1. *Thermal:* Heat created from intentional explosions or fires, or cold generated by cryogenic liquids.
   Radiological: Radioactive contamination from dirty bombs; alpha, beta, and gamma radiation.
   Asphyxiation: Oxygen deprivation caused by materials such as nitrogen; tissue asphyxiation from blood agents.
   Chemical: Injury and death caused by the intentional release of toxic industrial chemicals, nerve agents, vesicants, poisons, or other chemicals.
   Etiological: Illness and death resulting form biohazards such as anthrax, plague, and smallpox; hazards posed by bloodborne pathogens.
   Mechanical: Property damage and injury caused by explosion, falling debris, shrapnel, firearms, explosives, and slips, trips, and falls.
   Psychogenic: The mental harm from being potentially exposed to, contaminated by, and even just being in close proximity to an incident of this nature. (page 796)
2. S: Salivation
   L: Lachrymation (tearing)
   U: Urination
   D: Defecation
   G: Gastric disturbance
   E: Emesis (vomiting)
   M: Miosis (constriction of the pupil) (page 796)
3. (1) Inhalation: Through the lungs. (2) Absorption: By permeating the skin. (3) Ingestion: Via the gastrointestinal tract. (4) Injection: Through cuts or other breaches in the skin. (page 798)
4. (1) The amount of radiation absorbed by the body has a direct relationship to the degree of damage done.
   (2) The amount of exposure time ultimately determines the extent of the injury. (page 793)
5. H: Hydrogen
   A: Ammonia
   H: Helium
   A: Acetylene
   M: Methane
   I: Illuminating gas (neon and hydrogen cyanide)
   N: Nitrogen
   C: Carbon monoxide
   E: Ethylene (page 790)

Word Fun

Crossword solution with the following entries:
- 7 across: ALPHA
- 8 across: CARCINOGEN
- 12 across: ACID
- 13 across: INHALATION
- 14 across: INGESTION
- 15 across: CONTAMINATION
- 1 down: VAPOR
- 2 down: SARIN
- 3 down: BASE
- 4 down: GAMMA
- 5 down: INJECTION
- 6 down: SECONDARY
- 9 down: NERVE AGENT
- 10 down: BETA
- 11 down: NEUTRONS
- 12 down: ADSORPTION

Fire Alarms

1. Make every effort to reduce or eliminate the ability of the substance to enter your body, and keep the duration of the exposure to an absolute minimum. This requires you to reduce the time you are exposed to the material, to stay far enough away so that you are not directly exposed, and/or to shield yourself with personal protective equipment or solid objects. Time, distance, and shielding are methods used to protect fire fighters from the adverse effects of radiation. If you suspect a radiation incident at a fixed facility, you should initially ask for the radiation safety officer of the facility. This person is responsible for maintaining the use, handling, and storage procedures for all the radioactive material at the site. This person will likely be a tremendous resource to you and will know exactly what is being used at the facility.

2. From a terrorism perspective, irritants may be employed to incapacitate rescuers or to drive a group of people into another area where a more dangerous substance can be released. Irritants pose the least amount of danger in terms of toxicity of all the potential WMD agents a fire fighter may encounter. Exposed patients can be decontaminated with clean water, and the residual effects of the exposure should not be significant.

Chapter 29: Hazardous Materials: Recognizing and Identifying the Hazards

Matching

1. H (page 820)
2. F (page 820)
3. G (page 812)
4. C (page 810)
5. J (page 812)
6. A (page 820)
7. E (page 811)
8. B (page 811)
9. I (page 816)
10. D (page 811)

Answer Key

Multiple Choice

1. D (page 810)
2. D (page 810)
3. B (page 811)
4. D (page 811)
5. A (page 811)
6. C (page 812)
7. A (page 812)
8. C (page 813)
9. D (page 813)
10. C (page 813)
11. B (page 815)
12. A (page 815)
13. C (page 816)
14. D (page 818)
15. A (page 818)
16. B (page 820)
17. A (page 820)
18. C (page 820)
19. A (page 821)
20. B (page 822)

Labeling

1. Chemical transport vehicles.

Figure 29-4 A. An MC-306 flammable liquid tanker.

Figure 29-4 B. An MC-307 chemical hauler.

Figure 29-4 C. An MC-312 corrosives tanker.

Figure 29-4 D. An MC-331 pressure-cargo tanker.

Figure 29-4 E. An MC-338 cryogenic tanker.

Figure 29-4 F. A tube trailer.

Figure 29-4 G. A dry bulk cargo tank.

Figure 29-4 H. Intermodal tanks.

Vocabulary

1. **Shipping papers:** A shipping order, bill of lading, manifest, or other shipping document serving a similar purpose and usually including the names and addresses of both the shipper and the receiver as well as a list of shipped materials along with their quantity and weight. (page 820)
2. **Secondary containment:** Any device or structure that prevents environmental contamination when the primary container or its appurtenances fail. Examples of secondary containment include dikes, curbing, and double-walled tanks. (page 810)
3. **Hazardous materials:** Any materials or substances that pose an unreasonable risk of damage or injury to persons, property, or the environment if not properly controlled during handling, storage, manufacture, processing, packaging, use and disposal, or transportation. (page 809)
4. **Pipeline right-of-way:** An area, patch, or roadway that extends a certain number of feet on either side of the pipe itself and that may contain warning and informational signs about hazardous materials carried in the pipeline. (page 815)
5. **Placards and labels:** Placards are diamond-shaped indicators (10 inches on each side) that must be placed on all four sides of highway transport vehicles, railroad tank cars, and other forms of transportation carrying hazardous materials. Labels are smaller versions (4-inch diamond-shaped indicators) of placards and are used on the four sides of individual boxes and smaller packages being transported. Placards and labels are intended to give fire fighters a general idea of the hazard inside a particular container. A placard may identify the broad hazard class (e.g., flammable, poison, corrosive) of material that a tanker contains, while the label on a box inside a delivery truck relates only to the potential hazard inside that package. (page 816)

Fill-in

1. size-up (page 809)
2. 102 (page 811)
3. 312 (page 813)
4. tube (page 814)
5. pipelines (page 815)
6. W (page 818)
7. Hazardous Materials Information System (HMIS) (page 819)
8. material safety data sheet (MSDS) (page 819)
9. military (page 819)
10. waybills, consist (page 820)

True/False

1. T (page 809)
2. T (page 811)
3. F (page 811)
4. T (page 811)
5. F (page 816)
6. F (page 818)
7. T (page 818)
8. T (page 819)
9. T (page 819)
10. F (page 820)

Short Answer

1. The following items are included on a pesticide bag label:
   - Name of the product
   - Statement of ingredients
   - Total amount of product in the container
   - Manufacturer's name and address
   - U.S. Environmental Protection Agency (EPA) registration number, which provides proof that the product was registered with the EPA
   - The EPA establishment number, which shows where the product was manufactured
   - Signal words to indicate the relative toxicity of the material:
   Danger: Poison: Highly toxic by all routes of entry.
   Danger: Severe eye damage or skin irritation.
   Warning: Moderately toxic.
   Caution: Minor toxicity and minor eye damage or skin irritation.
   - Practical first-aid treatment description
   - Directions for use
   - Agricultural use requirements
   - Precautionary statements such as mixing directions or potential environmental hazards
   - Storage and disposal information
   - Classification statement on who may use the product

   In addition, every pesticide label must carry the statement, "Keep out of reach of children." (page 812)

2. (1) DOT Class 1: Explosives. (2) DOT Class 2: Gases. (3) DOT Class 3: Flammable combustible liquids. (4) DOT Class 4: Flammable solids. (5) DOT Class 5: Oxidizers. (6) DOT Class 6: Poisons (including blood agents and choking agents). (7) DOT Class 7: Radioactive materials. (8) DOT Class 8: Corrosives. (9) DOT Class 9: Other regulated materials. (page 816)

3. *Yellow section:* Chemicals in this section are listed numerically by their four-digit UN identification number. Entry number 1017, for example, identifies chlorine. Use the yellow section when the UN number is known or can be identified. The entries include the name of the chemical and the emergency action guide number.

   Blue section: Chemicals in the blue section are listed alphabetically by name. The entry will include the emergency action guide number and the identification number. The same information, organized differently, appears in both the blue and yellow sections.

   Orange section: This section contains the emergency action guides. Guide numbers are organized by general hazard class and indicate what basic emergency actions should be taken, based on hazard class.

   Green section: This section is organized numerically by UN identification number and provides the initial isolation distances for specific materials. Chemicals included in this section are highlighted in the blue or yellow sections. Any materials listed in the green section are always extremely hazardous. This section also directs the reader to consult the tables listing toxic inhalation hazard materials (TIH). These gases or volatile liquids are extremely toxic to humans and pose a hazard to health during transportation. (page 817)

4. The NFPA 704 hazard identification system uses a diamond-shaped symbol of any size, which is itself broken into four smaller diamonds, each representing a particular property or characteristic. The blue diamond at the nine o'clock position indicates the health hazard posed by the material. The top red diamond indicates flammability. The yellow diamond at the three o'clock position indicates reactivity. The bottom white diamond is used for special symbols and handling instructions. The blue, red, and yellow diamonds will each contain a numerical rating of 0 to 4, with 0 being the least hazardous and 4 being the most hazardous for that type of hazard. The white quadrant will not have a number but may contain special symbols. Among the symbols used are a burning O (oxidizing capability), a three-bladed fan (radioactivity), and a W with a slash through it (water reactive). (page 818)

5.
   - Physical and chemical characteristics
   - Physical hazards of the material
   - Health hazards of the material
   - Signs and symptoms of exposure
   - Routes of entry
   - Permissible exposure limits
   - Responsible-party contact

- Precautions for safe handling (including hygiene practices, protective measures, and procedures for cleaning up spills or leaks)
- Applicable control measures, including personal protective equipment;
- Emergency and first-aid procedures
- Appropriate waste disposal (page 819)

Word Fun

Across:
1. CARBOYS
2. VENTPIPES
4. LABELS
9. SECONDARYCONTAINMENT
11. DEWARCONTAINERS
12. CONSIST
13. AIRBILL
16. SHIPPINGPAPERS
17. TOTES
18. FREIGHTBILLS

Down:
1. CYLINDERS
3. PIPELINE
5. BILLOFLADING
6. SIGN
7. DRUM
8. INTERMODAL
10. TUBETRAILERS
13. AYBILL (waybill region)
14. WAYBILL
15. BUG
16. S

Fire Alarms

1. (a) Name of the caller and callback telephone number; location of the actual incident or problem; name of the chemical involved in the incident (if known); shipper or manufacturer of the chemical (if known); container type; railcar or vehicle markings or numbers; shipping carrier's name; recipient of material; local conditions and exact description of the situation

   (b) 1-800-424-9300

2. (a) A secondary device is a second explosive device placed in a location with the intent to kill emergency responders after an initial explosion has taken place.

   (b) Indicators of potential secondary devices may include trip devices such as timers, wires, or switches. They may also include common concealment containers, such as briefcases, backpacks, boxes, or other common packages, and uncommon concealment containers, such as pressurized vessels (e.g., propane tanks) or industrial chemical containers (e.g., chlorine storage containers). Attackers may be watching the site of the primary devices and waiting to manually activate the secondary devices.

Chapter 30: Hazardous Materials: Implementing a Response

Matching

1. B (page 836)
2. A (page 831)
3. D (page 829)
4. E (page 829)
5. C (page 829)

Answer Key

Multiple Choice

1. D (page 830)
2. B (page 830)
3. A (page 830)
4. A (page 831)
5. B (page 835)
6. A (page 835)
7. D (page 833)
8. B (page 833)
9. C (page 833)
10. D (page 831)

Vocabulary

1. **Defensive objectives:** Actions that do not involve the actual stopping of the leak or release of a hazardous material, or contact of responders with the material. These actions include preventing further injury and controlling or containing the spread of the hazardous material. (page 831)
2. **Material safety data sheet (MSDS):** A form, provided by manufacturers and compounders (blenders) of chemicals, containing information about chemical composition, physical and chemical properties, health and safety hazards, emergency response, and waste disposal of the material. (page 833)
3. **Hazardous materials safety officer:** A second safety officer dedicated to the safety needs of the Hazardous Materials Branch. (page 835)
4. **Hot zone entry team:** The team of fire fighters assigned to the entry into the designated hot zone. (page 836)
5. **Decontamination team:** The team responsible for reducing and preventing the spread of contaminants from persons and equipment used at a hazardous materials incident. Members of this team establish the decontamination corridor and conduct all phases of decontamination. (page 836)

Fill-in

1. resources (page 829)
2. offensive (page 830)
3. characteristics (page 830)
4. defensive (page 831)
5. hazardous materials (page 833)
6. Litmus (page 833)
7. command, operations, logistics, planning, finance, administration (page 835)
8. maximum (page 835)
9. responders (page 835)
10. incident commander (IC) (page 833)

True/False

1. F (page 829)
2. T (page 830)
3. F (page 830)
4. T (page 833)
5. T (page 833)

Short Answer

1.
   - The exact address and specific location of the leak or spill
   - Identification of indicators and markers of hazardous materials
   - All color or class information obtained from placards
   - Four-digit United Nations/North American Hazardous Materials Code numbers for the hazardous materials
   - Hazardous material identification obtained from shipping papers or MSDS and the potential quantities of hazardous materials involved
   - Description of the container, including its size, capacity, type, and shape
   - The amount of chemical that could leak and the amount that has already leaked
   - Exposures of people and the presence of special populations (children or elderly)
   - The environment in the immediate area
   - Current weather conditions, including wind direction and speed
   - A contact or callback telephone number and two-way radio frequency or channel. (page 830)

2. (1) Isolate the area affected by the leak or spill, and evacuate victims who could become exposed to the hazardous material if the leak or spill were to progress. (2) Control where the spill or release is spreading. (3) Contain the spill to a specific area. (page 831)

3. *Level I:* Lowest level of threat. A small amount of hazardous material is involved, and the incident can usually be handled by the local fire department. Fire fighters must wear turnout gear and SCBA. Example: A small gasoline spill occurs as a result of a motor vehicle accident.

   Level II: A hazardous materials team is needed at this level. Fire fighters only support the hazardous materials team. Additional PPE required will be specialized and carried only by the hazardous materials team. Civilian evacuations may be required. Decontamination may need to be performed. Example: A gasoline tanker overturns in a tunnel and spills gasoline onto the highway.

   Level III: The highest level of threat. Large-scale evacuations may be needed. Federal agencies will be called. Example: A ship in a highly populated harbor catches fire and begins to release chlorine vapors from its cargo area. (page 829)

4. The special technical group may include a second safety officer reporting directly to the hazardous materials officer. This hazardous materials safety officer is responsible for the hazardous materials team's safety only. The group may also include a hot zone entry team, a decontamination team, a backup entry team (rapid intervention team), and a hazardous materials information research team. (page 835)

Fire Alarms

1. Contact the engineer; he or she may have some information and the waybill that you need. Once you have gathered your on-scene information, call CHEMTREC and the National Response Center.
2. The diamond marking will give you a general idea of the flammability, toxicity, reactivity, and any special hazards associated with the materials that are stored in the chemical storage shed. Collect the information and limit access to the building until the hazardous materials team arrives.

Chapter 31: Hazardous Materials: Personal Protective Equipment, Scene Safety, and Scene Control

Matching

1. C (page 859)
2. E (page 844)
3. B (page 842)
4. G (page 859)
5. D (page 842)
6. A (page 859)
7. F (page 846)
8. I (page 846)
9. J (page 847)
10. H (page 851)

Multiple Choice

1. A (page 843)
2. A (page 844)
3. B (page 844)
4. D (page 844)
5. B (page 846)
6. C (page 847)
7. A (page 855)
8. A (page 857)
9. D (page 858)
10. C (page 858)
11. B (page 859)
12. A (page 859)
13. B (page 859)
14. C (page 859)
15. D (page 859)

524 Answer Key

Vocabulary

1. **Immediately dangerous to life and health (IDLH):** A designation indicating that an atmospheric concentration of a toxic, corrosive, or asphyxiant substance poses an immediate threat to life or could cause irreversible or delayed adverse health effects. (page 842)
2. **Heat stroke:** A severe and sometimes fatal condition resulting from the failure of the temperature-regulating capacity of the body. It is caused by prolonged exposure to the sun or high temperatures. Reduction or cessation of sweating is an early symptom; body temperature of 105°F or higher, rapid pulse, hot and dry skin, headache, confusion, unconsciousness, and convulsions may also occur. Heat stroke is a true medical emergency requiring immediate transport to a medical facility. (page 856)
3. **Heat exhaustion:** A mild form of shock caused when the circulatory system begins to fail as a result of the body's inadequate effort to give off excessive heat. (page 856)
4. **Backup personnel:** Individuals who remove or rescue those working in the hot zone in the event of an emergency. (page 860)
5. **High temperature–protective clothing:** Protective clothing designed to shield the wearer during short-term exposures to high temperatures. (page 843)

Fill-in

1. least (page 843)
2. Liquids (page 844)
3. Chemical (page 844)
4. single (page 845)
5. eyes (page 856)
6. Alkaline (page 856)
7. 240 (page 857)
8. socks (page 857)
9. 8 to 16 ounces (page 857)
10. Skin absorption (page 855)

True/False

1. T (page 843)
2. T (page 843)
3. F (page 845)
4. F (page 860)
5. F (page 860)
6. T (page 860)
7. T (page 860)
8. F (page 860)
9. T (page 859)
10. T (page 859)

Short Answer

1. *Safe atmosphere:* No harmful hazardous materials effects exist, so personnel can handle routine emergencies without donning specialized PPE.
   Unsafe atmosphere: A hazardous material that is no longer contained has created an unsafe condition or atmosphere. A person who is exposed to the material for long enough will probably experience some form of acute or chronic injury.
   Dangerous atmosphere: Serious, irreversible injury or death may occur in the environment. (page 843)
2. *Level A protection:* Personal protective equipment that provides protection against vapors, gases, mists, and even dusts. The highest level of protection, it requires a totally encapsulating suit that includes self-contained breathing apparatus.
   Level B protection: Personal protective equipment that is used when the type and atmospheric concentration of substances have been identified. It generally requires a high level of respiratory protection but less skin protection: chemical-protective coveralls and clothing, chemical protection for shoes, gloves, and self-contained breathing apparatus outside of a nonencapsulating chemical-protective suit.
   Level C protection: Personal protective equipment that is used when the type of airborne substance is known, the concentration is measured, the criteria for using air-purifying respirators are met, and skin and eye exposure is unlikely. It consists of standard work clothing with the addition of chemical-protective clothing, chemically resistant gloves, and a form of respirator protection.
   Level D protection: Personal protective equipment that is used when the atmosphere contains no known hazard, and work functions preclude splashes, immersion, or the potential for unexpected inhalation of or contact with hazardous levels of chemicals. It is primarily a work uniform that includes coveralls and affords minimal protection. (page 861)

3. *Hot zone:* The area immediately surrounding a hazardous materials spill/incident site that is directly dangerous to life and health. All personnel working in the hot zone must wear complete, appropriate protective clothing and equipment. Entry requires approval by the IC or a designated hazardous materials officer. Complete backup, rescue, and decontamination teams must be in place at the perimeter before operations begin.

Warm zone: The area located between the hot zone and the cold zone at the incident. Personal protective equipment is required when working in this zone. The decontamination corridor is located in the warm zone, which is also called the contamination reduction zone.

Cold zone: A safe area at a hazardous materials incident for those agencies involved in the operations. The incident commander, incident command post, EMS providers, and other support functions necessary to control the incident should be located in the cold zone, which is also called the clean zone or support zone. (page 859)

Word Fun

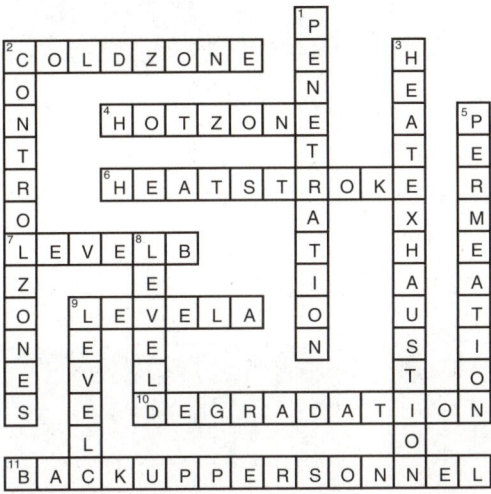

Fire Alarms

1. The recommended PPE for Level B protection includes the following components:
   - SCBA or SAR
   - Chemical-resistant clothing
   - Inner and outer chemical-resistant gloves
   - Chemical-resistant safety boots/shoes
   - Hard hat
   - Two-way radio

   Optional PPE for Level B protection includes the following components:
   - Coveralls
   - Long cotton underwear
   - Disposable gloves and boot covers

2. (a) Heat exhaustion is a mild form of shock that occurs when the circulatory system begins to fail because the body is unable to dissipate excessive heat and becomes overheated.

   (b) Although heat exhaustion is not an immediately life-threatening condition, the affected individual should be removed at once from the source of heat, rehydrated with electrolyte solutions, and kept cool. If not properly treated, heat exhaustion may progress to heat stroke.

Answer Key

Skill Drills

Skill Drill 31-1: Donning a Level B Encapsulated Chemical-Protective Clothing Ensemble

1. Conduct a pre-entry briefing, medical monitoring, and equipment inspection. While seated, pull on the suit to waist level; pull on the chemical boots over the top of the chemical suit. Pull the suit boot covers over the tops of the boots.

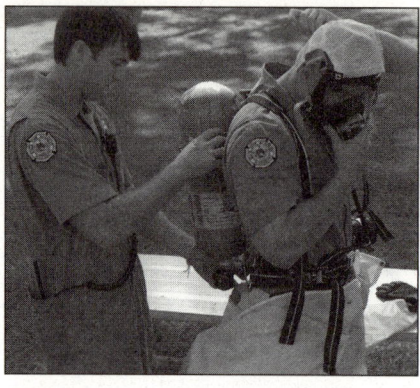

2. Stand up and don SCBA and the SCBA face piece, but do not connect the regulator to the face piece.

3. Place the helmet on the head.

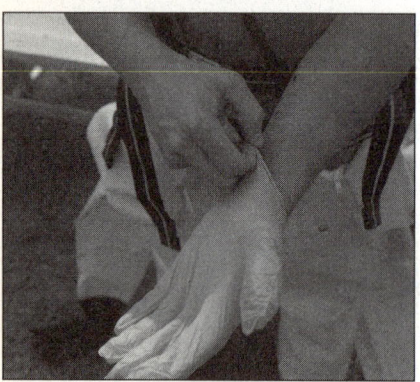

4. Don the inner gloves.

5. With assistance, complete donning the suit by placing both arms in the suit, pulling the expanded back piece over the SCBA, and placing the chemical suit over the head. Instruct the assistant to connect the regulator to the SCBA face piece and ensure air flow.

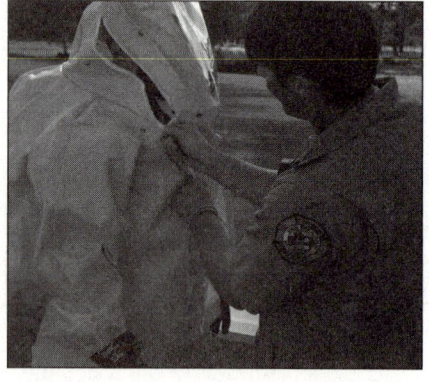

6. Instruct the assistant to close the chemical suit by closing the zipper and sealing the splash flap.

7. Review hand signals and indicate that you are okay.

Skill Drill 31-2: Doffing a Level B Encapsulated Chemical-Protective Clothing Ensemble

1. After completing decontamination, proceed to the **clean area**. Remove the hands and arms from the suit gloves and sleeves, and cross the arms in front inside the suit.
2. Instruct the assistant to open the **chemical splash flap** and open the suit zipper.
3. Instruct the assistant to begin at the head and roll the suit **down** and **away** from you until the suit is below waist level.
4. Sit and instruct the assistant to complete rolling down the suit and remove the outer boots and suit. Rotate on the bench to the direction that will allow you to place your **feet** on a dry, clean area.
5. Stand and doff the SCBA using the **quick-release** method. Keep the face piece in place while the SCBA frame is placed on the ground.
6. Take a deep breath, doff the SCBA mask, and walk away from the clean area. Go to rehabilitation area for medical monitoring, rehydration, and **personal decontamination shower**.

Skill Drill 31-5: Donning a Level C Chemical-Protective Clothing Ensemble

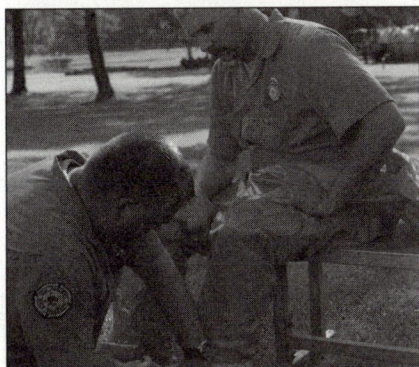

1. Conduct a pre-entry briefing, medical monitoring, and equipment inspection. While seated, pull on the suit to waist level; pull on the chemical boots over the top of the chemical suit. Pull the suit boot covers over the tops of the boots.

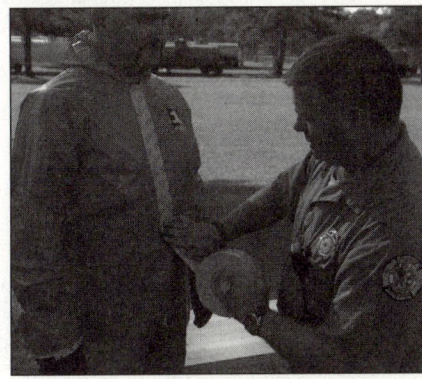

2. With assistance, complete donning the suit by placing both arms in the suit and pulling the suit over the shoulders. The assistant closes the chemical suit by closing the zipper and sealing the splash flap.

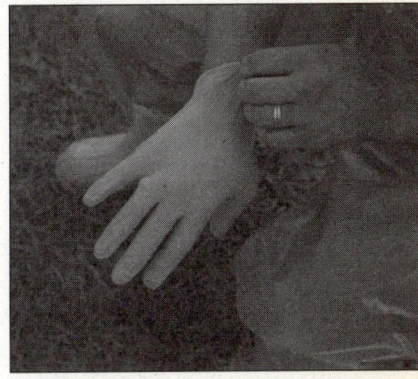

3. Don the inner glove.

4. Stand up and don the APR/PAPR and APR/PAPR face piece. With assistance, pull the hood over the head and the APR/PAPR face piece. Place the helmet on the head.

Skill Drill 31-6: Doffing a Level C Chemical-Protective Clothing Ensemble

1. After completing **decontamination**, proceed to the clean area. The assistant opens the chemical splash flap and suit zipper. Remove the hands and arms from the gloves (except the inner gloves) and sleeves, and cross the arms in front **inside** the suit. You are seated and the assistant completes rolling down the suit and takes the **outer boots** and suit away. The assistant helps remove the inner gloves. **Rotate** on the bench to the direction that will allow you to place your feet on a dry, clean area. Remove APR/PAPR. Go to the **rehabilitation** area for medical monitoring, rehydration, and personal decontamination shower.

Chapter 32: Hazardous Materials: Response Priorities and Actions

Matching

1. C (page 876)
2. A (page 870)
3. B (page 870)
4. F (page 873)
5. E (page 876)
6. D (page 876)
7. J (page 873)
8. I (page 872)
9. H (page 872)
10. G (page 871)

Multiple Choice

1. B (page 867)
2. D (page 868)
3. A (page 868)
4. C (page 868)
5. B (page 867)
6. A (page 870)
7. C (page 870)
8. C (page 871)
9. D (page 871)
10. A (page 873)
11. B (page 876)
12. D (page 876)
13. C (page 878)
14. B (page 878)
15. B (page 878)

Vocabulary

1. **Exposures (hazardous materials):** People, property, structures, or parts of the environment that are subject to influence, damage, or injury as a result of contact with a hazardous material. The amount of exposures that remain is determined both by the location of the incident and by the amount of progress that has been made in protecting those exposures via isolation and other indirect responses. Incidents in urban areas will likely have more exposures and, therefore, will likely need more resources to protect those exposures from the hazardous materials. (page 867)
2. **Shelter-in-place:** A method of safeguarding people in a hazardous area by keeping them in a safe atmosphere, usually inside structures. (page 868)
3. **Recovery phase:** The stage of a hazardous materials incident after the imminent danger has passed, when clean-up and the return to normalcy have begun. (page 878)

Fill-in

1. life (page 869)
2. safe area (page 867)
3. toxicity (page 868)
4. unnecessary (page 867)
5. technician (page 870)
6. Retention (page 873)
7. complete dam (page 872)
8. cylinders (page 871)
9. liquid fires (page 870)
10. not necessary (page 870)

True/False

1. T (page 869)
2. T (page 868)
3. F (page 870)
4. F (page 870)
5. T (page 878)
6. T (page 878)
7. F (page 878)
8. T (page 877)
9. T (page 876)
10. F (page 873)

Short Answer

1. Aqueous film-forming-foam (AFFF); fluoroprotein foam; protein foam; high-expansion foam (page 870)

Word Fun

Across: 2. CONTAINMENT, 4. DIVERSION, 9. RECOVERY PHASE, 10. DILUTION
Down: 1. DIKE, 3. CONFINE, 5. RETENTION, 6. ABSORPTION, 7. EXOMMSRE (crossword grid), 8. DAMMING

Fire Alarms

1. Collect the basic materials: absorbents, adsorbents, absorbent pads, and absorbent booms. Decide which material is best suited for use with the spilled product, and assess the location of the spill. Stay clear of any spilled product. Apply the appropriate material to control and contain the spilled material. Maintain materials and take appropriate steps for their disposal.

2. (a) AFFF, protein foam, and fluoroprotein foam

   (b) Foam should be gently applied or bounced off another adjacent object so that it flows down across the liquid and does not directly upset the burning surface. Foam can also be applied in a rain-down method by directing the stream into the air over the material and letting the foam fall gently, much as rain would.

Chapter 33: Hazardous Materials: Decontamination Techniques

Matching

1. F (page 884)
2. D (page 888)
3. G (page 888)
4. B (page 887)
5. C (page 888)
6. A (page 888)
7. E (page 887)
8. I (page 887)
9. J (page 888)
10. H (page 888)

Multiple Choice

1. A (page 884)
2. B (page 887)
3. D (page 886)
4. A (page 887)
5. C (page 888)
6. B (page 890)
7. A (page 890)
8. B (page 890)
9. C (page 890)
10. D (page 888)

Vocabulary

1. **Decontamination team:** The team responsible for reducing and preventing the spread of contaminants from persons and equipment used at a hazardous materials incident. Members of this team establish the decontamination corridor and conduct all phases of decontamination. (page 890)
2. **Contamination:** The process of transferring a hazardous material from its source to people, animals, the environment, or equipment, any of which may act as a carrier. (page 884)
3. **Adsorption:** The process of adding a material such as sand or activated carbon to a contaminant, which then adheres to the surface of the material and allows for collection of the contaminated material. (page 887)
4. **Solidification:** The process of chemically treating a hazardous liquid so as to turn it into a solid material, thereby making the material easier to handle. (page 888)
5. **Emulsification:** The process of changing the chemical properties of a hazardous material so as to reduce its harmful effects. (page 888)

Fill-in

1. decontamination corridor (page 884)
2. contaminants (page 886)
3. vapor dispersion (page 888)
4. removal (page 888)
5. dilution (page 888)
6. disperse (page 888)
7. adsorption (page 887)
8. Technical decontamination (page 886)
9. drains, streams, ponds (page 885)
10. identify (page 885)

True/False

1. F (page 884)
2. F (page 885)
3. T (page 888)
4. T (page 890)
5. T (page 890)

Short Answer

1. (1) Emergency decontamination is used in potentially life-threatening situations to rapidly remove the bulk of the contamination from an individual, regardless of the presence or absence of a technical decontamination corridor.

    (2) Gross decontamination, like emergency decontamination, aims to significantly reduce the amount of surface contaminant by delivery of a continuous shower of water and removal of outer clothing or PPE. It differs from emergency decontamination, however, in that gross decontamination is controlled through the decontamination corridor.

    (3) Technical decontamination is performed after gross decontamination and is a more thorough cleaning process. Technical decontamination may involve several stations or steps. During this type of decontamination, multiple personnel (the decontamination team) typically use brushes to scrub and wash the person or object to remove contaminants.

    (4) Mass decontamination is often used in incidents involving unknown agents or in the case of a contamination of large groups of people. It takes place in the field and is a way of quickly performing gross decontamination on a large number of victims who have escaped from a hazardous materials incident. (pages 885–886)

Word Fun

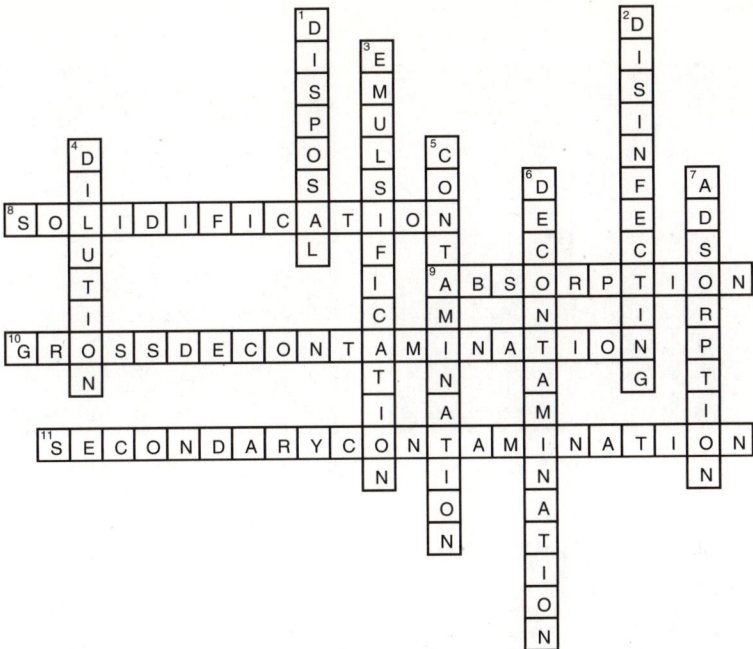

Fire Alarms

1. Ensure that you have the appropriate PPE to protect against the chemical threat. Stay clear of the product, and do not make physical contact with it. Make an effort to contain runoff by directing victims out of the hazard zone and into a suitable location for decontamination. Flush the victim to remove the product from the victim's clothing. Instruct and assist the victim in removing contaminated clothing. Flush the contaminated victim. Assist or obtain medical treatment for the victim, and arrange for the victim's transport.

2. Alternative decontamination procedures include the following techniques:
   - Absorption
   - Adsorption
   - Dilution
   - Disinfection
   - Disposal
   - Solidification
   - Emulsification
   - Vapor dispersion
   - Removal
   - Vacuuming

Skill Drills

Skill Drill 33-3: Performing Responder Decontamination

1. Drop any tools and equipment.

2. Perform gross decontamination.

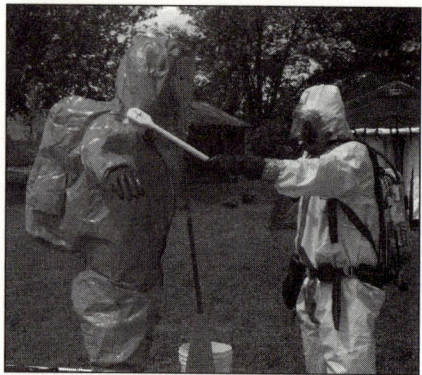

3. Wash and rinse the fire fighter one to three times.

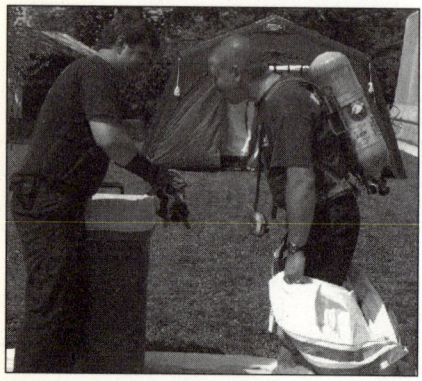

4. Remove outer hazardous materials protective clothing and isolate PPE.

5. Remove SCBA.

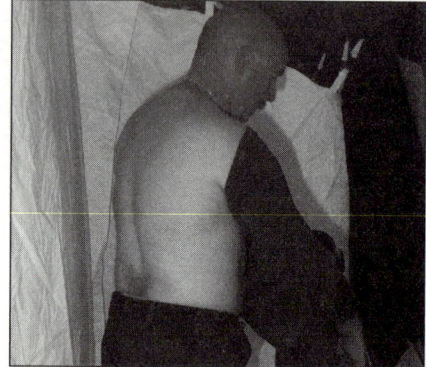

6. Remove personal clothing. Bag and tag all personal clothing.

7. Shower and wash the body. Dry off the body and put on clean clothing.

Chapter 34: Terrorism Awareness

Matching

1. E (page 906)
2. M (page 901)
3. A (page 914)
4. B (page 910)
5. K (page 900)
6. N (page 906)
7. O (page 906)
8. J (page 908)
9. G (page 903)
10. F (page 913)
11. C (page 902)
12. L (page 914)
13. D (page 914)
14. H (page 908)
15. I (page 896)

Multiple Choice

1. D (page 896)
2. A (page 899)
3. B (page 899)
4. C (page 899)
5. A (page 900)
6. C (page 901)
7. B (page 902)
8. D (page 902)
9. B (page 902)
10. C (page 908)
11. B (page 908)
12. C (page 908)
13. D (page 909)
14. A (page 910)
15. C (page 914)

Vocabulary

1. **V-agent:** A nerve agent, principally a contact hazard; an oily liquid that can persist for several weeks. (page 904)
2. **Plague:** An infectious disease caused by the bacterium *Yersinia pestis*, which is commonly found on rodents. (page 907)
3. **Smallpox:** A highly infectious disease caused by the virus *Variola*. (page 907)
4. **Tabun:** A nerve gas that is both a contact hazard and a vapor hazard. It disables the chemical connection between the nerves and their target organs. (page 914)
5. **Universal precautions:** Procedures for infection control that treat blood and certain bodily fluids as capable of transmitting bloodborne diseases. (page 908)
6. **Forward staging area:** A strategically placed area, close to the incident site, where personnel and equipment can be held in readiness for rapid response to an emergency event. (page 902)
7. **Radiation dispersal device:** Any device that causes the purposeful dissemination of radioactive material without a nuclear detonation; a dirty bomb. (page 909)

Fill-in

1. Terrorism (page 896)
2. pipe bomb (page 901)
3. Nerve agents (903)
4. SLUDGE (page 904)
5. air flow (page 903)
6. incubation period (page 908)
7. Radioactive materials (page 908)
8. absorption (page 911)
9. mass decontamination (page 911)
10. defined perimeter (page 911)

Answer Key

True/False

1. F (page 906)
2. F (page 910)
3. F (page 908)
4. F (page 908)
5. T (page 908)
6. F (page 914)
7. F (page 899)
8. T (page 897)
9. T (page 909)
10. T (page 908)

Short Answer

1. Terrorists are motivated by a cause and choose targets they believe will help them achieve their goals and objectives. Terrorist incidents aim to instill fear and panic among the general population and to disrupt daily ways of life. Given this goal, terrorists tend to choose symbolic targets, such as places of worship, embassies, monuments, or prominent government buildings. Sometimes the objective is sabotage—that is, to destroy or disable a facility that is significant to the terrorist cause. The ultimate goal could be to cause economic turmoil by interfering with transportation, trade, or commerce. (page 897)

2. Ecoterrorism refers to illegal acts committed by groups supporting environmental or related causes. Examples include spiking trees to sabotage logging operations, vandalizing a university research laboratory that is conducting experiments on animals, or firebombing a store that sells fur coats.

    Cyberterrorism refers to electronically attacking government or private computer systems. This type of terrorism would disrupt many day-to-day activities in our society, because the use of computers is woven into most things we do as part of contemporary life.

    Agroterrorism includes the use of chemical or biological agents to attack the agricultural industry or the food supply. The deliberate introduction of an animal disease such as foot-and-mouth disease to the livestock population could result in major losses to the food industry and produce fear among members of the general population. (page 899)

3. Your first priority should be to ensure the safety of the scene. During the initial stages of an incident, you will not know whether the event was caused by an intentional act or by accidental circumstances. In any incident involving an explosion, follow departmental procedures to ensure the safety of rescuers, victims, and bystanders. Consider the possibility that a secondary device may be in the vicinity. Quickly survey the area for any suspicious bags, packages, or other items.

    It is also possible that chemical, biological, or radiological agents may be involved in a terrorist bombing. Qualified personnel with monitoring instruments should be assigned to check the area for potential contaminants. The initial size-up should also include an assessment of hazards and dangerous situations. The stability of any building involved in the explosion must be evaluated before anyone is permitted to enter it, because entering an unstable area without proper training and equipment may complicate rescue and recovery efforts. (page 902)

4. Responding to a terrorist incident puts fire fighters and other emergency personnel at risk. Although responders must ensure their own safety at every incident, a terrorist incident may carry an extra dimension of risk. Because the terrorist's objective is to cause as much harm as possible, emergency responders are just as likely to be targets as are ordinary civilians.

    In most cases, the first emergency units will not be dispatched for a known WMD or terrorist incident. Rather, the initial dispatch might be for an explosion, for a possible hazardous materials incident, for a single person with difficulty breathing, or for multiple victims with similar symptoms. Emergency responders will usually not know that a terrorist incident has occurred until personnel on the scene begin to piece together information gained from their own observations and from witnesses. (page 909)

5. (1) Keep the time of exposure as short as possible. (2) Stay as far away from the source of the radiation as possible. (3) Use shielding to limit the amount of radiation absorbed by the body. (page 908).

Word Fun

Across:
1. BIOLOGICAGENTS
5. ECOTERRORISM
8. LEWISITE
10. ANTHRAX
11. TRIAGE
14. PERSONALDOSIMETERS

Down:
1. BETARADIATION (partial)
2. CHEMICALAGENTS
3. NERVEAGENTS
4. SARIN (partial)
6. INCUBATION
7. CYBERTERROR
9. PLAGUE
12. PHOSGENE
13. GMMA (partial)

Fire Alarms

1. The fire service role includes emergency medical services (EMS), hazardous materials mitigation, and technical rescue as well as fire suppression. All of these functions will probably be needed when a terrorist incident occurs. Terrorism presents challenges to the fire service on a scale that has never previously been experienced in North America; it also presents an unparalleled threat to the lives of fire fighters and emergency responders. The terrorist threat requires fire fighters to work closely with local, state, and federal law enforcement agencies; emergency management agencies; allied health agencies; and the military. It is critical that all of these agencies work together in a coordinated and cooperative manner. All emergency responders and law enforcement agencies must be prepared to face a wide range of potential situations.

2. Unless the cause of an explosion is known to be accidental, fire fighters at the scene should always consider the possibility that an explosive device was detonated. The first priority should be to ensure the safety of the scene. Fire fighters should also consider the possibility that a secondary device may be in the vicinity. Responders should quickly survey the area for any suspicious bags, packages, or other items. It is also possible that chemical, biologic, or radiological agents may be involved in a terrorist bombing. Qualified personnel with monitoring instruments should be assigned to check the area for potential contaminants. These precautions should be implemented immediately. The initial size-up should also include an assessment of hazards and dangerous situations.

The stability of any building involved in the explosion must be evaluated before anyone is permitted to enter. Entering an unstable area without the proper training and equipment may complicate rescue and recovery efforts.

Chapter 35: Fire Prevention and Public Education

Matching

1. B (page 918)
2. C (page 918)
3. A (page 919)
4. E (page 920)
5. D (page 921)

Multiple Choice

1. A (page 918)
2. B (page 918)
3. D (page 918)
4. A (page 919)
5. D (page 923)
6. C (page 919)
7. B (page 922)
8. C (page 924)
9. B (page 926)
10. A (page 926)

Vocabulary

1. **Fire code:** A set of legally adopted rules and regulations designed to prevent fires and protect lives and property in the event of a fire. (page 918)
2. **Fire prevention:** Activities conducted to prevent fires and protect lives and property in the event of a fire. These activities include the enactment and enforcement of fire codes, the inspection of properties, the presentation of fire safety education programs, and the investigation of the causes of fires. (page 918)

Fill-in

1. building (page 919)
2. legal (page 919)
3. interior (page 919)
4. systematic (page 924)
5. bedrooms (page 926)
6. prevent (page 918)
7. home fire (page 919)
8. buildings (page 919)
9. bare hand (page 921)
10. smoke alarms (page 922)

True/False

1. T (page 918)
2. T (page 919)
3. F (page 923)
4. T (page 928)
5. T (page 922)
6. F (page 922)
7. T (page 922)
8. F (page 921)
9. T (page 926)
10. T (page 926)

Short Answer

1. (1) Stop, Drop, and Roll program; (2) Exit Drills in the Home (EDITH); (3) installation and maintenance of smoke alarms; (4) Learn Not to Burn; (5) Change Your Clock—Change Your Battery; (6) fire safety for babysitters; (7) fire safety for seniors; (8) fire safety for college students; (9) wildland fire prevention programs (page 919)
2. (1) Explain the importance of properly installing and maintaining smoke alarms. (2) Stress the importance of installing a smoke detector on each floor and outside each sleeping area. (3) Remind students to avoid installing smoke alarms in kitchens and garages, or near fireplaces, windows, and exit doors. (4) Mount smoke alarms on the ceiling or as high as possible on walls. (5) Test smoke alarms once a month using the test button. (6) Change the smoke detector batteries twice each year. (7) Clean alarms regularly to prevent false alarms. (page 922)
3.
   - Do not leave anything cooking unattended on a stove.
   - Keep all flammable materials, cleaning supplies, cooking oils, and aerosols away from the stove. Do not place anything that could ignite on a cooking surface, even when it is turned off and cold. Do not place towel racks near the stove.
   - Store smoking materials out of the reach of children.
   - Do not overload electrical outlets or extension cords.
   - Keep electric cords properly maintained; replace any frayed cords.
   - Check that the extinguisher is visible and properly charged. (page 926)

Fire Alarms

1. Teenagers are ready for lessons that they can apply to everyday life. Many teenagers work as babysitters, so prevention messages might be geared toward cooking safety and evacuating children in the event of a fire. Teenagers are also a high-risk group for motor vehicle collisions, so messages about safe driving, the dangers of drinking and driving, and action to take if they see or hear an emergency vehicle while driving are appropriate. Some teenagers might be interested in becoming fire fighters or joining a fire fighter explorer group. This age group can be a good source of recruits for both career and volunteer departments.
2. Tell the home occupant that storage of gasoline and other flammable substances is a major concern, because an open flame or pilot light can easily ignite leaking vapors. Gasoline and other flammable liquids should be stored only in approved containers and in outside storage areas or outbuildings. Small quantities of flammable and combustible liquids (such as paint, thinners, varnishes, and cleaning fluids) should be stored in closed metal containers away from heat sources.

Chapter 36: Fire Detection, Protection, and Suppression Systems

Matching

1. I (page 953)
2. J (page 959)
3. C (page 944)
4. G (page 958)
5. D (page 959)
6. K (page 944)
7. A (page 945)
8. L (page 953)
9. N (page 945)
10. E (page 958)
11. M (page 945)
12. B (page 954)
13. H (page 954)
14. F (page 945)

Multiple Choice

1. C (page 939)
2. A (page 939)
3. C (page 937)
4. C (page 937)
5. B (page 942)
6. A (page 942)
7. D (page 944)
8. D (page 944)
9. A (page 944)
10. C (page 945)
11. B (page 945)
12. D (page 950)
13. A (page 950)
14. C (page 954)
15. D (page 957)
16. B (page 957)
17. A (page 960)
18. C (page 961)
19. B (page 961)
20. B (page 965)

Vocabulary

1. **Verification system:** A fire alarm system that does not immediately initiate an alarm condition when a smoke detector is activated. The system will wait for a preset interval, generally 30 to 60 seconds, before checking the detector again. If the condition is clear, the system returns to normal status. If the detector is still sensing smoke, the system activates the fire alarm. (page 945)
2. **Zoned system:** A fire alarm system design that divides a building or facility into zones so that the area where an alarm originated can be identified. (page 946)
3. **Deluge sprinkler system:** A sprinkler system in which all sprinkler heads are open. When an initiation device, such as a smoke detector or heat detector, is activated, the deluge valve opens and water discharges from all of the open sprinkler heads simultaneously. (page 959)

4. **Fire department connection:** A fire hose connection through which the fire department can pump water into a sprinkler system or standpipe system. (page 957)
5. **Outside stem and yoke valve:** A sprinkler control valve with a valve stem that moves in and out as the valve is opened or closed. (page 954)
6. **Post indicator valve:** A sprinkler control valve with an indicator that reads either open or shut depending on its position. (page 954)

Fill-in

1. Photoelectric (page 939)
2. brain (page 937)
3. faults (page 937)
4. beam (page 942)
5. Rate-of-rise (page 943)
6. unwanted (page 945)
7. remote station (page 948)
8. Sidewall (page 954)
9. piping (page 954)
10. I (page 961)

True/False

1. T (page 939)
2. F (page 939)
3. F (page 938)
4. F (page 943)
5. T (page 943)
6. F (page 945)
7. T (page 946)
8. F (page 949)
9. T (page 950)
10. T (page 957)

Short Answer

1. (1) Alarm initiation device; (2) alarm notification device; (3) control panel (page 937)
2. (1) Local alarm; (2) remote station; (3) auxiliary system; (4) proprietary system; (5) central station (page 948)
3. (1) Automatic sprinkler systems; (2) standpipe systems; (3) specialized extinguishing agents (page 949)
4. (1) Wet sprinkler system: The piping in a wet system is always filled with water. When activated, water is immediately discharged.
   (2) Dry sprinkler system: The pipes are filled with pressurized air, which keeps water out until the air pressure is released. The system utilizes accelerators and exhausters.
   (3) Preaction sprinkler system: This system is similar to a dry sprinkler system except that a secondary device (pull alarm, smoke detector) must be activated before the water is released.
   (4) Deluge sprinkler system: In this type of dry sprinkler system, all of the sprinkler heads are activated as soon as the system is activated. The sprinkler heads are always open. (page 958)
5. (1) A Class I standpipe is designed for use by fire department personnel only. Each outlet has a 2½-inch male coupling and a valve to open the water supply after the attack line is connected. Often the connection is located inside a cabinet, which may or may not be locked. Responding fire personnel carry the hose into the building with them, usually in some sort of roll, bag, or backpack. A Class I standpipe system must be able to supply an adequate volume of water with sufficient pressure to operate fire department attack lines.
   (2) A Class II standpipe is designed for use by the building occupants. The outlets are generally equipped with a length of 1½-inch single-jacket hose preconnected to the system. These systems are intended to enable occupants to attack a fire before the fire department arrives, but their safety and effectiveness are questionable.
   (3) A Class III standpipe has the features of both Class I and Class II standpipes in a single system. This kind of system has 2½-inch outlets for fire department use as well as smaller outlets with attached hoses for occupant use. (pages 961–962)

Word Fun

```
 U N W A N T E D A L A R M
                 O       T
       H     E C F       A
       A     X A L       M
   C   L   D H L A       P
 N O N C O D E D A L A R M
   D   N   L U L E       E
   E   1   U S A D       R
   D   3   G T R E       S
   S   0   E E M T       W
   Y   1   V R   E       I
   S       A     C       T
   T     F L O W S W I T C H
   E       V     O
 S M O K E D E T E C T O R
```

Fire Alarms

1. From a safety standpoint, fire fighters need to understand the operations and limitations of fire detection and suppression systems. A building with a fire protection system will have very different working conditions during a fire than will an unprotected building. Fire fighters need to know how to fight a fire in a building with a working fire suppression system, and how to shut down the system after the fire is extinguished.

   From a customer service standpoint, fire fighters who understand how fire protection systems work can help dispel misconceptions about these systems and advise building owners and occupants after an alarm is sounded.

2. Carbon dioxide extinguishes a fire by displacing the oxygen and creates a dangerous situation. Responding personnel should wear SCBA at all times when entering the server room.

Chapter 37: Fire Cause Determination

Matching

1. B (page 985)
2. E (page 985)
3. A (page 977)
4. F (page 977)
5. D (page 977)
6. C (page 984)
7. J (page 975)
8. I (page 977)
9. H (page 977)
10. G (page 973)

Multiple Choice

1. B (page 972)
2. D (page 973)
3. D (page 973)
4. A (page 973)
5. C (page 974)
6. B (page 975)
7. A (page 975)
8. D (page 975)
9. C (page 977)
10. A (page 979)
11. B (page 981)
12. D (page 982)
13. C (page 983)
14. A (page 984)
15. B (page 984)
16. D (page 985)
17. C (page 985)
18. A (page 986)
19. B (page 986)
20. C (page 986)

Vocabulary

1. **Trailers:** Combustible materials (such as rolled rags, blankets, and newspapers or ignitable liquid) that are used to spread fire from one point or area to other points or areas, often used in conjunction with an incendiary device. (page 982)
2. **Competent ignition source:** An ignition source that can ignite a fuel under the existing conditions at the time of the fire. It must have sufficient heat and be in close enough proximity to the fuel for a sufficient amount of time to ignite the fuel. (page 973)
3. **Depth of char:** The thickness of the layer of a material that has been consumed by a fire. The depth of char on wood can be used to help determine the intensity of a fire at a specific location. (page 974)
4. **Chain of custody:** A legal term used to describe the paperwork or documentation describing the movement, storage, and custody of evidence, such as the record of possession of a gas can from a fire scene. (page 979)

Fill-in

1. validity (page 973)
2. lessons (page 985)
3. fuel supply (page 973)
4. exterior (page 974)
5. burn (page 975)
6. demonstrative (page 977)
7. contaminated (page 977)
8. not (page 980)
9. observes (page 980)
10. location (page 983)
11. obstacles (page 982)
12. Clothing (page 983)

True/False

1. T (page 973)
2. T (page 973)
3. F (page 973)
4. F (page 974)
5. F (page 977)
6. T (page 979)
7. T (page 980)
8. T (page 980)
9. T (page 981)
10. F (page 981)
11. T (page 982)
12. F (page 983)

Short Answer

1. (1) Suspend salvage and overhaul, and secure the scene. Keep nonessential personnel out of the area. Deny entry to all unauthorized and unnecessary persons.

   (2) Photograph the fire scene extensively. Start with the area with the least amount of damage and work your way toward the area of possible origin. Take several pictures of the point of origin from various angles. Photograph any incendiary devices on the premises exactly where they were found.

   (3) If weather, traffic, or other factors could destroy the evidence, take steps to preserve it in the best way possible. Protect tire tracks or footprints by placing boxes over them to prevent dust accumulation. Use barricades to block off the area to further traffic. Rope off areas surrounding plants, trailers, and devices, and post a guard at the site. (page 984)

2. - Fires are set in easily accessible locations, such as immediately inside entrances, on basement stairs, in trash bins, or on porches.
   - Fires are set in structures such as occupied residences of all types, barns, and vacant buildings.
   - Accelerants are rarely used. The pyromaniac is impulsive, so materials readily at hand are used.
   - Each pyromaniac follows a unique pattern—for example, setting fires at the same time of day or night, using the same method, and setting fires in similar locations. (page 985)

3. (1) Vandalism; (2) excitement; (3) revenge; (4) crime concealment; (5) profit; (6) extremism (page 986)

Word Fun

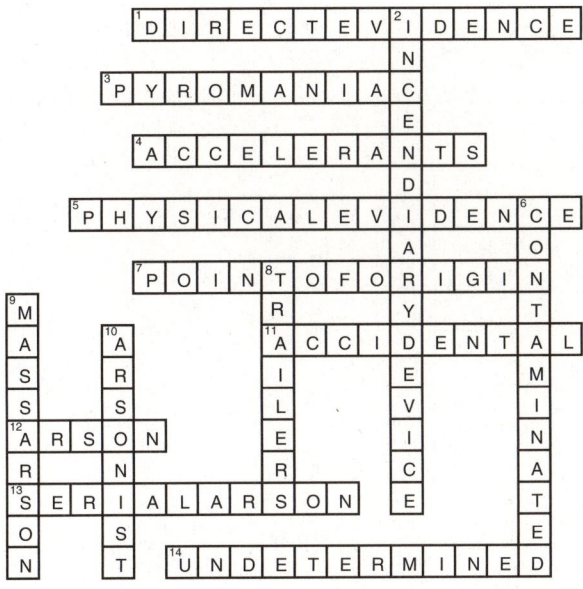

Fire Alarms

1. Take photographs of each piece of evidence as it is found and collected. If possible, photograph the item as it was found, before it is moved or disturbed. On the fire scene, sketch, mark, and label the location of the evidence.

   Place evidence in appropriate containers to ensure its safety and prevent contamination. Unused paint cans with lids that seal automatically when closed are the best containers for transporting evidence. Glass mason jars sealed with a sturdy sealing tape are appropriate for transporting smaller quantities of materials. Soak up small quantities of liquids with either a cellulose sponge or cotton batting. Protect partially burned paper and ash by placing them between layers of glass.

   Tag all evidence at the fire scene. Evidence being transported to the laboratory should include a label with the date, time, location, discoverer's name, and witnesses' names. Record the time the evidence was found, the location where it was found, and the name of the person who found it. Keep a record of each person who handled the evidence.

   Keep a constant watch on the evidence until it can be stored in a secure location. Evidence that must be moved temporarily should be put in a secure place accessible only to authorized personnel. Preserve the chain of custody in handling all the evidence. A broken chain of custody may result in a court ruling that the evidence is inadmissible.

2. After the investigator identifies the area of origin, fire fighters could be asked to assist in digging out the fire scene. "Digging out" is a term used to describe the process of carefully looking for evidence within the debris. Sometimes the entire fire scene must be closely examined to determine the cause of the fire and gather evidence. Removing and inspecting the layers of debris enables the investigator to determine in what order items burned, whether an item burned from the top down or from the bottom up, and how long it burned. Systematically digging through the debris often can uncover the exact point of origin and cause of both accidental and deliberate fires. Common search methods include the "grid" search (also known as the "double strip" search) and the "strip" search. The investigator will explain generally what to look for, how to search, and what to do with any potential evidence.

Photo Credits

Chapter 7

7–19 Courtesy of Amerex Corporation; SD 7–3 © 2003 Berta A. Daniels; SD 7–7 Courtesy of Ansul Incorporated

Chapter 9

9–9A, 9–9B Courtesy of Yale Cordage

Chapter 11

11–8 © AbleStock

Chapter 16

16–7 © 2003 Berta A. Daniels; 16–19A, 16–19B, 16–19C Courtesy of Akron Brass Company

Chapter 29

29–5A, 29–B Courtesy of Polar Tank Trailer, LLC; 29–5C Courtesy of National Tank Truck Carriers Association; 29–5D, 29–5E, 29–5F Courtesy of Jack B. Kelley, Inc.; 29–5G Courtesy of Polar Tank Trailer, LLC; 29–5H Courtesy of Eurotainer